The Scientific Spirit of American Humanism

MEDICINE, SCIENCE, AND RELIGION IN HISTORICAL CONTEXT
Ronald L. Numbers, Consulting Editor

The Scientific Spirit
of American Humanism

STEPHEN P. WELDON

Johns Hopkins University Press

Baltimore

© 2020 Johns Hopkins University Press
All rights reserved. Published 2020
Printed in the United States of America on acid-free paper
9 8 7 6 5 4 3 2 1

Johns Hopkins University Press
2715 North Charles Street
Baltimore, Maryland 21218-4363
www.press.jhu.edu

Library of Congress Cataloging-in-Publication Data

Names: Weldon, Stephen P., 1962– author.
Title: The scientific spirit of American humanism / Stephen P. Weldon.
Description: Baltimore : Johns Hopkins University Press, [2020] | Series:
    Medicine, science, and religion in historical context | Includes
    bibliographical references and index.
Identifiers: LCCN 2019052322 | ISBN 9781421438580 (hardcover) | ISBN
    9781421438597 (ebook)
Subjects: LCSH: Humanism—United States—History—20th century. |
    Humanism, Religious—History—20th century. | Science—Philosophy—
    History—20th century. | Religion and science—United States—History—
    20th century. | United States—Intellectual life—20th century.
Classification: LCC B821 .W434 2020 | DDC 211/.609730904—dc23
LC record available at https://lccn.loc.gov/2019052322

A catalog record for this book is available from the British Library.

*Special discounts are available for bulk purchases of this book. For more
information, please contact Special Sales at specialsales@press.jhu.edu.*

Johns Hopkins University Press uses environmentally friendly book
materials, including recycled text paper that is composed of at least 30
percent post-consumer waste, whenever possible.

For Tomoko, Karen, and Julia

Preface ix

Introduction. The Scientific Spirit of American Humanism 1

Humanism between Religion and Science 4 / The Scientific Spirit of American Humanism 8

1. Liberal Christianity and the Frontiers of American Belief 13

Populist Radicals: Free Thinkers and Atheists 16 / Denominational Radicals: Unitarians 17 / Academic Radicals: Protestant Modernists 23 / Understanding Radical Religion and Unbelief in the Nineteenth Century 29

2. The Birth of Religious Humanism 33

Heretics Find a Home 34 / The Proliferation of Religious Humanism 37 / Content of Humanism: Religion, Science, and Democracy 39 / Unitarian Controversy 43 / Humanism's Place in American Theology 45

3. Manifesto for an Age of Science 47

The Fellowship of Religious Humanists 48 / The Humanist Manifesto 53 / Unpacking the Manifesto 55 / The Humanist Movement in the 1930s 60

4. Philosophers in the Pulpit 63

The Professionalization of American Philosophy 64 / The University of Columbia Philosophy Department 65 / Roy Wood Sellars: The Architect of Evolutionary Naturalism 74 / Max Otto: Advocate for a Personal Pragmatism 77

5. Humanists at War 81

The Conference on Science, Philosophy and Religion 83 / Historical Shifts in American Theology 87 / Humanist Response 89 / The Scientific Spirit Conferences 92 / The Problem of Religion 95

6. Scientists on the World Stage 99

The International Humanist and Ethical Union 100 / The Scientists of the Chicago Circle 103 / The UNESCO Affair 106 / The Pugwash Movement 110

7. Eugenics and the Question of Race 115

Birth Control and the Population Problem 117 / The Question of Eugenics 120 / The Jensen IQ Controversy 123 / Race in the Humanist Movement 127

8. Inside the Humanist Counterculture 131

Science and Human Love 132 / Toward a Countercultural
Humanism 134 / Humanism and the Third Force 137 / San Francisco and
Buffalo 141 / Humanism beyond Freedom and Dignity 145

9. Skeptics in the Age of Aquarius 151

The New Cults Symposium: Debating the New Age 153 / Stars Aligned
against Astrology 157 / Organizing Skeptics 159 / The Mars Effect
Controversy 164 / CSICOP and Religion 167

10. The Fundamentalist Challenge 171

Church and State in Humanist History 173 / The Evangelical Antihumanist
Crusade 176 / The Humanist Response 179 / A Manifesto and a Declaration 183

11. Battling Creationism and Christian Pseudoscience 191

Early Battles over Evolution 192 / The Fundamentalists' Antievolution
Agenda 196 / Defending Evolution: Two Strategies 198 / Beyond Antievolution:
Attacking Fundamentalist Pseudoscience 203

12. The Humanist Ethos of Science in Modern America 209

Two Figureheads of Scientific Humanism 210 / Scientists and
Humanism 214 / Science and Atheism 219 / Secular Critics at the End of the
Century 221 / The American Foundations of Humanism 226

Epilogue. Science and Millennial Humanism 231

Notes 235
Archival Sources and Personal Papers 271
Index 273

Photographs appear following pages 80 and 150.

Though many people talk about the nature of our secular world, few have come to grips with the very slippery boundaries that define it. In this volume, I try to do just this by tracing the history of a small but influential social movement. That story is especially revealing in uncovering some of the unique ways in which the American experience in the twentieth century shaped institutions and gave impetus to different modes of dialogue, debate, and conflict.

Some projects take years to figure out. This one took much longer than I ever could have imagined. It started as an idea for a short paper that would be finished in the months before I began working on a dissertation at the University of Wisconsin–Madison. Before long, that paper had *become* the dissertation. The scope of the project touched on widely diverse areas of American intellectual and cultural history, and researching and writing the dissertation was challenging, but it was never onerous. I finished my research, discovered the main narrative, wrote it up, and completed my degree in history of science.

But a thesis is not a book, and I had a vision for the book that required a more comprehensive assessment of the topics I had covered. The issues I grappled with shed light on a number of areas that students of American intellectual history and history of science are deeply engaged in, and a book-length treatment of humanism needed to do justice to those issues. Moreover, I recognized that the broad topic area of science and religion and the cast of well-known characters that were integral to the humanist story would put the book in a category that might make it especially attractive to humanists themselves and their detractors. In addition, it was clear that it could say something to the nonacademic reader, someone who wanted to know more about the place of science and religion in American life and how our current fraught cultural climate came into being. The book, I decided, should reach out to nonspecialists.

It is not unusual for historians to write for a broader audience. The nature of our discipline is especially open in that respect. Indeed, I have been guided over the years by many excellent histories that accomplish this task. So I embarked on what proved to be a very long journey to create a book that I hoped would capture the spirit and importance of the events I had uncovered, both for colleagues and future researchers in the academy and for the curious and engaged readers who might find this narrative enlightening.

As the years passed, my ideas also changed and matured. I came to see that the meaning of the narrative was not exactly what I had set down in my thesis many years earlier. Time provided perspective, and it became clearer to me that the humanist story ramified more widely than I had believed, and I found myself exploring areas, like the interaction between race and science in the humanist movement, that had been entirely off my radar at the beginning. Along the way, many conversations with colleagues, students, friends, and family transformed my thinking. I periodically used drafts of chapters in my classes, which helped me see how differently a new generation of Americans thought about the world, and how this story spoke to them in different ways than it did to me.

The manuscript turned into an experimental space, and gradually I discovered a historical voice that I felt was proper for the story I wanted to tell. What I have ended up with, the book you now hold in your hands, accomplishes this task to the best of my ability. Now it is up to you, the reader, to decide whether it truly succeeds.

Because so much of my text discusses how other academics have engaged religion, I think it only proper to give a short autobiographical note here to put my own experience and motivations into perspective. I am a scholar by training and profession. I teach courses, write reviews, attend conferences, and do research in the field of history of science. During my training as a historian, I spent a lot of time trying to work out the particularities of religious doctrines in Catholic, Anglican, and other Protestant sects in the West in order to better grasp the motivations of scientists in their study of nature. What I have learned I now try to convey in an undergraduate class that I call Science in a Religious World. That said, I came to my career by way of a personal fascination with the nature of science and especially its relationship to religion. Over the years I have discovered that many of my colleagues in history of science had similar motives for entering the discipline. As a result, the field is replete with people who started their studies with a strong personal desire to dig deeply into the nature of our current world and how and why we know it as we do.

Growing up in central Texas, I was quite familiar with Bible Belt controversies over secular humanism and evolution. The issues were frequently in the news and were often discussed among high school classmates at lunch and in other forums. My family was secular in many respects but not dogmatically so. Moreover, various counterculture influences periodically found their way into our world. Over the years, I've studied or participated in several nontraditional religious and quasi-religious organizations, including Transcendental Meditation, Anthroposophy, Unitarianism, Buddhism, and even humanism itself.

Through marriage, I have become connected to Jodo Shinshu Buddhism. My father-in-law was a priest in a small village in western Japan, and it has been from my experience as part of this old temple family that I have gained something of an insider's perspective on what it means to be religious in that country. This experience, more than anything else, exploded all of my expectations about institutions, beliefs, and practices that we in the West associate with the concept *religion*.

I hope that this book, too, will help explode some of the myths about religion and science that tend to be commonly held in this country. Even more, I hope it will contribute to the discussion of how intellectual movements operate and how the ideas behind them are shaped and in turn how they shape and motivate people and institutions. In the end, the book offers my interpretation of an important swath of American culture. Writing it has required that I leave out many people and stories, but I have done my best to be true to the historical record. My narrative is premised on the view that the history of humanism really does provide valuable insight into one aspect of the American experience, a rich and instructive place to explore American ideas and ideology.

I am indebted to many people and institutions that have helped me greatly over the years. The staff at the archives listed at the end of the book were exceptionally helpful. Though it has been quite long since I have visited many of those places, I must acknowledge the valuable help that I was given. Financially, I could never have made all of those visits unless I had had extra support, which I received from a National Science Foundation Dissertation Improvement Grant (no. 56438) at the outset of my study. Financial support was also provided from the Offices of the Vice President for Research, the Provost, and the Dean of the College of Arts and Sciences, University of Oklahoma.

Many people generously gave me their time as I sought to better understand humanism. Michael Schuler, Tom McIver, Marcello Truzzi, and Will Provine were very helpful in getting me started with this research. I was openly welcomed into their offices, homes, and collections. The staff at the American Humanist Association, the Council for Democratic and Secular Humanism (now Center for Inquiry), the Humanist Institute, and the

Committee for the Scientific Investigation of Claims of the Paranormal (now Committee for Skeptical Inquiry) were all very generous in allowing me to look through their extensive files and to obtain photographs. In my search for photographs, I must especially thank Timothy Binga, Jennifer Bardi, David Breeden, Jan Sharar, Henry P. S. Robertson, and Jim Lippard.

Other people I interviewed by phone, in person, or via email were Bette Chambers, Jean Kotkin, Fred Edwords, Janet Asimov, Paul Kurtz, Carl Sagan, Edward O. Wilson, Corliss and Beth Lamont, Arthur Jackson, Edd Doerr, Lloyd Morain, and Tolbert McCarroll. Khoren Arisian spoke to me at length and was very kind to invite me to a Young Humanist Conference in Minneapolis, where I was able to talk with humanists at the turn of the new millennium.

At the outset, several faculty members at the University of Wisconsin read and commented on the dissertation; Lynn Nyhart, Paul Boyer, and Ronald Numbers were most important in this respect. In the years between that early study and the book's completion, I talked with many people and gained a great deal of help. In preparing the manuscript for press, I was fortunate to receive very useful comments from a number of excellent scholars, including Jon Roberts, Matt Stanley, Paul Murphy, and Adam Shapiro. Murphy and Shapiro both spent a lot of time with the manuscript, and I am indebted to their help. Ronald Numbers has been a guiding voice and strong supporter throughout, displaying unflagging confidence in my efforts. An advisor turned friend and colleague, I hear his consistent call to produce "more truth!" echoing in the background during all of my historical work.

Working at the University of Oklahoma for the past seventeen years has been enormously rewarding. Though we are an out-of-the-way school in many respects, the community of scholars in the department has helped me a lot. The faculty both in and outside the department are generous, the graduate students are sharp and engaging, and the administration has been consistently supportive of my work on this and other projects. I must recognize Steven Livesey, Hunter Heyck, Kerry Magruder, JoAnn Palmeri, Piers Hale, and Rienk Vermij especially for their reading, listening, and prodding. The University Libraries have been remarkable allies for me. Students and colleagues have given advice on various parts of this book, and for that I'm truly grateful. I must especially thank Caitlin Beasley and Aja Tolman for helping me develop a website for some of my data.

I've had a chance to talk about my project with a number of religious men and women who range across a wide spectrum of belief. Smart and thought-

The Scientific Spirit of American Humanism

The Scientific Spirit of American Humanism

I would suggest that science is, at least in part, informed worship.

Carl Sagan (1985)

In April of 1981, Cornell astronomer Carl Sagan received the Humanist of the Year award at the annual convention of the American Humanist Association in San Diego. Sagan was one of those few scientists who could achieve a level of stardom usually reserved for actors and politicians. His fame was due in large part to the extraordinarily popular science show *Cosmos* that was then being broadcast on TV stations around the world. The show's cautious yet optimistic vision of the future was grounded in Sagan's own enthusiasm for modern science, which he considered to be the pinnacle of human culture. He imbued the show and his story of science with an almost religious reverence. The humanists who gave him their award found his message particularly appealing—to many, it seemed to illustrate precisely those aspects of humanism that the group had promoted in sermons, lectures, books, and articles for many decades.[1]

American humanism had originated in the early twentieth century as a form of liberal religion, established by Unitarian ministers who had abandoned what they considered unfounded myths and superstitions—including the idea of God. In place of these outdated views, the humanists had turned to a modern outlook based on scientific knowledge and democratic values. But that modern outlook was being seriously attacked in America in the latter half of the twentieth century. Cold War tensions were at a peak, Islamic fundamentalism had challenged American supremacy and ended the political career of President Jimmy Carter, conservative Christianity had gained significant political influence at all levels of government, and nuclear-armed nations threatened the very future of life on earth. In this tense climate, humanists had been singled out by fundamentalists as the cause of America's moral and religious decline, and the US House of Representatives had even debated banning humanism in the public schools. On top of this, humanists looked around and saw large swaths of the culture succumbing to cult-like thinking and New Age mysticism that encouraged belief in pseudoscience and irrationality: UFOs, sea monsters, conspiracy theories, psychic powers, and a myriad of invalid health scams pointed to a retreat from the principles of the Enlightenment on which humanism was based.

That the humanists selected Sagan to be a figurehead of their movement at this time says a lot. Sagan was undoubtedly the most well-known living scientist in the country, if not the world, at this time. Not only did he study fascinating topics—space exploration and the possibility of extraterrestrial life—but he also wrote engagingly for a general audience. His prose was articulate and visionary. More impressive for a scientist, his genial manner, good looks, and captivating public persona brought him widespread media attention. A few appearances on Johnny Carson's *Tonight Show* had sealed his place as a spokesperson for the scientific community as a whole.[2]

And now, he had produced this thirteen-part public television series *Cosmos*, which would become one of the most watched science documentaries of all time. In the series, Sagan gave his personal account of science and its place in civilization through the ages. He linked that picture of science with an ethical perspective that embraced the values of freedom, democracy, and peace. The show also reverberated with religious imagery that reinforced Sagan's message. It was a creative blend of science and spirituality—not traditional religion by any means, but rather one that gave satisfaction by recalling human achievements and the grandeur of nature. The new sort of spirituality was suggested in the very design of the "spaceship of the imagination" that Sagan used to take his viewers to far-off places and times. The walls and ceiling of the ship were like that of a futuristic cathedral. We humans, it seemed to suggest, could gain not just knowledge but also spiritual and moral sustenance from a worldview centered on our connection to nature and embrace of scientific values.[3]

The American Humanist Association had given the award to Sagan for his skill in popularizing science in a humanist mode. For Sagan and the humanists, science was much more than simply the discovery of new knowledge. It was also one of the best examples of how mankind's innate curiosity could make us better people. He encouraged us to work together to overcome ignorance and superstition, and he made connections between the pursuit of knowledge, ethical behavior, and emotional fulfillment. The morals that he drew from the story of science highlighted the values that humanists had discussed for nearly seventy years. By emphasizing the importance of science to civilization, Sagan's popular appeal did a lot of work for humanists. He channeled an enduring element of the humanist movement—a faith in the scientific enterprise in its many aspects: as a way of critical thinking, as a means of discovery, as a form of control, and (not insignificantly) as a form of worshipful devotion to human ingenuity and the wonders of nature. This moralistic and visionary use of science was central to humanist thought, and it gave rise to controversies both within and without the humanist movement. This book explores both.

For one, it tells the story of humanism's battles with orthodox Christians—a conflict that is simply one chapter of a much larger history of the pervasive and often vicious struggles between conservative and liberal forces in America.[4] Humanists saw themselves as defenders of secular Western civilization against threats posed by religious zealotry. Their attack on religion took many forms and included political, ethical, and philosophical objections, and the humanists' ideas regarding science played a major role. Supernaturalistic religions disparaged science, the humanists believed, and refused to accept its hard-won truths. The humanists' metaphysical beliefs were based on scientific naturalism—a stance that gave credence only to natural phenomena and that denied the existence of supernatural beings and forces—and as a result, their ideas challenged the religious and ethical beliefs of the great majority of Americans. This metaphysical conflict pitted humanists against religious conservatives, and fights between them flared up again and again over the decades in different but very recognizable forms. The most intense battles, which occurred between humanists and fundamentalists, peaked late in the century. This book sheds light on that story, helping to explain why and how it arose, and what is underpinning these widely different temperaments in American thought.

Science was also at the center of a major struggle *within* the movement. While the fundamentalist-humanist split is well known, this internal conflict is more easily missed, yet it also reflects existing, deep-seated differences in American thought. Science has been very important to people on the left. Liberals and progressives have often looked to science for solutions to human problems, but they have also turned to it to justify their ideas about mankind, their social ideals, and even their politics.[5] Humanists, who represent the rationalist wing of progressive American thought, embrace science with an even greater passion than other liberals, so the divisions within humanism are especially revealing. The core of these differences has to do with the way that ideas about science collided with assumptions about human nature and human society. Once again, scientific naturalism played a key role.

The challenge of naturalism concerned especially its relationship to human nature and human values.[6] For humanists, it was not so much a question whether naturalism per se was a threat to humanism—since they asserted that it was not—but rather whether one or another specific naturalistic philosophy was a problem. Did some scientific philosophies undermine the qualities that humanists considered to be fundamental to their own moral outlook? Could a particular understanding of science threaten the humanistic value system in the same way that it threatened a Christian value system? For example, many humanists believed that a narrow mechanical philosophy was incompatible with such humanistic ideals as freedom, dignity, equality, and the well-being of all people.

These issues lay at the heart of many of the conflicts that humanists waged in the second half of the century. The internal battles were fought over two fundamentally different science-based outlooks: pragmatism—most especially the version that was widely disseminated by John Dewey and his circle—and positivism and the dominant Anglo-American analytic tradition that succeeded it.[7] Personal and intellectual conflicts intensified the differences between these two schools, and the struggles took various forms, touching on questions of mechanism, determinism and reductionism, the limits of scientific knowledge, and the justification of values. These philosophical arguments extended far beyond the technical problems of humanist philosophers in the academy, shaping the social conflicts that the movement as a whole encountered. The visionary framing of science played a major role in how humanists reacted in both the academy and the popular realm.

Humanism between Religion and Science

The word *humanism* has been used in many ways over the years. In European history, we talk about Renaissance humanism, referring to the movement of art and culture in that era. In more recent times, people have used the term *humanism* as a synonym for the study of the humanities in contrast to the sciences. Some will hear the word *humanism* and think about humanitarian projects to help the poor and underprivileged. Some twentieth-century Christians even adopted the term to emphasize their focus on the specific human aspects of their theology. All of these are quite valid uses of the word, but they do not refer to the idea that forms the focus of this book.

The humanism discussed in these pages, by contrast, is an outgrowth of liberal modernist thought, a view of the world that emphasizes human dignity, democracy as the ideal form of government, universal education, and scientific rationality. As a broad philosophical perspective, it is hard to overestimate its influence on the modern world. Intellectuals and activists for more than two centuries have embraced this outlook, which can be traced back to the Enlightenment. In this usage, much of what we think of as fundamentally modern is also humanistic; the two seem inseparable. To one extent or another, most of us in the West are deeply influenced by these ideas, even if we would not consider ourselves humanists.

The stories told here follow the activities of a relatively small group of self-declared humanists, people who wholeheartedly embraced this general Enlightenment outlook and who became its defenders and promoters. This group originated in the early decades of the twentieth century as a clearly defined movement within American liberal religion. It flourished within Unitarian congregations and Ethical Culture Societies. There were humanist assemblies scattered throughout the country—in New York City, in Chicago, in Minneapolis, in Los Angeles, and in smaller cities in the South and Midwest. And in each of these places, leaders gave sermons on Sunday and ministered to their congregations the rest of the week. The most distinctive aspect of these meetings was that the sermons never referred to the Lord, the Heavenly Father, or any other supernatural deity. God was not relevant to the message that these men were preaching. We were living in a modern scientific age. Religion, they thought, should focus on people and their lives here and now, not on mythical ideas believed in ages past. God was merely a distraction from the things that mattered most. The idea may have seemed radical for a majority of the population, but it was not marginal. It thrived in some of the wealthiest and most well-educated communities in the country. Those who came out of the liberal religious tradition understood humanism as "the next step in religion," which was the title of one of the earliest apologies for religious humanism by the philosopher Roy Wood Sellars, published in 1918.[8]

Although these early adherents considered their outlook "a religion without God," that way of thinking about humanism did not satisfy all humanists. Many humanists explicitly rejected that line of thinking. Indeed, the idea that humanism was simply another religion became ever more controversial as the century wore on and sharply divided members of the movement. The ardent New York socialist Corliss Lamont—one of the American Humanist Association's wealthiest and most loyal patrons—was in this camp, arguing that the word *religion* was so strongly identified with supernaturalism that humanists should avoid it altogether.[9]

The secularists described humanism as a philosophy or a way of life rather than a religion, but the practical distinction between the two viewpoints was sometimes obscure. Lamont, for example, wrote wedding vows and funeral orations for unbelievers, yet he still objected to using the word *religion* to describe his brand of humanism. In other words, the division was ideological, not based on practice. Eventually the question assumed legal significance, and a number of landmark US Supreme Court cases strongly influenced how humanists saw themselves, as well as how others characterized them. When fundamentalist Christians attacked "secular humanism" in the 1970s, they called it a religion, drawing on some of these legal decisions to make their case that the state was unequally favoring humanist religion over traditional Christianity.[10]

In short, the humanist movement had both religious and nonreligious characteristics, but in all cases their worldview was more than a political philosophy or a scientific outlook. It was broad and touched on all aspects of life—the personal, psychological, and ethical—and it was associated with the building of institutions and with activism and outreach of many kinds. This comprehensive outlook grounded humanists, performing that typically religious function of offering adherents a "sacred canopy"—in the terminology of sociologist Peter Berger—that is, an integrated picture of the world.[11]

That the organized movement grew out of American liberal religiosity is deeply significant as we shall see, but the religious question is not my central concern here. Instead, it is the role of science in humanism that drives the narrative. This, I argue, gets closer to the core of the humanist tradition than their various debates and disagreements about religion. Looking at the ways in which science became aspirational for humanists helps us understand what humanism was about and how it was able to embody such a multifaceted place in American culture.

I treat humanism as both an intellectual and a social movement, and the narrative of this book follows the trajectories of a few major humanist organizations. The men and women who have built and operated these institutions tell us about a world that has not been well studied. This is partly because it is so complex and crosses so many boundaries—disciplinary boundaries within the academy, professional boundaries that divide the religious and the secular, social boundaries that separate lay organizations from academic and religious ones. Several institutions formed the scaffolding that enabled humanism to survive and flourish: the Unitarian Church (which is now called Unitarian Universalism); the Fellowship of Religious Humanists and its successor, the American Humanist Association; the Committee for the Scientific Investigation of Claims of the Paranormal (which is now called the Committee for Skeptical Inquiry); and the Council for Democratic and Secular Humanism (now simply the Council for Secular Humanism). Other groups that were also important, both formal and informal, are introduced in due course.

This volume goes beyond much previous work on the history of humanism in a few different ways. First, it places the humanist story into the larger context of American intellectual history. To understand humanism fully, we need to see how it has related historically to the many other philosophical and religious movements of its time, how humanists have engaged and responded to major concerns of the day, and indeed, why humanism matters so much, as I think it does. Second, unlike many histories of the American humanist movement, the book documents the institutional changes beyond the first decades of its founding, taking us into the second half of the twentieth century, which witnessed the movement's growing cultural influence.[12] Third, while there are many histories of freethought, atheism, and secularism in the United States, few pay much attention to organized humanism. In these works, humanism has been treated sporadically and most often as a footnote. Though humanism's outreach has been less aggressive than many other more in-your-face atheist and freethought groups, I believe it to have been more effective and influential in its long-term impact on the culture.[13] This can be seen, in large part, in the diverse, well-established membership and coterie of fellow travelers—people like Carl Sagan, Ernst Nagel, Margaret Sanger, John Dewey, B. F. Skinner, Abraham Maslow, and A. Philip Randolph. These prominent men and women have brought to humanism goals and ideals that transcend a strict secularist agenda, giving the movement as a whole a broader area of influence and connecting it more closely with the country's long liberal and progressive traditions. Finally, by focusing on science and philosophy, this book explores a different series of questions than other works that focus primarily on the issues of religion and secularism, and it highlights concerns that are especially close to those of the humanists themselves.[14]

This study has a deep payoff for students of American culture, and especially those interested in the role of the academy and the authority of science. The social relationships that unfolded over the century help to explain how and why humanism thrived. The humanist movement crosses so many boundaries and involves so many prominent American thinkers that its story reveals many tacit social relationships that undergird our world. Indeed, I argue that it was largely because of the intertwined relationships that humanist ideas and outlooks gained such traction in different areas. The social status of both the academy and the church helped to elevate humanism's cultural significance.

The school and the church, two of the foundational institutions of the country's civil society, have given rise to a national conversation that shapes our culture—albeit one that sometimes seems recalcitrantly fixed in place— and this book helps to show how that development has occurred. In particular, it explores some of the surprising ways that academic and religious cultures have intersected. As other historians have shown, an extraordinary synergy developed between liberal religion and academia early in the century that had far-reaching ramifications, shaping the way we talk about religion and science even today. This small group of prominent scholars and scientists grappled seriously with the nature and role of religion and its ties to modern scholarship. Even readers who are not interested in the humanist movement per se should recognize the wider relevance of this story to the history of higher learning and faith in America.[15]

The Scientific Spirit of American Humanism

My narrative is guided by the ways in which the humanists used and thought about science. I frame the history in three distinct periods. The first period, which gave birth to humanism, was dominated by ministers and philosophers, and it was they, radical Unitarians and protégés of the influential pragmatist John Dewey, who most influenced the movement, establishing the philosophical foundations of humanism during the high tide of liberal American Protestantism. The second period, which spanned the middle of the century— the World War II and early Cold War eras—witnessed a transition from a primarily religious movement to a more secular one. As the leadership secularized, a number of well-known scientists began to replace clergymen and philosophers as the movement's figureheads. As the number of ministers declined, so too did the Deweyans, which opened the way for new philosophical influences. The third period, from the early sixties to the late eighties, ushered in an even more diverse leadership in which lawyers and secular activists vied for influence alongside a new cohort of philosophers from the positivist and analytic traditions, as well as an ever-increasing number of prominent scientists. It was during this period that the cultural revolutions of the 1960s were sweeping the country—the youth movement, the counterculture, civil rights marches, and antiwar protests, as well as the less noticed, but equally important, rise of politically engaged conservative Christians. The cultural changes shocked humanists like everyone else in this period, and one result was a boisterous struggle over the nature and role of science. Many of the foundational assumptions that had undergirded humanism at its outset were being called into question, and this forced individual humanists to take sides in the wider culture wars.

Intellectual shifts in the academy were especially important in the way that the humanist outlook changed over the years, for it was scholars—especially those in philosophy, psychology, and the natural sciences—who provided humanism with authoritative worldviews that gave their faith a metaphysical and ethical foundation. The humanists were especially indebted to the two schools of naturalistic philosophy mentioned earlier—pragmatism and positivism—and the changing fortunes of these very different outlooks affected the way humanists responded to cultural shifts around them. In humanist hands, these two outlooks were much broader than the philosophical schools that go by those labels, and their proponents were often not philosophers at all. Out of Deweyan pragmatism came the idea that science was a natural outgrowth of the human quest for understanding, an evolutionary result of mankind's complex mind. For humanists steeped in this tradition, the scientific enterprise was simply the most rigorous and institutionalized form of common, everyday thinking and problem-solving. At the same time, Deweyans also found science to be a noble enterprise, insofar as it was tied to the highest human values, qualities like courage, individual responsibility, concern for others, and the democratic frame of mind. In subsequent years, humanistic psychologists, drawing on a different set of scholarly traditions, picked up on this outlook and embraced the idea that the human sciences were paths to understanding and affirming the complex and multifaceted lived experience. And, as we see in chapter 8, one humanist writer influenced by them even considered science to be an outgrowth of motherly love.

By contrast, the positivist outlook that became so influential in the second half of the century considered scientific practice to be fundamentally different from other human forms of thought, which were deeply fallible. Rigorous scientific method was not immune to error, but it set up various kinds of checks to guard against the cognitive and perceptual mistakes that were so common in most areas of human thought. Science's success was tied to its adherence to math and logic and its rigid experimental and observational procedures, practices that set it apart from the ordinary human endeavors. For positivists, science was utilitarian and mechanical and offered a means of escape from the organic and primitively human. Though positivism has often been associated with the idea that knowledge is value neutral and divorced from political and social beliefs, in fact these positivistic humanists were no less visionary than the pragmatists. Science was humanistic for them precisely because it could transcend our human faults and help us rationalize our world. They were just as adamant as the pragmatists about the importance of science for a more humane and orderly world.

The positivists used science very differently than the pragmatists, and those differences help us to understand why humanists responded in such diverse ways to changes in American culture. When humanism emerged in the early years of the twentieth century, the dominant liberal religious environment of the period was in many ways quite receptive to their ideas. The aspirational rhetoric that was common among the pragmatists gave them a means of attracting intellectuals and other liberals in search of new ways to express their religiosity. Later in the century, the liberal climate gave way to more orthodox religious views. American institutions throughout the country began to become ideological battlegrounds where conservatives and liberals fought for control. In that climate, humanism was forced to play a much more defensive role than it had before. Positivism, I argue, provided humanists with better tools for combatting these opponents than did pragmatism. In the hands of the positivists, science could be employed as a weapon in the culture wars to attack the superstitious, mystical, and antiscientific ideas that were being promoted by conservative religion. It was a tactic that turned out to be especially useful after the late 1960s, when humanists all of a sudden were threatened by aggressive and vocal opponents.

The changing tactics proved to be a source of tension within the movement itself. The positivist outlook often supported a reductionistic view of human nature that was anathema to many humanists. At issue in these internal differences were a host of different issues: the propriety of eugenics, the acceptability of behavioristic psychology, the legitimacy of claims about the biological inferiority of African Americans, and the value of attacking astrology and other popular pseudoscientific ideas. Those philosophical differences influenced the very character of the movement. Humanists faced the world very differently based on how they perceived the fundamental nature of mankind, in either generous or pessimistic terms. Those who had trust in the fundamental goodness of human nature spoke with a rhetoric that elevated the aspirational elements of science and humanistic values, whereas those who saw humanity in pessimistic terms—who believed we needed the strong arm of rationality to pound us into good sense—promoted a hard-edged and belligerent response to their opponents. Neither force prevailed entirely. Each one retained a place in the humanist movement. Moreover, those differences in outlook were often outweighed by the humanists' overall agreement on the importance of science and the authority of scientists in helping to shape a new and better world for humanity.

We see elements of humanism as far back as Thomas Jefferson. Reading Jefferson, one can sometimes forget that he was writing two centuries ago. He was effusive about the importance of science to the new nation in so many different ways. Science was not simply a utilitarian tool for Jefferson; he believed that the philosophy of science was integral to public welfare. He praised "the value of science to a republican people" because it gave us security "by enlightening the minds of its citizens." Far from being a disinterested, rigid, and elite endeavor, science was closely linked to the public good: it inculcated virtue and was linked to "power, morals, order, and happiness."[16]

The central thread of this book traces the trajectory of the visionary outlook I have called the scientific spirit of American humanism, the notion that there is more to science than simply the knowledge it provides us. The very phrase *scientific spirit*, which was used by humanists themselves, suggests this point of view. It indicates that science and scientific ideas are tied to broad human concerns. Science is not just descriptive; it can have prescriptive aspects to it as well. There are moral codes and ideals embedded in the very core of what it means to look at the world from a scientific perspective, and these can give us guidance and suggestions about who we are and how we might live our lives.

The scientific spirit not only describes a centerpiece of humanistic thought but also reveals aspects of American culture. In this respect, I hope the book opens up new ways of understanding the larger role of science in the twentieth century, since it helps us to see some of the contours of the broad tradition of American liberalism—in which science plays such a prominent role—and illuminates ideas and events that continue to inform our world.

CHAPTER ONE

Liberal Christianity and the Frontiers of American Belief

The Almighty Lecturer, by displaying the principles of science in the structure of the universe, has invited man to study and to imitation.

Thomas Paine (1794)

Icon of the American Revolution Thomas Paine held radical ideas on religion just as he did on politics. Although he was more iconoclastic than many Americans, his general outlook, deism, was not uncommon at the time, especially among the well-read. Several of the country's founding fathers, including Benjamin Franklin and Thomas Jefferson, echoed his views on religion. Deism was a radically stripped-down version of theism, and it offered believers a faith in a philosopher's God: the world was created by a deity who was purposeful, benevolent, and rational. The cosmos he created was governed according to logical and mathematical principles. Science—"natural philosophy" as it was frequently called—was the primary means for understanding God's Creation, and it was sufficient. Deists did not believe in special revelations or miracles.[1]

Paine's writings are among the most famous expositions of American deism. Writing in France during the darkest hours of the Terror, he set down his ideas on God and religion in a book that would become a classic for free-thinkers and religious radicals from that time on. Rhapsodic in some places and bellicose in others, *The Age of Reason* established a tone that has been admired and emulated by his followers. In it, Paine expounds the moral, philosophical, and scientific principles of the deist worldview.[2]

His positive and inspirational message revolves around the idea of God's benevolence and the beauty and perfection of nature. It is an expression of religion that is called *natural theology*: essentially, the drawing of theological and religious ideas from the study of nature under the assumption that God's design of the natural world is the best means of understanding and appreciating God himself. This is an idea that was common among devout Christians of his time as well—indeed, the concept goes back to ancient Greek times and is a central idea that appears again and again in Western religious thought. What makes the deist form of natural theology so different and radical with respect to Christianity is that the deists believed that this was the primary path to God's truth. While Christians believed that revealed truth as found in the Bible was the core of their faith, deists worshiped God by using reason. Paine emphasized this when he stated that "natural philosophy is the true theology." It was an obvious idea if you accepted science as "the study of the works of God, and of the power and wisdom of God in his works."[3]

Behind this positive message in *The Age of Reason*, however, was a deep disdain for Christianity, and the feeling was mutual—contemporary Christians considered deists anathema to their faith. Deism sprang up as a protest movement against Christianity. Nowhere was this more evident than in their different understandings of mankind and his purpose on earth. According to the deists, God created human beings with minds capable of understanding the rational order of the world he had also created. The human mind is thus an integral part of his Creation, for it allows us to recognize its beauty and rationality. By contrast, the Christians Paine criticized tended to be suspicious of the intellect, believing that mankind needed an external hand to guide them. The deists were appalled by such thinking and were not afraid to say so.

A second aspect of deist theology that challenged the Christian understanding of man was its insistence on the absolute transcendence of God and his utter dependence on natural law. The deist God did not perform miracles. Paine thought that the very idea of having God perform miracles was evidence of the impoverishment of Christianity: claiming that God performs miracles on earth, Paine stated, "is degrading the Almighty into the character of a show-man, playing tricks to amuse and make the people stare and wonder."[4] The ethical corollary to the rejection of God as a miracle maker is that human beings were on their own. We were created to help ourselves, not to pray for special dispensations. It was simply pitiful, the deists thought, that Christians had to seek external relief from suffering.

The deists also had an intense disdain of scripture, which they thought could not be trusted. Paine claimed that the Bible was simply a collection of made-up tales. The story of Christ's resurrection, for example, is described by Paine as baseless: "The story, so far as relates to the supernatural part, has every mark of fraud and imposition stamped upon the face of it." How can one trust a book that provides such "fabulous" tales and does not provide any critical evidence? Religious faith, the deists argued, must be based on reasonable ideas and verifiable evidence from the world itself. It must also follow from man's own abilities to reason about right and wrong.[5]

In the end, this mix of an iconoclastic rhetoric against Christianity with inspirational ideas about nature, reason, and science created an outlook that captured the imaginations of several generations of American men and women who had a rebellious spirit. Thomas Paine, with his knack for capturing the essence of this sentiment in writing, remains to this day a popular author for those who find the ideas engaging. Today, one can still find small groups of deists around the country who embrace this set of religious beliefs, people who accept the existence of God as a rational and benevolent being and who believe that scientific thought can illuminate the order of God's Creation.

I begin this chapter with Paine and deism because this story captures several key elements that have characterized many of the radical religious movements in America since its inception. In this respect, the deism of Paine's time has a great deal in common with the humanist movement of the twentieth century. From the attack on Christianity to the reverence for science, from the disavowal of miracles to the assertion that people must not rely on God to fix their problems—in all of these ways, deism reflects ideas that would recur again and again in America in both religious and secular forms. Humanists would adopt these ideas in their own way. In this chapter, I will chart a course through nineteenth-century radical religion by surveying several of the movements and institutions that proved instrumental in laying a foundation for American humanism. Overall, one can discern three major movements over the course of the nineteenth and early twentieth centuries: populist radicalism (of which deism is a prominent example), denominational radicalism, and academic radicalism.

This radicalism was an explicit attack on religious orthodoxy, and its many separate movements, a few of which I discuss below, illuminate America's noisy and vibrant radical religious tradition. The point to keep in mind throughout is that the majority of conflicts were not between science and religion, or between atheists and believers, but rather between radicals and conservatives, all of whom thought they were defending religion. The catch was that each side thought that their religion was the true or best one. In other words, religion was used to agitate for radical and liberal ideas as often as it was used to defend conservative and orthodox ones. Religion for the liberals expressed a faith in progress and transformation rather than in tradition and dogma. Moreover, although science and atheism were not irrelevant to the debates between these different religious positions, they played subsidiary roles that were part of larger arguments about the nature of the world and the nature of God.

Populist Radicals: Free Thinkers and Atheists

The most widely quoted statement from Paine's book—"My own mind is my own church"—aptly set the tone for what came to be called freethought, whose advocates, aptly called freethinkers, aggressively attacked everything from "Romanism" (that is, Roman Catholicism) to slavery and obscenity laws.[6] They attacked religious orthodoxy either because it was irrational and unscientific or because it seemed to them to be distinctly immoral, advocating and praising behavior that they deemed unacceptable. A large group of freethinkers in America came from the lower classes and embodied a rather fierce and iconoclastic attitude toward the pretensions of the church. They also tended to advocate an arch-individualist attitude that sanctified freedom of conscience in all forms. Many of these freethinkers were activists who held meetings, published newspapers, or agitated against the oppressive behavior of religion. However, although there were active communities of atheists, there was no major national organization, and so these groups had virtually no social clout.[7]

As the United States expanded westward, freethought societies spread with it. German immigrants, in particular, promoted this type of iconoclasm. Men and women who had been radicalized in their native country formed groups called *Freie Gemeinde* leagues, which published radical newspapers that regularly ranted against conservative political and religious beliefs. Many of these immigrants settled in frontier communities in places like Wisconsin and Missouri and established their own societies in the years following the Civil War.

The ornery dispositions of many of these freethinkers made them ideal activists, and many of them joined radical social and political movements that were not necessarily tied to unbelief or religious unorthodoxy. The abolitionist and women's rights movements, for example, attracted freethinkers who blamed the social cancers of slavery and sexual inequality on the dogmas of the church. In these groups we find leaders like William Lloyd Garrison, Lucretia Mott, and Elizabeth Cady Stanton. Garrison, who was the nation's most notorious abolitionist, held quite unorthodox religious ideas. He had been deeply influenced by Paine's views on rational religion and on the importance of individual conscience; he besieged the organized churches with the same venom that he attacked the institution of slavery.[8]

Stanton was another widely known freethinker of her era. An agnostic, she opposed traditional religion for its extreme conservatism. She was a founder of the woman's suffrage movement and, like Garrison, attached her social agenda to her religious iconoclasm, sometimes to the dismay of fellow activists. Her religious heterodoxy, for example, led her to announce in a famous statement her reluctance to participate in religious gatherings: in 1878, on the eve of the first vote in the US Senate on women's suffrage, she commented to a friend, "I did not attend the prayer meeting [at the close of the convention] for, as Jehovah has never taken a very active part in the suffrage movement, I thought I would stay at home."[9] Stanton's unbelief was most fully developed in her book entitled *The Women's Bible*, which was designed to correct and eliminate misogynistic passages that permeated the Christian Bible.[10] Her religious radicalism was strengthened by her adoption of Darwinism, where she found more ammunition for her attack on the male-dominated religious and social establishment.[11]

These freethinkers were not, for the most part, atheists. Some were deists who accepted the existence of a supreme Creator. Some, like Garrison, thought of themselves as evangelical Christians. And, of course, there were a few who gave up on the supernatural altogether: Stanton, for example, simply called herself an agnostic. In general, these freethinkers were social and political radicals who sometimes merged science-based beliefs with their fight against injustice, but on the whole, scientific ideas played only a minor role. Most were driven by feelings of moral injustice that were elicited by their understanding of the church and its dogmas.

Denominational Radicals: Unitarians

Not all religious radicals were as anticlerical as the freethought and allied groups. Indeed, the churches themselves often fostered a bourgeois radicalism among the educated and professional classes. Recent historical literature on the origins of unbelief illuminates how some of this occurred. New ideas developing within religious groups—rather than in opposition to them—were crucial in establishing this bigger framework. The setting, in this case, made all the difference. That a well-connected, educated, upper-middle-class gentlemanly culture adopted these radical new ideas meant that they now had a home in a stable institutional environment. Thus, the atmosphere of the church often made the radicalism quite different than in the more individualistic freethought movement.

One of the most prominent and influential of America's denominations was the ultraliberal Unitarian Church. As with deism and freethought, the Unitarian emphasis on the authority of reason over dogma shaped its character. Even the name of the denomination is a reminder of this rationalist legacy. It comes from the term *unitarian*, which refers to the rejection of the Christian doctrine of the Trinity. That doctrine had long been considered by the orthodox as one of the most essential concepts from early Christian theology: the central mystery of God's existence as a single divine being composed of three distinct persons. That mystery set the Godhead apart from the world as something unintelligible to man by human reason alone. Those who placed a high value on rationality in religion found the Trinity doctrine unacceptable. All of these anti-Trinitarians were considered heterodox, and in many areas of the Old World, going back to the Middle Ages, they were persecuted. By the nineteenth century, a more pluralistic and legally tolerant religious environment in Europe and America made it possible for this theological position to flourish. In particular, in America it survived and thrived, becoming the centerpiece of a separate denomination. These Unitarians believed in a rational God and in the comprehensibility of the Creation. Like the deists, they characterized the religious mysteries as obfuscations and corruptions of religion.[12]

William Ellery Channing, the spiritual father of American Unitarianism, expressed the idea in these terms: "God has given us a rational nature and will call us to account for it."[13] Channing made the same point in a sermon in 1819, where he argued that the only path to the understanding of the scriptures was through reason. Scripture was written in ordinary language with ordinary words, and the only way to understand it was through careful study of that language with the rational faculties that God gave us, for "we honor revelation too highly to make it the antagonist of reason or to believe that it calls us to renounce our highest powers."[14] The sermon went on to set forth two key principles of Unitarian belief: the unity and moral perfection of God and the ethical necessity of the individual's freedom of conscience. His main point, though, was about the intellect. Reason, Channing argued, was the proper means of understanding God's Word; reason and science were the most important paths to the Christian God.

American Unitarianism emerged in the early nineteenth century, breaking away from the Massachusetts Congregational Church. Since that time, it has been at the vanguard of American religious change. Unitarians have always composed only a tiny fraction of America's churchgoing population, yet they have been one of the country's most elite and prestigious denominations.[15]

Polls of religious groups have shown that Unitarians have ranked high among American denominations in income, education, and professional status, a characteristic of the group since its formative period among the wealthy elite of Boston, Cambridge, and Harvard. This has given Unitarianism a great deal of cultural clout despite its relatively small membership. Moreover, notwithstanding their radical theology, Unitarians have retained their place as respected members of America's church-going population.[16]

Unitarian theology arose as a reaction to conservative Calvinism in Congregational churches in colonial New England. The conservatives believed that all authority and all salvation rested in God's hands. They depicted God as a severe judge whose thoughts were incomprehensible to the human mind. However, by the early years of the nineteenth century, liberal ministers were portraying God in very different terms: as a gentle, wise, and benevolent deity who worked according to reasons that were accessible to human understanding.[17] These liberal ideas gradually gained ground in Massachusetts, and the movement eventually succeeded in getting one of their own appointed to the influential Hollis Professorship of Divinity at Harvard College. The liberal movement spread out from there, and in 1825 liberal ministers officially separated from the Congregationalists by forming the American Unitarian Association. This was a key moment in the history of religious liberalism in America, for within the space of a decade, well over a hundred Congregational churches had become Unitarian.[18]

The Unitarians had clearly won the battle. They took the lion's share of the Congregationalists' property, money, and educational resources. The value of their church property alone placed them far ahead of all other denominations in America at that time. In fact, the value of properties owned by the average Unitarian church outstripped that of the second-ranked Episcopalians by almost two to one. Moreover, the Boston Unitarian churches claimed the city's most important merchants, lawyers, and men of affairs as members. All that, as well as control of religious education at Harvard College, gave Unitarianism a degree of prestige unrivaled by any other denomination.[19]

The other distinctive, and especially significant, characteristic of Unitarianism was its rejection of creed as a means of determining the parameters of the group. While most other Christian denominations had a fixed statement of belief that enabled them to enforce orthodoxy, Unitarians gave that up. However, unlike rationalism, which was a basic principle of Unitarian theology from the outset, anticreedalism was established only gradually and contentiously after a series of confrontations over the course of the nineteenth century. The radicals continuously wrangled with Unitarian conservatives, and—after many years of schism in the church—all groups finally came to the table to negotiate a reintegration, arriving at the understanding that there would be no single dogmatic statement of faith required for fellowship. This accommodation changed the face of the denomination by opening it up to a wide variety of theological and spiritual forms—including, ultimately, humanism. The story of that decades-long conflict and its resolution is instructive.

The conservative-radical divide showed up quite early in the denomination's history. In the 1830s, the iconoclastic Ralph Waldo Emerson tested the limits of Congregationalism's tolerance. After giving a sermon at the Harvard Divinity School that was unorthodox even by Unitarian standards, Emerson was expelled from the ministry and denied further engagements at the Divinity School, where he had been ordained. Less rationalistic than Channing, Emerson's more romantic conception of religion was nonetheless pathbreaking insofar as he questioned the personal God of Christianity and adopted a pantheistic-type reverence for nature as divine in and of itself. This position was a far cry from the rational Godhead that Channing had pronounced, and one that many Unitarians considered to be beyond the pale. What resulted was the Transcendentalist movement, which followed the more romantic and individualistic thinking of Emerson and other New England visionaries. Emerson's thought matured over the next few years, becoming influenced by Hinduism and other forms of Eastern religion, and he ended up on the margins of American faith and could no longer be considered a Christian by any standard sense of the term.[20]

Emerson was not alone in his thinking. The ferment of ideas in the first half of the nineteenth century had given rise to non-Christian religiosity of many kinds, and as the Civil War was coming to a close, conflicts within Unitarianism flared up yet again. In April of 1865, just days before the Confederate forces surrendered, the Unitarian clergy met to establish the National Conference of Unitarian Churches—the first coordinated administrative body for the denomination. This conference became the site of the first major struggle between the radicals and traditionalists.[21]

Though all Unitarians were radical in the sense that they rejected Trinitarianism, the conservatives among them remained committed to many aspects of Christian belief, and all accepted a belief in a personal God. Given Emerson's example, the conservatives remained concerned about the slippery slope of radicalism within the denomination, which helps to explain their long-standing and adamant defense of a creed. The devoutly Christian Unitarians believed that a biblically based theology must remain central to the faith. Step beyond this, as Emerson had done, and one would enter the realm of heresy. Christianity remained the bedrock of their faith; accordingly, they could not sanction non-Christian beliefs.

By contrast, the radicals lobbied for non-Christian fellowship. They didn't mind affiliating with Unitarians who held a biblically based theology—indeed, many of them were close friends and colleagues—but they wanted alternative religious expressions to be tolerated and accorded the same respect they gave the conservatives. Unitarianism should be a big tent, they thought, open to the vibrant liberal religious faiths that flourished in churches and towns across the country. Yet when they proposed reforms to the conference's constitution that would open up the denomination, they were defeated. The final wording of the charter at the conference identified Unitarians as "disciples of the Lord Jesus Christ" who worked "in the service of God" toward "the building up of the kingdom of his Son."[22] A second conference the following year saw a similar defeat. Frustrated and with no other path, the radicals walked out.[23]

The following May, these radical Unitarians joined with an eclectic group of religious dissenters including Universalists, Quakers, liberal Jews, and even spiritualists. By the end of the meeting, the Free Religious Association had been born. The notion of *freedom* was a rallying cry for the assembly. It had been only two years since the conclusion of the Civil War, and the political associations of that word continued to reverberate. They considered the spiritual battle they were currently fighting to be a continuation of the fight for emancipation: one contemporary writer even characterized the group as "a spiritual anti-slavery society."[24] The main goals of the Free Religious Association were the abolition of religious servitude to doctrinal Christianity and the right to speak freely within the group. The members demanded complete tolerance of religious opinion, no matter how unorthodox. Before long, a second Unitarian association, the Western Unitarian Conference, which was free of the powerful conservative voices from the older churches in New England, embraced the radicals and adopted a noncreedal constitution.

By the end of the century, the radicals finally overcame conservative resistance in the National Conference as well. The radicals who had walked out in Syracuse remained active in their congregations—indeed, many of them had never formally left—and year by year, death claimed the conservative opposition.[25] Finally, in 1894 the National Conference amended its charter to make all Unitarianism welcoming to non-Christians. Even though their new credo still affirmed the "Kingdom of God" and the "religion of Jesus," it also stated that "nothing in this constitution is to be construed as an authoritative test," which was the crucial point.[26]

The rejection of creedalism within Unitarianism would end up becoming one of the denomination's most important characteristics, for while other denominations could be called liberal, the Unitarians explicitly rejected affirmation of a creed as a criterion for fellowship. They would ultimately adopt such an open, tolerant attitude that it became impossible to point to any single viewpoint as "Unitarian." Thus, by the early twenty-first century, membership in the Unitarian Universalist church has come to include people whose outlooks range from liberal Christianity, to eco-spirituality, to scientific humanism, and even to self-proclaimed Buddhists and pagans.

One of the more influential radicals in nineteenth-century Unitarianism was Francis Ellingwood Abbot. Rationality and science played a dominant role in his thinking. In contrast to Emerson, who was more of an intuitionist in his theology, Abbot praised science rather than nature. Science, he claimed, was "destined to be the world's true Messiah." Indeed, in Abbot's hands the very word *science* became as much a slogan as a descriptive term. After completing a degree at Harvard in 1859, he attended the Unitarians' Meadville Theological School in Pennsylvania, where he discovered evolution and began to develop a science-based theology. Abbot promoted what he called "scientific theism" and argued that God was to be found through the concrete and objective methods of science. "If faith in God is good for anything,—if it is based on <u>truth</u>,—I fear no harm to it from the broad daylight of science." But he went much further. Not only should modern religion accept the newest discoveries, but it also ought to courageously "throw itself into the arms of science."[27]

The year he graduated, 1859, was the year in which Darwin's *Origin of Species* was published. Abbot became an immediate devotee of evolution and based his theology on the ideas of British philosopher Herbert Spencer and the German scientist-popularizer Ernst Haeckel, both of whom considered evolution not simply a biological process but a universal law of matter. Evolution, said Abbot, "is the glory of science." For him, the centrality of evolution meant that the universe was "dynamical" and in flux. Like Emerson, he made little distinction between the deity and nature as a whole; his God was an immanent being that was *in* the world. Also, like Emerson, his romanticism shaped his depiction of nature, which he believed was infused with spirit: he spoke of "a universal Intelligence in Nature of which the most resplendent souls of men are but tiny sparks."[28]

From Channing to Emerson to Abbot, Unitarianism broke boundaries of many kinds, and all this occurred within the confines of an established American church. Unlike the deists and freethinkers, the men who broke new ground within Unitarianism were professionals, well-educated ministers who tended congregations of relatively well-to-do, cultured families. No matter how radical the theology, the setting was a respectable one composed of churchgoers and men of the cloth. The theology was strongly influenced, as we saw, by the internal politics of the denomination. Without the final removal of creedal tests for Unitarians in the Western and National Conferences, the more far-flung theologies of ministers like Abbot would have remained marginalized or been expelled altogether as heretical and against the creed. Instead, the denomination came to be recognized for its liberal tolerance and acceptance of extreme variety in American religiosity.

Although the Unitarians used the terms *rational, nature, evolution,* and *science,* these words took on special meanings in a theological context. Channing's "rationality" was not simply a means of understanding things in the world or an approach to mathematical truths; it was a way of understanding God's mind. Rationalism was the key to his theology. Similarly, when Emerson and Abbot spoke about nature, they were referring to a deeply spiritual concept that was nothing like how materialist philosophers used the term. This was not naturalism, but instead Nature with a capital N, an immanence to be revered. When Abbot wrote of "evolution," the term took on a significance far beyond the simple fact of biological change. Like nature, his evolution was a teleological and optimistic concept. And as for Abbot's use of "science," in his hands this became a term that grounded his theology. It pointed to a new authority that trumped the Bible, and therefore a new way of coming to know God. His use of the word was not analytical or rigorous, but it was extremely important in capturing the essential foundation of his faith. Even though the way these ministers used these words differed from that of the philosophers and scientists, it would be foolish to think that actual science was unimportant to the theologians. The increasingly authoritative world of science and academic study had encouraged their use of these notions. Without the ongoing scientific quest to understand the world, the Unitarian theologians would have had nothing to draw on.[29]

Academic Radicals: Protestant Modernists

There was one other major source of radicalism within American mainline churches: the seminaries. The schools that trained clergy were just as deeply influenced by naturalism as was the secular academy, and the result was what came to be known as Protestant modernism. Most of the scholars and clergy that espoused modernist theology were not nearly as heterodox as the radicals among the Unitarians and freethinkers, but they did form an intellectual bloc that produced unorthodox interpretations that created ever-greater challenges for traditional Christian belief. Their ideas were heavily influenced by the philosophical and scientific concepts of the period. Also, it was important that these ideas came to be situated in the academy because just as the denomination gave critical prestige and support to Unitarian radicals, so too did the academy support modernist theology, giving it similar cultural stature.

The move toward modernism happened in the second half of the nineteenth century, when Protestant theologians across the denominational spectrum nurtured new interpretations of scripture based on scientific and historical findings. American seminary students, many of whom studied in Germany, found the new discoveries in history and natural science exciting; they saw in this scholarship the word of God as written for an age of science and progress. In particular, a new way of studying religion, called higher criticism, was being taught in the German schools. The scholars of higher criticism were naturalists, meaning that they eschewed supernaturalism and relied on new techniques for studying language and history that departed from older philological approaches that had been the foundation of biblical studies in the early part of the century. It is important to emphasize here that this scientific and naturalistic methodology was designed as a means for better understanding, not eliminating, God. In other words, it was a limited form of naturalism, very different from strict philosophical naturalism that implied a wholly material cosmology. Most theologians of this school were confident that modern naturalistic methodology gave evidence for God's existence.[30]

Many of the most well-known preachers in the country at this time—Washington Gladden, Lyman Abbott, and Herbert Willett—proselytized a modernist Christianity. Despite its naturalistic foundations, this scholarship was not a threat to these men's faith; instead, they found it to be an important and intellectually exciting means of supporting their Christian apologetics. Modernism as a theological movement would, once established, remain influential in American theology going forward, but its heyday lasted for about forty years: from roughly 1880 to 1920. During that time, it transformed Christian theology in most of the mainline American denominations. Although in many denominations the most radical ideas seldom saw their way into Sunday morning sermons, the faculties of the seminaries and the students that they turned out were often much more liberal than the families in the pews, and young preachers quickly learned where to draw the line.[31]

Baptist theologian William Rainey Harper became one of the most important popularizers of modernism, not only because of his effectiveness as a speaker and writer but also because of his fortuitous access to the deep pockets of American billionaire industrialist John D. Rockefeller. Harper came to fame as a prized speaker. His enthusiastic championing of modern biblical studies made him popular on the Chautauqua Literary and Scientific Circle lecture circuit. From there, he was able to enroll over sixty thousand people in his Bible study course, where he defended higher criticism and rejected biblical literalism.[32]

Harper explained that all religions, including Christianity, changed over time as they were absorbed into new cultures. Despite the general view that religion was a static tradition with an unchanging dogma, Harper asserted that all religions *evolved* through the action of God's spirit in his people. His word choice was not accidental; like Abbot, he was excited by the new dynamic universe that had been uncovered by Darwin, and so he adopted the term and the concept into his defense of the Christian faith. Unlike many fundamentalists, who would reject evolution and hold it responsible for the erosion of the Christian faith, Harper and other modernists welcomed it and placed it at the center of their theology. Harper embraced the two areas of modern knowledge that many Christians found most threatening: evolution and higher criticism. Conservatives were wrong, he thought—the new scholarship did not undermine Christianity; it bolstered it.[33]

Harper became instrumental in the rise of modernism in America when he was hired as president of the newly built University of Chicago. Based on his success and popularity as a lecturer in, and later director of, the Chautauqua movement, Harper was the man whom Rockefeller turned to for help in establishing his new school in the Windy City. Drawing on his own religious views, Harper molded the university's Baptist Theological Union Seminary into one of the country's preeminent modernist institutions, hiring noted and able theologians like Shailer Mathews, George Burman Foster, and Edward Scribner Ames. All of the seminary faculty were active outside the classroom: preaching sermons, proselytizing, and authoring numerous tracts and handbooks for the general reader. Rockefeller's architectural tastes were medieval, but the school that Harper built was anything but that. The neo-Gothic campus belied the extreme modernism taught there, and it would play a major role in the rise of humanism.[34]

So, what were the specific principles that Protestant modernism was founded on? Primarily, it was a theology immersed in the scholarship of higher criticism, which so many young American scholars had learned when they traveled to Germany for advanced study in theology. The German scholars treated the history of Christianity as being no different from the history of any other human endeavor, and in particular, they did not resort to supernatural explanations. Friedrich Strauss's two-volume *The Life of Jesus*, published in 1835 and 1836, was the seminal work in this tradition. Avoiding supernaturalism in all its forms, Strauss retold the life of Christ. Stripping the Christian story of its miracles, he radically transformed the traditional Christian narrative. It was a bold and dangerous move at that time, one that cost Strauss his professorship, but it revolutionized the discipline, and several decades later, when Harper was picking faculty for the University of Chicago, his ideas had become widespread throughout Europe and America—although they were still extremely controversial in many churches and seminaries. To this naturalistic historical account was joined modern linguistics, which demonstrated that the Bible was composed of confusing and intertwined narratives that frequently contradicted each other. The picture that emerged from all of this was of a patchwork of stories, each with its own internal purposes and based on its own culturally and linguistically distinct sources. The resulting scholarship tended to demythologize Christianity by explaining biblical stories as the product of prescientific minds.[35]

Modernists were not atheists, however, and their revisionism had limits. While this scholarship seemed to undercut traditional areas of theological certainty, its reliance on naturalistic methodology and scientific reason did not prove destructive of faith itself. Even as the Protestant modernists ceased to talk about miracles in any but the most restricted sense, they remained convinced that God's presence and action in the world were evident all around them. Theologians had long acknowledged that God could work his will in the world through secondary causes—that is, he could use the ordinary laws of nature rather than miraculous intervention to accomplish his purpose. A shift in this view had taken place in the nineteenth century, however, and many modernists came to believe that this was the *only* way in which God worked. This view mirrored that of the deists and Unitarian rationalists, and these people believed that naturalism in this form could reinforce the Christian faith.

It should be clear by now that the way that these theologians thought about the world was crucially tied to the rising authority of natural science. Science was important insofar as it encouraged these theologians to revere evidence from nature and the various methods of interrogating it. The modernist theologian Shailer Mathews, whom Harper selected to become dean of Chicago's theology faculty, illustrates the point. He was one of the country's most well-known and influential modernists, a founder of the Northern Baptist Convention, and a president of the Federal Council of Churches.[36] Mathews argued that the Christian religion must accept the new authority of science. In past ages, scripture and other church dogmas were the primary paths to knowledge; however, now we lived in a modern, scientific world and had new means at our disposal for understanding. Christians needed to give primary authority to "the consensus of investigators," as he put it.[37] Empiricism and rationalism trumped revelation as sources of reliable knowledge.[38]

This thinking was especially apparent in his enthusiasm for evolution. Like Harper, Mathews believed that evolution demonstrated the dynamic nature of the world and how it was being transformed by God's action. Through evolution, we could understand better the purposeful unfolding of a divine plan. "Matter itself has ceased to be dead, and has become vibrant with activity," he said. "Our religion thus gets new support."[39]

That wholesale embrace of the evolutionary process reveals much about how modernism differed from more orthodox thinking. Christians across the spectrum responded with a great diversity of arguments to the way that modern science was reshaping the origin story. Evolution was just one aspect of it. Few geologists after 1830 believed that the world was as young as is implied in the Old Testament story of Genesis, and this fact forced theologians to respond. The most orthodox conservatives denied the validity of scientific findings altogether and reaffirmed the Bible's account of the seven-day Creation. More moderate conservatives believed that fidelity to the scriptural account did not depend on a literal reading, and they found ways of interpreting Genesis so that the written story did not contradict a multimillion-year geological record.[40] Liberals, by contrast, were able to talk about the story of Genesis as a literary and mythological text, and many of them considered it primarily a moral fable. Among the modernists like Harper and Mathews, the Genesis narrative was treated in precisely this way, and nothing about it was considered to be relevant to modern scientific knowledge. To them, it was senseless to attempt to reconcile the words of scripture with scientific facts because the Bible was not a science textbook. Natural history should be left to the scientists, they thought.[41]

Protestant modernism was liberal both theologically and politically, and adherents tended to be strong advocates for social action to benefit the poor and downtrodden. Reform efforts of all sorts were embraced: temperance societies, settlement houses, and wage and hour laws were all championed. Ministers and religious activists preached what they called the Social Gospel, justifying it by referring to Christ's teachings regarding the brotherhood of mankind and the care of the poor. Although the Social Gospel was not specifically tied to the liberal theological agenda, many of the modernists embraced it and adopted quite radical—often explicitly socialist—ideals, which tended to make them outliers in capitalist America's political environment.[42]

Mathews, once again, is representative here. His popular book from 1897, *The Social Teachings of Jesus*, illustrates this liberal version of the Social Gospel. In it, he argues that Christians have a duty to work toward the Kingdom of God on earth—which, in Mathews's case, looked very little like a kingdom: God's preferred political system was not monarchy but democracy. This political shift had theological implications: the emphasis now was on people and human problems, not on metaphysical questions or the hereafter. In this respect, the Social Gospel theology tended to turn away conservatives who held that human beings were inherently weak, sinful, and impotent and who believed that too great a focus on the here and now led to a devaluing of God, our ultimate savior.[43]

Modernism and the Social Gospel were squarely within the Christian fold. Unlike the freethinkers, or even the Unitarians, these liberals were working within well-established church traditions. Mathews, for example, was a liberal Baptist, and this made a difference in a couple of ways. First, he was a mainstream preacher in a denomination that had a much greater influence over many more people than was the case for the Unitarians. Also, liberalism was closely tied to the centers of power and prestige in the country. It was not unimportant that the Chicago Divinity School was supported by the deep pockets of Rockefeller. And, of course, Chicago was only one of many seminaries around the country that nurtured modernism in one form or another. This was a widespread academic movement in an age in which academia was growing in stature and importance. So, despite a periodic radicalism that included dalliances with socialism and a strong opposition to Christian orthodoxy, the modernists were part of an institutional framework that, although it was more cautious, was ultimately more socially influential.[44]

Understanding Radical Religion and Unbelief in the Nineteenth Century

The three strands of radicalism that we have looked at have one thing in common: they were all sources of American religious thought that took issue with orthodox Christianity, and in that sense, they all encouraged a change in American religious faith, specifically in its content. The story that I have told about these American radicals is not new, but it does frame the issues differently than other historians have in recent years. Whereas many of these scholars have been concerned about the direction or origin of intellectual change, I'm more interested in its texture.

Historians of agnosticism, freethought, secularism, and radical religion have focused largely on the *origins* of unbelief—why and how atheism, agnosticism, and unbelief arose in a largely Christian culture—as the titles of their histories suggest. The middle of the nineteenth century was a transformative period in Anglo-American culture, in which radically new beliefs developed and much of the old faith was discarded. As one author has recently indicated, it ushered in "a secular age." The question these authors ask, then, is why? What happened to create it? How did it become possible to disbelieve in God?[45] In their answer to these questions, these historians have documented, again and again, how the conditions for atheism and unbelief arose not out of *opposition* to religion but rather as a result of the *defense* of religion, in the face of changing ideas about nature, philosophy, history, ethics, and society. These scholars argue that concerned men and women looked at these titanic changes going on around them and sought means to preserve what, in their view, was the core of religion.

The historian Bernard Lightman, for example, deftly shows that agnosticism, the belief in the unknowability of God's existence, was abetted by the perceived need to purify and reform Christianity. His ideas about agnosticism in Victorian Britain are echoed in works about America. James Turner has shown that when the Protestant theologians attempted to accommodate themselves to modern knowledge, they replaced tradition and dogma with science and rationalism, unwittingly undermining Christianity in the process.[46] In a recent variation of this literature, Matthew Stanley has explored the rise of *naturalistic* science in this period, in contrast to the *theistic* science that had been practiced up until this time; he argues that there occurred a change in the assumptions about God and the natural world that encouraged scientists to set aside their religious beliefs at the laboratory door.[47]

My purpose differs from that of these historians. I want to show the vitality of American radical religion and highlight its institutional foundations. In all of the origin studies, the historians are focused primarily on the changing content of belief, and they make a sharp distinction between belief and unbelief even as they show a gradual path from one to the other. While the content of belief is important in my story, it is the activities of the institutions and people promoting this new thought that help us understand its social status and its persistence. It also helps to demonstrate its normality—radical religion of one type or another was part of the American landscape from the outset. The rise of radical religion was as much a social movement as an intellectual one, and in that social transformation we find a huge mix of ideas. Although many of those ideas fall into categories that we today associate with irreligion and secularity, they were not at the time so considered. The point is that the interplay of the religious and the secular in this period provided the motive force for the religious change that forms the central core of American radical religion.

My characterization of nineteenth-century American religious culture also puts my narrative in direct conflict with some writers who want to simplify American history for polemical purposes. Two forms of this polemical history stand out. Conservatives will sometimes claim that America is a Christian nation founded on traditional Christian values, and that its laws and its moral foundations emerge from pious principles drawn from our Judeo-Christian heritage. Christian teachings, they say, are fundamental to our identity as a nation. The assumptions behind this belief are, I believe, unsupportable, as the evidence I provide in this chapter helps to show. Although traditional Christian ideas, people, and institutions have made enormous contributions to this country, it is a mistake to consider traditional Christianity as the country's primary foundation. Just as important were liberal and radical forms of religion. Unorthodox, and even anti-Christian, ideas were there at the beginning and have played instrumental roles in shaping American values.

Likewise, my narrative opposes a strictly secularist interpretation of American history. There are some writers who will claim that America was, from the outset, a fundamentally secular country; I consider them mistaken, and for similar reasons. There is an oversimplification on this side as well, one that ignores the extraordinary religious motives, both liberal and conservative, that were behind so much of America's ideals and ideology. The misconceptions and oversimplifications on both sides have made for a recurring situation where both secularists and fundamentalists are astonished at the tenacity of their opponents. Neither can quite come to acknowledge that the American fabric contains both strands intricately woven together, and it is that weave that has produced the complex country we live in. From my perspective as an American historian, our nation's past is one of continual dialog and conflict between competing ideologies and religious views; consequently, its history cannot be adequately reduced to one or another dominant theme.

There is one other misconception that I hope this chapter helps to dispel, one crucial to the argument of the rest of the book: namely, the tendency to consider religion an essentially conservative force. I emphasize this because one finds that both secularists and fundamentalists associate religion with orthodoxy. This is reinforced by popular usage, in which the term *religious* is used as a synonym for dogmatism and rigid adherence to doctrine. Yet, as I have shown here, frequently religion is radical. When we think about religion as conservative, we neglect the truly revolutionary role that liberal and radical religion has played in this country. It was not secularism that drove so many of the changes in thinking during the nineteenth century; it was liberal and radical religion. Both churches and people motivated by religious belief have influenced reform and progressivism throughout the country. This chapter has highlighted that liberal and radical heritage and has shown how it set the stage for another truly radical development in the twentieth century: religious humanism.

CHAPTER TWO

The Birth of Religious Humanism

On the one hand, [religion] must accept the conclusions of modern science in every department of learning, and on the other . . . it must use the principles and standards of the scientific method in its own particular inquiries and work.

John Dietrich (1927)

In 1911, St. Mark's Reformed Church in Pittsburgh tried its young minister for heresy. Thirty-three-year-old John Dietrich was charged with preaching doctrines contrary to the Heidelberg Catechism, even though he didn't advocate anything that wasn't already commonly taught in the nation's seminaries. This liberal modernist theology, unfortunately, frequently did not reach far beyond the classroom in many denominations, and in Dietrich's case, it upset enough influential parishioners to get him into trouble. But Dietrich refused to back down, and a trial was set. He was sadly familiar with the situation, having witnessed a similar scene earlier as a seminary student, when professors whom he respected came under fire for holding ideas that the school's wealthy donors objected to. He watched those professors buckle under when their jobs were threatened. Demand for orthodoxy overruled freedom of conscience at the school, and this had deeply disappointed Dietrich. His actions at St. Mark's reflected the lessons he learned from his teachers; unlike his professors, however, he preferred to be defrocked rather than to compromise his integrity.[1]

To the dismay of his many supporters—he was, by all accounts, well liked by most of the parishioners—he made no attempt to capitulate, offering no defense and admitting that his views were contrary to the catechism.[2] That act of will and, as he saw it, daring honesty would define his self-image. In a few years, his story would shape how he understood humanism: that it was a courageous act to renounce tradition and accept the findings of modern scholarship.

It did not take long for Dietrich to find a new church. He was intelligent, talented, likable, and willing to move. Traveling cross-country to the state of Washington, he took a post at a Unitarian church in Spokane. With its values of tolerance and noncreedalism, Unitarianism fit Dietrich well and brought him into contact with a network of liberal scholars and ministers.

His heresy trial had launched him on a theological journey that took him into uncharted territory. It was a journey shaped as much by his study of new scholarship as by his experience as a minster, and it was also deeply informed by his understanding of what it meant to be part of a religious community. After several years of reading and preaching in this liberal denomination, he emerged with ideas more radical than anything he had studied at seminary. His Christianity fell by the wayside, and he eventually abandoned talk of God altogether. At the same time, he continued to believe more than ever in the importance of the ministry.

Dietrich's story was unique in its particulars, but he was not alone in his journey out of orthodoxy and into radicalism. Other nonconformist preachers had also left their home denominations and, like Dietrich, ended up in Unitarian pulpits, each one seeking a way to extend and develop the American liberal religious tradition. Even as these ministers abandoned "God talk," they continued to view their work as essentially religious. Churches, they recognized, were important institutions in American life, and Unitarianism, with its open and tolerant attitude, gave them freedom to explore and preach new ideas. By the end of the second decade of the twentieth century, these men had come together in a loose-knit organization: they called themselves *humanists*.[3]

What these preachers maintained in their sermons was that modern man must base his religion on the best knowledge available—on the ideas and discoveries of modern science—and that religion must live up to the American democratic and progressive spirit. The most characteristic elements of their humanist religion were its emphasis on freedom, self-reliance, and scientific method, three qualities that were closely intertwined in their thinking. This chapter charts the activities of several humanists over the course of the second decade of the twentieth century as they laid the framework for this new religion.

Heretics Find a Home

Two years after Dietrich was expelled from his pulpit, Curtis Reese, a young Southern Baptist preacher, left his church to become a Unitarian minister. Like Dietrich, he eventually abandoned all talk of God in his sermons and began promoting what he called the "religion of democracy." Reese was the son of a Southern Baptist deacon, and his family was full of preachers, from his great-grandfather and grandfather to his uncles and brothers. His own journey away from orthodoxy was, like Dietrich's, quite traumatic, although in a more personal way. When he became a liberal Unitarian, his mother went into a rage: "My mother said very sincerely that she would rather have seen me dead," Reese noted sardonically. "This is understandable, for had she heard of my death she would have had the satisfaction of knowing that I was flying around with angels in heaven. But now she was sure that if and when I died, I would burn in hellfire and brimstone forever and ever."[4]

Having been disowned by his family, Reese moved to a small Illinois town near St. Louis, where he accepted a ministerial post at a Unitarian church. He combined his liberal theology with the aggressive moralizing of his Baptist heritage and crusaded against gambling, prostitution, and pornography; he also joined a campaign to better the housing conditions of the poor. Later, working in Des Moines, Iowa, he sponsored chaperoned parties for the enlisted men at Fort Hood in order to save them from "moral decay."[5] As with Dietrich, Reese took a daunting intellectual leap when he left orthodoxy, but his conservative temperament remained, both in his morally driven activism and in his belief in the crucial importance of institutionalized religion.

For those ministers who adopted, like Dietrich and Reese, a naturalistic perspective, an entirely new path in the realm of religion had to be forged. Although the concept of a religion that had no supernatural being or transcendent figurehead was not new, it was mostly foreign to American soil. In the nineteenth century, the French social scientist Auguste Comte had invented an atheistic religion, what he called the Religion of Humanity, but it never caught on in the United States. This was largely because its liturgy was modeled on Roman Catholicism, which was diametrically opposed to the liberal wing of American religious activity as exemplified by Unitarianism and freethought. In the Orient, of course, one could point to Buddhism and Taoism as ancient traditions without anything like the theism of the West, but these Eastern forms of thought were still foreign and poorly understood in America. People like Emerson drew on some of the religious ideas of the East, but few non–Asian Americans paid much attention to them. So, all in all, the idea of a religion without God was both strange and repellent to most Americans. However, in Dietrich's hands, as well as those of his fellow humanists who sermonized in front of liberal well-educated congregants, the naturalistic project succeeded. The humanist movement that resulted thrived in pockets of liberalism around the country.[6]

Of course, success did not come without a struggle. As we saw in the previous chapter, the new knowledge and an increasing appreciation of scientific method invigorated Protestant theology by secularizing and naturalizing it. The evidence forced these modernists to conclude that some of Christianity's most sacred myths had to be abandoned: Christ's miracles, and—among the more radical thinkers—even the Resurrection. Although young theologians seldom adopted an all-out naturalism, their ideas were disorienting and, to the orthodox mind, utterly heretical. While the modernists believed that it was not just proper but also imperative that Christianity adopt these radical new theologies, the orthodox and the ultraconservative fundamentalists saw only misguided and dangerous ideas. It was because of this sharp division between two visions of Christianity that a turbulence arose within the American churches, one that manifested itself in charges of heresy that drove congregations and families apart. It also provided the catalyst that drove men like Dietrich and Reese out of the Christian orbit altogether.[7]

During the two decades spanning the turn of the twentieth century, heresy trials similar to Dietrich's sprang up across the country. Battles flared in American seminaries and churches as conservatives fought with liberals over who should define Christian theology for the new century. Both sides vied for control of their denominations. The crux of the debates was whether modern knowledge or ancient dogma was key to the survival of Christianity. While liberals proposed a religion that privileged modern science and historical research, conservatives held that dogma and divine revelation should retain their traditional authority.[8]

The intellectual path that Dietrich and Reese traversed—from orthodoxy to modernism, and then to outright humanism—was supported by the work of other radical theologians who taught in institutions that fostered the kind of academic freedom that allowed truly radical ideas to develop, even within an otherwise conventional religious space. At the University of Chicago, the theologian George Burman Foster was one such scholar, and his work deeply influenced Reese. Foster's controversial book *The Finality of the Christian Religion* (1906) referred to traditional Christianity as an "authority-religion," which he contrasted with the modern American ideals of freedom and individuality. By equating Christianity with an authoritarian mind-set, Foster compared it unfavorably to modern American democratic society. He argued that Christians had consistently looked up to God as a divine king, making monotheism not much different from an earthly monarchy. But now we lived in a democracy, and since our values about governance were radically different than in times past, so must our religious ideas develop to keep up.[9] Going beyond Foster, Reese took the idea to its logical conclusion: we must dethrone God altogether.

Dietrich and Reese met for the first time in 1917, at the annual Western Unitarian Conference. There they found that they were both thinking along the same lines, preaching sermons devoid of any supernaturalism.[10] The two men concluded that "God talk" was no longer useful. More than that, they considered that the traditional faith in miracles, angels, and an afterlife was actually detrimental because it distracted people from what really mattered— the world we lived in here and now. The difficulties we faced, as well as the successes we achieved in our personal and social lives, deserved our full attention. There were many problems to solve in the world, but Christianity was unsuitable for helping us solve them; the comfortable belief in God's beneficence and omnipotence weakened our resolve: we could always be sure that someone else—God—would come to our rescue.

Even as they let go of God, they continued to value the church, believing that strong and stable religious institutions fulfilled a social purpose. In fact, they occasionally expressed concern when they noticed signs that organized religion was in decline. It was important that religion endure, according to Dietrich, "not as a survival of ancient custom, but as a living force in the development of society."[11] He believed that "the prime task of religion" was to contemplate and revere human life and to help improve it. Religion, he said, could "unfold the personality of men and women" and help them achieve "the fullest enjoyment of the natural world and the human society around them."[12] This aspect of American humanism was probably the most significant insofar as it contributed to the sustainability of the movement over the long term. By allying themselves with institutionalized religion at this time, humanists were able to maintain their community and take advantage of the support structures that the American churches of this period offered. That Dietrich and the other humanist ministers adopted this way of thinking is evidence less of tactical maneuvering than their belief in the social necessity of churches—they were ministers, after all. Tactically, though, their decision to remain in fellowship with the Unitarian denomination had far-reaching consequences.

The Proliferation of Religious Humanism

Throughout this period, more and more people who considered themselves modernists were coming to similar conclusions, and many of them were neither ministers nor theologians. In 1918, Roy Wood Sellars, a prominent professor of philosophy at the University of Michigan and a loyal Unitarian, published a book entitled *The Next Step in Religion*. Sellars spoke of the coming humanist reformation: "Every age," he said, "must possess its own religion," and whereas supernaturalism was appropriate for mankind's childhood, humanism would be the religion for "an adult and aspiring democracy."[13] Sellars compared the progress of human civilization to the maturing of a boy into a man. Intellectually and ethically, as well as scientifically and politically, Western man's youthful beliefs must give way to mature thought. Thus, the profound transformations in science, politics, and religion over the past century made it inevitable that religion would have to change as well. It was natural, he thought, that religious humanism would pave the way for a new synthesis of science, religion, and progressive politics.

Sellars had reason to be optimistic. Humanism seemed to be on the rise among the Unitarians. Spearheaded by Dietrich and Reese, nontheistic religiosity was being promoted not just by a number of ministers but also by laymen like Sellars. In 1916, the Unitarian minister Charles H. Lyttle delivered a lecture entitled "Humanism, America's Real Religion" while serving at the pulpit of the Second Unitarian Church in Brooklyn, New York.[14] The sermons, articles, and books that were produced by these first-generation religious humanists helped it spread quickly throughout the denomination in the second decade of the twentieth century. It continued to expand and flourish in the years that followed as well. One of the most prominent religious liberals in America, Charles Potter, a showman who would make his name battling Christian fundamentalists over evolution, would establish his own quite popular independent humanist church in New York.[15]

Nearly all of the men leading the movement were in their thirties: old enough to be established in their careers and settled intellectually into a consistent, well-developed worldview, but young enough to exude great optimism and press their cause energetically. They were ambitious men, anxious to make their mark on the world, and this led to real success within the denomination. Not long after his meeting with Dietrich, Reese assumed the influential post of secretary of the Western Unitarian Conference, a position that he held for eleven years. As the chief administrator of the association, Reese supported the movement with strategic decisions as organizational head of the conference. He was responsible for supplying churches in the West with ministers; he sat on the board of trustees of the Unitarians' Meadville seminary, the school that vied with Harvard in producing ministers for the church; and he was a trustee on the board of the American Unitarian Association. Without

Reese, it is hard to imagine that the humanist movement would have grown so quickly within the denomination.[16]

On a day-to-day level, humanist ministries looked pretty much like those of other Unitarian preachers: there were weddings and funerals to officiate, parishioners to be counseled, and Sunday services to write. The ministers had to manage the affairs of the congregation as a whole and oversee the business side of the organization as well. In other words, Unitarian humanists kept to the American church model, and most of them were very good at their profession. Some became locally or nationally known because of their speaking skills—outstanding preachers, by all accounts. Dietrich, for example, became widely recognized in the Minneapolis area while building up the First Unitarian Church there. Not only did the church grow phenomenally under his leadership, but his sermons also became popular enough to be broadcast weekly on the radio.[17]

Content of Humanism: Religion, Science, and Democracy

So, what were the religious humanists preaching in their sermons? After all, if they did not have recourse to God or the afterlife, and they were no longer focused on the Christian Bible, then what was this new outlook, and how was it either consistent or inspiring? Looking at some of Dietrich's sermons, the answers to these questions become clearer, and they reveal how integral science and scientific method were in his thinking. "The Quest is the thing," he said in one sermon, suggesting that it was in our human nature to be curious and to be seekers of truth.[18] We were problem-solvers, and the most important enterprise in this regard was science, which he invested with enormous epistemological and moral authority. "To the scientific mind," he wrote, "truth is something to be earnestly pursued and patiently sought, and knowledge is something that is built up by slow and laborious steps and much painful effort and struggle." This was a description of science that explained its purpose as Dietrich understood it, but more than that, it told a story of how scientists persevered in the face of challenges and conquered them. It was essentially a morality tale used to encourage and inspire his listeners. And if it was unclear just how he believed science should be integrated into the religious mission, he explained in a 1927 sermon, as quoted in the epigraph to this chapter, that religion "must accept the conclusions of modern science in every department of learning" and apply "the principles and standards of the scientific method" to its work.[19] Dietrich was drawing on the long scholarly tradition of treating the study of religion as a scientific practice with a naturalistic method, the paramount consequence of which is that we must never assume the existence of a supernatural agency, which is by definition beyond the ability of science to affirm. In other words, for Dietrich, religion was entirely dependent on science as a source of knowing about the world.

This was a very special reading of what science was and how it could be used, and it drew heavily on the way that the liberals characterized and used science. By and large, the humanist ministers' version of science came from their reading of liberal theologians, not usually from natural scientists themselves. As a result, their discussion of science, its method, and its role in the world was guided by a desire to understand its religious import along lines not entirely different from the Protestant modernists. In this way, they were closely tied intellectually to the liberal Protestant mission, and much of their work as humanists was to challenge the liberals and differentiate their own thinking from them.

Opponents of humanism often considered it to be a stark and gloomy outlook because of its denial of God, but humanists demurred. Such an outlook, they claimed, was just the reverse; it was optimistic and hopeful, and their respect for rationalism and science explained why. The human ability to learn, as characterized by the scientific method, made it possible for us to better ourselves. Knowledge and science were the keys to our future. What science took away with one hand it gave back with the other. Although there was no supernatural cavalry that would come down to save us or guarantee our survival, the magnificent qualities that we had *in ourselves*—our intelligence, our resourcefulness, our minds—all gave us power to control our future. Without any divine intervention, we could accomplish most things that our predecessors had consigned to God. Rational thought and scientific knowledge were the means to our success and salvation.[20]

The humanist ministers were philosophical naturalists: they believed that there was nothing outside of nature—nothing supernatural, no divine beings, no transcendent principles, and no Platonic ideas underlying the cosmos. As we have seen, Protestant modernists of the nineteenth century and contemporary religious scholars at places like the University of Chicago frequently espoused a naturalistic epistemology when it came to looking at and thinking about the world, and much of their theology arose out of the desire to better understand the nature of Christianity using the tools of scientific and historical analysis. Although some of the philosophers and students of theology followed the methodological naturalism of the academy into a full-blown rejection of supernaturalism, the main thrust of Protestant modernism was to use naturalistic methods to understand a more-than-natural God. Many of these men, like Mathews, still talked about God and imagined God as immanent in nature—not the transcendent deity of old, but a being that was integral to the natural order. In nearly all cases, they envisioned this world-plus-God as purposeful and welcoming to human beings, providing the ground of our ethical nature.[21] In this respect, the humanist form of naturalism was different from that used by most of their fellow liberal theologians. Although they all agreed on the importance of science and scientific method, and they generally agreed that there was no supernatural causation, the humanists did not talk about God and did not believe that nature had any inherent purpose.

It was this last notion—the rejection of teleology (i.e., the idea of a purposeful cosmos)—that especially distinguished humanists from other modernists. Most modernists, even the most radical, tended to see a purpose in the world. Indeed, purpose was often fundamental to their belief. It is not hard to see why. This kind of theology in which some form of human-like meaning is woven into the fabric of the cosmos is a comforting vision that appealed to many modernists, especially when they were giving up on the idea of a personal God. It seemed to make the universe friendlier by providing a cosmic guarantee that our lives were worthwhile and important. The notion of teleology has been remarkably resilient in modern Western thought going back to Greek times. There is a widespread, but mistaken, belief that Darwinism utterly defeated teleological thinking in biology and, indeed, in science in general. However, even after Darwin, many scientists and philosophers found ways to bring teleology back in. Many people interpreted evolution in a teleological light, and the modernist theologians followed suit. As a case in point, Mathews's theology, as we saw, was based on a teleological evolutionary view of the world.[22]

Humanists, by contrast, rejected teleology on scientific, philosophical, and moral grounds. They agreed with those evolutionists and philosophers who adopted a contingent universe: in other words, a directionless world, one that worked according to the random and chaotic interaction of environmental forces. In this antiteleological interpretation of evolution, there was no need for a providential plan. Natural forces were sufficient. On moral grounds, the humanists were even more aggressively opposed to the teleological vision because it undermined their idea of human self-reliance. Dietrich fiercely defended this idea of self-sufficiency, which he contrasted with an absolute dependency on a supernatural being: "How different the world might be to-day," he speculated, "if religion, instead of teaching man to depend upon some supernatural power for wisdom and to recognize this power as the source of all his blessings, had boldly declared that all he had was the result of his own effort, and that the future is full with promise of endless blessings in proportion as he labors and strives."[23] All in all, both orthodox and progressive Christianity failed in this account, the conservative positing a divine being and the modernist imagining a progressive force. Both views jeopardized the self-sufficiency and freedom of human beings, according to the humanists.

Dietrich had a masculine idea of science that reinforced his rejection of the Christian outlook. Science for him was a source not only of knowledge but also of moral authority. Once again, Dietrich's sermons are exemplary. In them, he explained how science embodies the virtues of courage and self-reliance. He asserted that a modern "virile and human religion" would allow man to recognize and perfect the mastery "of himself and of his environment."[24] Reese echoed these sentiments. The religious liberal, he said, is "man at his best, sane in mind, healthy in body, dynamic in personality; honestly facing the hardest facts, conquering and not fleeing from his gravest troubles."[25]

In the end, Dietrich found humanism to be the most consistent position that was in keeping with the modern scientific worldview. In contrast to humanism, he argued, liberal modernists were too timid to give up the comfortable dependence on a supernatural overseer or a divine plan, and sometimes he showed more disdain for these liberals than for Christian fundamentalists. The fundamentalists were at least consistent, for when they appealed to the ancient authorities and denounced modern scientific ideas like evolution, they recognized the fact that Christian beliefs required a complete disavowal of modern knowledge. If one wanted to be a Christian, Dietrich explained, then one should be a fundamentalist.[26] By contrast, the position of Protestant modernism was illogical, and its contradictions compromised both science and religion. Modernists wanted religion to be scientific, but they had a sentimental attachment to certain outdated ideas. They tried to harmonize Christianity with modern science, but the result was what Dietrich characterized as a "half and half, middle of the road position."[27] It was simply wishful thinking, unscientific and unprovable.[28]

In addition to its reliance on science, religious humanism in this period had an intensely political side that was manifested in three ways. First, humanists promoted progressive, reformist agendas, and many were out-and-out socialists. In large part, they advocated an agenda based on a managerial social science that was typical of reformers during this period. Social workers provided expert management in urban centers by running settlement houses and similar projects for the poor; Reese, for example, was for many years the dean of the Unitarians' main settlement house in Chicago, the Abraham Lincoln Center. For the humanists, reformist politics was one of the means by which modern society could work toward a better world for humanity on earth.

Second, humanism was political insofar as democracy was so integral to the humanist ethos that Reese had initially referred to his new faith as the "religion of democracy." He meant that modern religion was an expression and embodiment of democratic ideals, an idea he latched onto after reading a book by the radical modernist theologian George Burman Foster. Foster had argued that traditional Christianity was built on an outmoded social structure—it was essentially a monarchy with a God-king ruling over his servile subjects. Reese considered humanism to be the radical democratic alternative to this. Human beings managed their own affairs. In America, men governed themselves, working together as equals, so why should religion be any different?[29]

Third, humanism was political with regard to denominational affairs because it stood for freedom of conscience. The humanists inherited the Unitarian opposition to creeds, which meant that they opposed all efforts to enforce theological doctrine. Dietrich's own heresy trial, which opened this chapter, demonstrates how central this notion was to the humanist identity.[30] Reese went even further. He argued that humanism was simply the natural outgrowth of the Unitarian ideal of freedom, and that when Unitarianism abolished its reliance on creeds as a condition of membership, it had established complete freedom of conscience as the *raison d'être* of liberal religion. Liberal religion was not about theology; it was about tolerance.[31]

Unitarian Controversy

It was this last point, however, that proved to be the most challenging—humanist acceptance within Unitarianism did not come without a struggle. Reese, as secretary of the Western Conference, did not run into any major difficulties because of his humanism; however, that group had long been given to theological radicalism. Despite Unitarianism's long nineteenth-century battle over freedom of conscience, the General Conference turned out to be especially resistant to the humanist eruption. This became very clear when Reese was invited to present an address at none other than the Harvard Divinity School during its summer program in 1920. Boston at the time was still dominated by Christian Unitarians whose theologies had not departed radically from the nineteenth-century Christian rationalism of Channing. Humanism, not yet a decade old, seemed like a threatening imposter to these old-style Unitarians. The talk turned out to be explosive, and one report even suggested that it brought some of the ladies in the audience to tears![32]

When the lecture was published in the denominational journal, the humanist movement became a major controversy. Although they were liberal by the standards of mainstream American churches, conservative Unitarians from around the country found humanism deeply troubling.[33] The following year, two leading ministers were so concerned that they tried to have humanism pushed out of the denomination by arranging a vote on the question at the next Unitarian General Conference. One of the ministers, former Catholic priest William Laurence Sullivan, wrote critical articles in the denomination's main journal in the months leading up to the conference that satirized Reese and humanism. Then, at the conference itself, a debate of sorts was held—actually two back-to-back lectures—in which Sullivan put forth his reasons that humanism should not be considered part of Unitarianism while Dietrich presented the case for its inclusion. The precise point at issue was a resolution that would deny unbelievers fellowship in the denomination. The evening concluded in a rout for the humanists: Sullivan withdrew his resolution, and the Unitarian constitution remained unchanged.

Worded as it was, Sullivan's resolution was a hard position to defend given the history of the denomination. Despite its apparent radicalness, humanism was simply another religious tent under which some modernists were finding a home, and the long history of Unitarianism in the nineteenth century pointed to a strong antipathy to creedal divisions and an even stronger embrace of freedom of conscience.[34] When Sullivan and other conservatives sought to ban humanism, Reese and Dietrich reminded them of this history. In one published piece, Reese pushed the point in the following way: the very core of liberal religion was not any specific idea, but rather the practice of spiritual freedom, and this freedom was essential to the life of an honest faith, for only "free and positive human souls" can speak truthfully.[35]

The humanist victory at the General Conference was significant for the movement, but it did not have the earthshaking ramifications within Unitarianism that the truly transformational events of the nineteenth century had had. The victory of the Unitarians over the Trinitarians at Harvard established the denomination and was foremost an economic victory that ensured that

Unitarianism would thrive in America as a wealthy, high-status, elite church. The resolution of the noncreedalism controversy, decades later, reunified the radical Western Conference with the national denomination and gave a stable home to free religionists. The victory of the still small humanist movement, by contrast, simply reaffirmed the denomination's gradual drift leftward.

The consequences for the humanist movement, however, cannot be over-estimated. The triumph consolidated humanism's place within the denomination. This gave institutional support at the national level that turned out to be crucial to the long-term stability of the movement. Humanists gained ecclesiastical status, the ability to influence seminary curricula for new ministers, and organizational strength that came through fellowship. Since it was a time when church affiliation mattered a great deal, this victory energized the movement, so that soon more humanist congregations emerged. It also strengthened the political influence of Reese in his capacity as secretary of the Western Conference. Finally, it introduced young seminarians to humanism, a development that turned out to be crucial to the movement's continued rise.

Humanism's Place in American Theology

Looking back at this period, it is tempting to try to figure out how significant this change from liberal modernism to religious humanism was. On the one hand, it seems very significant insofar as it was a profound departure from both Christianity and religious belief in nearly every sense. Religious humanism denied God, miracles, and the supernatural in all forms; it even abandoned the primacy of the Bible as the central religious text. In these respects, the change was dramatic and important. The political and religious struggles that followed in the wake of humanism were spawned in large part by the sharp and final break that divided it from the Christian worldview out of which it was born.

If one looks at it from the point of view of the history of ideas, however, the establishment of a religion without God was a relatively small step to take. The modernists had come to consider scripture as a historical document and dispensed with the idea that it was in any literal sense divinely inspired. The main thrust of modernism was naturalistic, and many contemporary versions of Protestant modernism and the Social Gospel already held ideas of God unlike the personal and human-like deity of orthodox Christianity. These abstract notions of God often bore no resemblance to this older conception. In this respect, then, the ideological shift of religious humanism shows it to be a small step away from the American Protestant modernist tradition.

These two ways of looking at humanism in relation to Christianity have consequences, and both help us understand its history. When considered as a group that gradually moved away from Protestant modernism, we can see clearly how these religious humanists first emerged, influenced as they were by ideas then current in American politics, liberal religion, and modern scholarship. When considered as a radical break from Christianity, it becomes possible to see why people reacted to it as they did. Considered as a godless religion, it provoked strong responses—both positive and negative—and that feature of its character would never disappear in America. The atheistic nature of humanism would drive most of the controversies that followed it in the twentieth century.

The story of the relationship of humanism to Protestant modernism reveals one of the curious paradoxes of religion in the modern world. Religious liberals were arguing that the churches would survive only if they accepted scientific knowledge and values, but by doing so, they were forcing the churches to become ever more secular. Religious traditionalists believed that this kind of accommodation would weaken religion, but the immediate effect of these ideas was to invigorate it. Most importantly, the liberal denominations gained status within the elite and educated classes when they decided to more fully incorporate modern secular learning. In this way, secularization led to an increase in the social influence and prestige of the more liberal denominations.

Within the spectrum of liberal religious groups, humanism distinguished itself by embracing science in distinct ways. First of all, it tied scientific knowledge to internal conflicts over church polity. Dietrich and many of the first-generation humanist ministers considered the embrace of science inseparable from their noncreedal stance because matters of scientific knowledge had to be independent of human creeds. Humanists also found science particularly important because it was an honest and unprejudiced account of the world that offered mankind improved means for survival, on both technological and political levels, through expert knowledge and social reform. Humanists also found in science praiseworthy values like courage and independence. For humanists, science could provide a principled ethical code, a metaphor, or even a guide to living. The religion of humanism thus served these ministers and their congregations as a space where science could be connected to broad political ideas like democracy and freedom.

CHAPTER THREE

Manifesto for an Age of Science

In every field of human activity, the vital movement is now in the direction of
a candid and explicit humanism.... To establish such a religion is a major
necessity of the present. It is a responsibility which rests upon this generation.

The Humanist Manifesto (1933)

In the spring of 1933, as the country lay in a crippling depression, newly
elected president Franklin D. Roosevelt launched a series of programs that
reshaped the role of the federal government in the life of the nation. While
the president and Congress tackled the economic plight of Americans, the
leaders of the religious humanist movement diagnosed another fundamental
problem. That same spring, they worked on a document that argued for the
reshaping of America's religious foundations. This document, entitled "A
Humanist Manifesto," announced the writers' ambitious program of action:
"The time has come for widespread recognition of the radical changes in religious beliefs throughout the modern world. The time is past for mere revision of traditional attitudes."[1]

The movement remained small, but it had grown modestly over the years
and was now more organized than it was when Reese gave his lecture to the
Harvard Divinity School more than a decade earlier. It consisted of a network
of liberal churches scattered across the country, with a thin bimonthly magazine, written primarily for church leaders, holding it together. Despite its
size, however, the group was more than just a collection of ministers; its leadership included scholars from some of America's top colleges, and the manifesto itself exemplified this. It was drafted not by a theologian or minister but
by a philosopher, and the document was signed by a number of intellectuals
who were not themselves religious men. Although fully half of the signatories
were ministers, the list also included a few journalists, a couple of doctors,
several philosophers, a few scholars in the humanities and social sciences,

and even a lawyer. Surprisingly, given the thrust of the religious humanist movement, practicing scientists were almost entirely absent.

The manifesto, containing fifteen terse propositions that filled just over two and a half pages, outlined the philosophical foundation for the religious humanist movement, covering topics such as democracy, human nature, ethics and responsibility, and, not surprisingly, the definition of religion. Humanists were by no means unanimous in their embrace of the document, and a few even refused to sign it, but it met overall approval and was considered to have captured the basic thrust of humanism at the time. The history of American humanism would be incomplete without an analysis of this document and its social and intellectual foundations. This chapter traces the movement up through the publication of the manifesto and explores four philosophical premises that helped define it.

The Fellowship of Religious Humanists

The Meadville Theological Seminary was one of only three Unitarian seminaries in the country in the first decades of the twentieth century. Since Unitarianism was a small denomination, so were its schools, and Meadville reflected this. It was a tiny institution. Located in a small town in western Pennsylvania, the seminary consisted of a student body of about fifteen mostly male students.

The eight permanent faculty members taught specialized classes on a number of subjects, including practical theology, church history, New Testament studies, philosophy, sociology, and homiletics. A look at the course descriptions in the early 1920s shows its continued ties to Judeo-Christian history and theology, though these were supplemented by a few classes dealing with non-Western religions.[2] It was certainly not humanistic in any sense, and Meadville's more progressive students found the curriculum stultifying and entirely inappropriate for training Unitarian ministers for a liberal church in modern America.

So, as engaged and assertive college students, they did the most obvious thing: about half a dozen of them formed a club to study the subjects their professors were ignoring. They called themselves the Radical Religious Research Society, and they thought of themselves as radical not just because they were talking about ideas that their professors found too extreme or subversive but also because they were the first student club in the history of the school allowed to convene without faculty participation and oversight. One of the topics that the group explored was humanism, the philosophy that soon would be debated on the floor of the General Conference between Sullivan and Dietrich.[3]

It would be a mistake to see the school as entirely closed to innovation, however; one of the most important programs that it organized was a yearly trek to Chicago. Every summer, according to the prospectus, "the entire student body and one or more professors become for the summer, members of the University of Chicago and are eligible to all its privileges."[4] That arrangement turned out to be especially fortuitous for the radicals at Meadville because it was at Chicago that these students discovered university faculty members who would talk about the more controversial material that they were interested in. The importance of Meadville's ties with Chicago cannot be overestimated since the Divinity School was so central to the Protestant modernist movement. This intersection of two extremely liberal institutions, pushing against orthodoxy in their different ways, injected energy into the latent radicalism of both places and provided the perfect incubator for the radical religious humanism that they fostered.

Meadville's summertime collaboration with Chicago eventually paved the way for a permanent move. In 1926 Meadville relocated to Hyde Park, which was adjacent to the University of Chicago. There, Meadville students were allowed to attend classes year round at the university to supplement their ministerial studies. All of this was made possible in large part by Curtis Reese, who used his influence as secretary of the Western Unitarian Conference to negotiate both the financing of the relocation and a working relationship with the Chicago Divinity School.[5] Meadville's shift from rural Pennsylvania to metropolitan Chicago was accompanied by a thorough modernization of the Meadville curriculum.

Edwin H. Wilson, who would later become editor of the *Humanist*, matriculated to Meadville two years before the relocation and witnessed these dramatic changes firsthand. When he arrived, Meadville ostensibly followed the Unitarian principle of noncreedalism. The charter asserted that "no doctrinal test shall ever be made a condition of enjoying any of the opportunities of instruction." Despite the charter, however, Wilson immediately found that humanists were not welcome.[6] Three students who Wilson ran into early in his freshman year were leaving because of the conservative atmosphere. Students who were humanistically inclined complained of limited resources and of being forced to use prepared materials that were meant for traditional services; "the bloody psalms," they called them.[7]

The fortunes of the student radicals changed quickly after Meadville's move to Chicago. Within a year, Meadville students joined with University of Chicago graduate students to form a new club, the Humanist Fellowship. This small group was the embryo of what would become one of America's most important humanist organizations. The fellowship, which sponsored lectures and distributed a mimeographed newsletter, both expanded interest in humanism and gave the movement a critical mass that widened the scope of humanist influence beyond the Meadville circle. It operated without any ties to the seminary other than having several Meadville students as members. The organization also acted as a bridge between the two schools, helping to cement relationships between the secular academy and the seminary.

Although the group soon became predominantly a Meadville club run by its students and graduates, its early years were centered at the university. The original leaders were graduate students unaffiliated with either Meadville or the Divinity School. The first president of the fellowship, Herrlee Glessner Creel, was a doctoral student in Chinese philosophy who went on to become an influential sinologist, and the editor of the newsletter was a graduate student in English who became an anthropologist. Professor A. Eustace Haydon, one of the University of Chicago's most popular instructors, provided the necessary faculty sponsorship and also gave moral and even financial support to the fledgling group. Haydon's high reputation across campus contributed to the diversity of the club's early membership.[8]

Haydon was probably one of the most important figures in the early movement, and he provides an excellent window into understanding how humanism wound its way through the different institutions in Hyde Park. Although Haydon was a religious studies scholar and worked with many of the Divinity School students, he was situated administratively in the university as chair of the department of comparative religion. Anyone familiar with University of Chicago politics of the day would have immediately understood why. Comparative religion had originally been part of the Divinity School, but it had been relocated administratively as a unit within the university proper in order to protect one of its faculty, the radical professor George Burman Foster, from expulsion. Foster's controversial book would have been his undoing if he had stayed in the Divinity School because the Baptist board of directors could not countenance such heretical ideas. President William Rainey Harper knew this and engineered the move. That was some years earlier, and Haydon, who had been Foster's student, ended up succeeding him as head of the department. Even more than Foster, Haydon's intellectual agenda took him far beyond the boundaries of Baptist theology. The comparative perspective that he adopted no doubt contributed strongly to his outlook, setting Christianity equally alongside all other historical religions. That kind of objectifying gave Haydon a much different perspective than many other religious scholars who confined themselves solely to the Christian tradition.

Haydon was a popular teacher who straddled the fence between the seminary and the university, teaching students from both divisions. Moreover, despite the administrative separation, Haydon never severed his ecclesiastical ties. He had been ordained and, like many of his colleagues at the Divinity School, remained active as a preacher. For six years he serviced the First Unitarian Society in Madison, Wisconsin, commuting there every weekend. He also gave sermons for other liberal churches, as well as for the local Reformed Jewish temple. After his retirement from the university, his religious service continued, and he became a leader of the Chicago Ethical Society.[9]

With an unorthodox professional career that placed him in both ecclesiastical and scholarly camps, Haydon had a unique view of religion in America. Despite his usual stance of academic objectivity, when reflecting on his fellow preachers, he did not refrain from speaking frankly. For example, he was disgusted by the way his contemporaries—even liberal ones—kowtowed to convention. Complaining to his wife at one point, he noted how sad it was that churches were feeding the people "stupid falsehoods which all students of history now understand, appreciate, and drop." Meanwhile, he said, "the preachers go on living in the middle ages."[10] Haydon's popular book *The Quest of the Ages* (1929) became a classic for humanists because it succinctly defined religion in a way that made sense of his unorthodox blending of the secular and the ecclesiastical. According to Haydon, religion was "a shared quest for the good life, . . . the age-old, heroic adventure of earth-born man wrestling for self-fulfillment on a tiny planet swung in the vast immensities of the stars."[11] This was a vision that romanticized the religious quest, while at the same time modernizing it and making it agreeable to the contemporary scientifically literate person. Just as Dietrich was doing in his ministry, Haydon elevated the questing spirit of humanity and gave voice to the desire of all people to find satisfaction in their lives through something beyond themselves.[12]

As faculty sponsor of the Humanist Fellowship, Haydon played a critical role in uniting the different students, and this made the club less ecclesiastical than the church-based humanism that developed in strictly Unitarian forums. The Humanist Fellowship turned out to be a much stabler group than one might have imagined, and one reason for this was that, unlike the Radical Religious Research Society that existed solely for the education and edification of its Meadville student membership, the fellowship was from the beginning an outward-looking group that sought to promote its cause, not just study it. The fellowship, as part of this mission, began issuing a monthly four-to seven-page mimeographed bulletin called the *New Humanist*. By the early 1930s, the *New Humanist* had evolved into a formal typeset journal.

Members of the Humanist Fellowship charted an ambitious course for the movement. In the first issue of the *New Humanist*, they proclaimed that "great and effective social movements" were not merely intellectual outgrowths of one or a few men. Humanism required collaboration, they said; to that end, one of their main goals was "bringing the Humanists of this country (and the world, if possible) into relations of mutual awareness and cooperation." When they spoke about "the cooperative quest for the satisfying life," and of the desire to build a society where everyone shall have "the greatest possible opportunity for the best possible life," Haydon's influence was clear.[13]

The students identified themselves strongly with scientific and democratic ideals. The Constitution of the Humanist Fellowship, written in late 1927, illustrates this. Nearly half of the four-paragraph document deals with science, scientific method, and democracy:

> We believe that religion, comprehensively defined is not less necessary in a world dominated by science, but more so. We believe that the vast energies which science places at the disposal of the human race, and the specialization indispensable to the efficiency of science, render imperative that synthetic function, and the recognition that the proper end of all human activity is the enrichment of human life and happiness. . . .
>
> We conceive our task, in the broadest sense, to be the promotion of cultural democracy. We believe the best means for the accomplishment of that end to be the scientific method of gathering facts and of basing the solution for any problem solely on those facts. We believe that it is essential to the success of our endeavors that the implications of the scientific method be clearly recognized.[14]

Science was crucial to their thinking, but it is important to note that despite their emphasis on science and the scientific method, none of these men were actually scientists. They were students of history, philosophy, and religion who, when they spoke about applying the scientific method, were mostly interested in its application to social and political problems and not to questions about the natural world. For them, it seemed that scientific advances would naturally result in political change and "cultural democracy."[15]

After a few years, the leadership of the Humanist Fellowship passed from graduate students and seminarians to young, working humanist ministers. This gave more stability to the organization and broadened its reach beyond the university community, but it also returned it to the Unitarian fold for a while. The various ministers active within it were in positions that gave it a national reach, and so the fellowship was refashioned as the Humanist Press Association, the primary goal of which was to publish the *New Humanist*. All fellows were Meadville graduates who were employed in ecclesiastical positions. One of them, the person most responsible for the survival of the *New Humanist* and, eventually, the long-term viability of the national humanist movement, was Edwin Wilson, who had moved to Dayton, Ohio, where he served that city's First Unitarian Church. A second fellow succeeded Reese as secretary of the Western Unitarian Conference. A third fellow began working at the Ethical Culture Society in New York City, probably the longest-lived nontheistic religious group in the country, and one that would become a close partner in the subsequent history of humanism.[16]

The Humanist Manifesto

In 1932, several members of the Humanist Press Association conceived of a project that would both define and elevate the status of humanism. They had been searching for a way to bring humanism to the attention of a wider public because—especially now that it was run mostly by Meadville graduates—the movement was in danger of becoming a club for a few radical Unitarian preachers. They wanted to reach out beyond this small group of radicals and announce humanism's transformative potential to the wider world. What they came up with was a document that would define the core spirit of humanism for the next forty years: the Humanist Manifesto.

The Humanist Manifesto was very much a team project. It was conceived by three clergymen, Leon Milton Birkhead, Raymond Bennett Bragg, and Charles Francis Potter, who then turned to University of Michigan philosopher Roy Wood Sellars to pen the first draft. Having already written two books expounding the principles of humanism, Sellars was a natural choice.[17] Whether these ministers explicitly wanted someone who was not a clergyman to write the statement is unclear, but incorporating Sellars's academic perspective was certainly in keeping with the earliest spirit of the Humanist Fellowship. Sellars's finished draft then went to four other men for extensive revision and editing. Among them were Haydon, Bragg, and Reese, again a mix of academics and ecclesiastics. The finished draft was sent to over fifty people with a request that they give it their formal support. The promoters called on their acquaintances—mostly fellow ministers—as well as prominent people known to be sympathetic to humanism. This group included people like publicist Harry Elmer Barnes, science popularizer Maynard Shipley, astronomer Harlow Shapley, lawyer Clarence Darrow, *New Republic* editor Walter Lippmann, historian James Harvey Robinson, and social historian Lewis Mumford. Not all of these men accepted, and some didn't even respond.[18] In the end, thirty-four men endorsed the manifesto.[19] Despite the dominance of ministers, the signatory list contained a relatively wide mix of professionals. The most notable name on the list was that of John Dewey, the University of Columbia philosopher who was by this time one of the country's most influential public intellectuals.

The Humanist Manifesto was published in the May/June issue of the *New Humanist*. Although it didn't make headlines in the newspapers, *Time* magazine carried a capsule summary of it, and a couple of religious periodicals also reported the event. The entire text of the manifesto was republished in the Unitarian's *Christian Register*, as well as in the liberal Protestant magazine *Christian Century*, where it received a long and angry rebuttal. The fact that it made the pages of the *Christian Century* and *Time* shows that the endorsers managed to achieve their goal of outreach beyond the community of liberal Unitarians.

Edwin Wilson, now editor of the *New Humanist*, published some critical responses by humanists who had refused to sign the manifesto. Some of them complained that the manifesto was a type of creed that adopted an absolutist position on the nonexistence of God. Some believed that it expressed "an unjustified cocksureness." Max Otto, a professor of philosophy at the University of Wisconsin, and otherwise a strong supporter of the movement, wrote that the manifesto was "an ineffectual gesture and a tactical error" that was "detached from the living experience."[20] However, the overall response in humanist circles was positive, and the manifesto marked a turning point in the movement, one that consolidated the decades of advocacy that led up to it.

Because the manifesto is so central to the core ideas of humanism and to its history in the twentieth century, it is important to understand what these foundational ideas were and how they were expressed. In the sections that follow, I look at the relationship between science and religion; the philosophical concerns about human nature, especially reductionism and determinism; the way in which ethics and human meaning were described; and, finally, the prominence of socialism in the movement. Each of these items led to complications down the road, as the movement attracted more people and differences of opinion multiplied. However, taken together, these ideas represent a coherent picture of the humanist worldview as it was accepted by its main supporters at the time.

Unpacking the Manifesto

The Humanist Manifesto presented a nuanced understanding of the relationship between science and religion, pointing to different ways in which these concepts were intertwined. Although it is ambiguous in its phrasing—or perhaps because of that—the sentence that best captures the writers' understanding of that relationship is the one that concludes the fifth proposition: "Religion must formulate its hopes and plans in the light of the scientific spirit and method."[21] The language was both evocative and clear about which endeavor held the upper hand: religion must bend to accommodate science. What science discovers about the world and how it comes to that knowledge are paramount and must guide what we do in our religious lives. The document used the term *science* and its cognates broadly, not in the fashion of a textbook definition. Indeed, no precise definition is provided in the document; instead, the phrase evokes meanings that are left implicit—thus "the light of the scientific spirit and method" alludes to enlightenment and uplift, something rising above the mundane results of a laboratory experiment. It indicates a way of thinking and experiencing the world, an almost romantic outlook illuminating the way we interact with the cosmos.

The term *religion* is used nearly as ambiguously. The traditional understanding of religion based on worship of a deity is changed by modernity: "Today, man's larger understanding of the universe, his scientific achievements, and the deeper appreciation of brotherhood, have created a situation which requires a new statement of the means and purposes of religion."[22] Although there is nothing new here—insofar as it follows in the time-honored footsteps of American liberal religion—the manifesto went further than most liberal religious documents by denying the existence of God and even presenting a countervailing mythos explicitly opposed to the one found in the Christian scriptures. Reading the opening sections of the manifesto, one can see an implicit rebuttal of the first passages of the book of Genesis and its theological claims insofar as it offers a parallel account of the origin of the universe based on modern science with naturalistic and evolutionary premises:

First: Religious humanists regard the universe as self-existing and not created.
Second: Humanism believes that man is a part of nature and that he has emerged as a result of a continuous process.

Third: Holding an organic view of life, humanists find that the traditional dualism of mind and body must be rejected.[23]

The writers of the manifesto wanted there to be no mistake about the humanist worldview. They did not call themselves atheists, but they could not be understood to be anything other than that. God was entirely missing. There was no supernatural, no Creator, and no act of Creation. Human beings were part of a material world that gradually changed and transformed on its own. Nature simply existed and evolved, and human beings arose within it. By making such a direct parallel to the biblical Creation story, the text replaced the ancient yet still widely accepted Christian narrative with a naturalistic account that functioned in much the same way. It told us where we were in the cosmic scheme, providing a foundation on which the rest of the manifesto's declarations were built.

The somewhat esoteric point about dualism makes sense when we consider the history of early twentieth-century philosophy and recognize that this was considered to be a crucial point in the larger academic debates over naturalism that Sellars was engaged in; the debate over dualism featured prominently in his 1922 book on evolutionary naturalism.[24] Dualism, the idea that mind and body are two separate kinds of reality, had long been invoked to defend supernaturalism on philosophical grounds. By rejecting dualism, the manifesto ruled out the possibility of a personal afterlife since, in this view, an individual's mind and personality could not exist without a material body. Similarly, the idea of a Heaven existing outside of and apart from the material world fell by the wayside. Human beings were in every respect, both physically and mentally, a part of nature. Taken in this context, as a religious statement based on the philosophy of naturalism, the manifesto implied that the scientific method was the only way of approaching the world. It had no competitors because nothing else could tell us about the world with any certainty.

Another set of ideas implicit in the manifesto concern the philosophical concepts of reductionism (i.e., describing human beings in terms of physical, chemical, or biological materials and processes) and determinism (i.e., seeing all human behavior as rigidly law bound). Here the philosophical issues were important for the humanists because they shaped the understanding of what it meant to be human. In particular, Sellars and many other humanists at that time opposed these notions, especially insofar as they were used to characterize human nature. These two ideas implied a very restrictive view of humanity that did not correspond to the religious humanist position.

Although the manifesto did not attack reductionism and determinism explicitly, as it had theism, supernaturalism, and dualism, it did issue a soft defense of holism and human freedom. It celebrated the richness of the human world, often using holistic and aspirational terms and concepts common to traditional religious discussion. When the manifesto addressed the purpose of humanism, for example, it spoke of hopes and feelings. In the eighth point, for example, it states that "religious humanism considers the complete realization of human personality to be the end of man's life and seeks its development and fulfillment in the here and now." This, it says, "is the explanation of the humanist's social passion."[25] Further, when describing what it means to be human, it lists "labor, art, science, philosophy, love, friendship, recreation," and it adopts a human-centered outlook.[26] The language of the manifesto also called into question determinism in favor of real, meaningful human choice, a position made explicit in the fourteenth point, which states that "the goal of humanism is a free and universal society in which people voluntarily and intelligently cooperate for the common good."[27]

Philosophical naturalism is a big tent, however, and according to some positions, a naturalistic outlook *requires* reductionism and determinism to a degree that humanistic language like that found in the manifesto appears superstitious or unscientific. If people are only material beings, a strong reductionist would argue, and there is no immortal or transcendent human soul, then we must be defined by our material composition, and in that sense we might be best understood as complex machines whose parts obey the deterministic laws of physics and chemistry. Human emotions, aspirations, and ideals would simply be the manifestations of interactions between different constructions of matter. The job of science for the reductionist is thus to understand the physical bases of life, so that we can understand human behavior as the result of more basic material processes. Instead of talking about passions, like love and friendship, we would be better off using a reductionist language that discusses human characteristics in terms of genetics, chemical hormones, neural synapses, and such.[28]

Sellars and the other humanist philosophers argued against this line of thought, contending that life, certainly human life, was not mechanical. Reductionism was not an inevitable consequence of naturalism. Instead, one could understand nature as producing complex and multifaceted beings that were not forced to behave in rigidly determined ways. This was an essential issue for humanists, who believed that the notions of freedom, goodness, and beauty were categories in and of themselves; accordingly, they believed that reductionistic accounts of human behavior were insufficient to explain them. This version of naturalism, which Sellars strongly defended, provided a strong philosophical justification for the sermons and writings of the humanist ministers.[29]

The celebration of science in the manifesto had a very particular character. These early humanists understood science broadly and did not privilege the natural sciences over the social sciences. If anything, the reverse was true. The fourth point, for example, explicitly mentions the human and social sciences as foundations of our understanding of humanity: "man's religious culture and civilization, as clearly depicted by anthropology and history, are the product of a gradual development due to his interaction with his natural environment and with his social heritage."[30] Evolution was not assumed to be a strictly biological process; our cultural and social beginnings were as important as our biological origins, if not more so. We are both material and social creatures.

In terms of values, the manifesto argued for a naturalistic ethics, a mode of thought quite different from that held by traditional religionists, both conservative and liberal. "The nature of the universe depicted by modern science," it said, "makes unacceptable any supernatural or cosmic guarantees of human values."[31] On the one hand, the document rejected outright the idea that moral rules were handed down from a supernatural deity. On the other hand, it also opposed the notion that an ethical system was in some way embedded in the fabric of the universe, what it called "cosmic guarantees of human values," a view that had become common among some secularists and religious liberals.

This second claim—concerning cosmic guarantees—was related to the notion of teleology, which, as we have seen, was something the humanists found especially distasteful in modernist thought. For precisely the reason that this idea was compelling to the modernist theologians, it was anathema to humanists: it offered a too-comfortable replacement for transcendent values. The transcendent source of values was simply replaced by nature and made immanent in the world itself. Ethics was not God given but instead nature given, and so our ethical beliefs were guaranteed. This was usually done through some kind of goal-driven evolutionary force that, according to the humanists, bypassed the truly radical nature of Darwinian thinking.

A more radical evolutionary outlook was the result of the humanists' different reading of evolution, one founded on contingency. Just as the biological world was a response to changing environmental conditions, our ethical system, too, was contingent and flexible, responding to the environment. Sellars's book *Evolutionary Naturalism* was a long philosophical defense of this outlook, and it was the basis for his understanding of both biological and social organization on earth. And Sellars was not alone: almost all of the philosophers who joined the humanist movement at this time adopted some form of nonteleological evolutionary thinking.

The ethical ramifications fell neatly into line with the ideas that humanists like Reese and Dietrich had developed based on the notion of human self-reliance. Nonteleological evolutionary naturalism did not offer people any assurance about their future or the future of mankind; it was up to humanity to develop its values on its own. The idea that people must consciously build the moral foundations of their worldview and take responsibility for their own lives was a sharp break from modernism. The humanists believed that the modernists were still holding onto a crutch, and that it was essential that mankind walk on its own. In the words that conclude the manifesto, "Man is at last becoming aware that he alone is responsible for the realization of the world of his dreams, that he has within himself the power for its achievement. He must set intelligence and will to the task."[32]

Despite the strong individualistic ethos expressed in those last statements on self-reliance, when it came to politics, the manifesto revealed the left-wing political ideologies of its writers. Sellars himself was a socialist, a stance he defended in his 1916 book *The Next Step in Democracy*. Socialism, he explained, was a type of democracy compatible with individualism.[33] A majority of early humanists were also either socialists or progressives in their politics. In 1932, the *New Humanist* even endorsed Norman Thomas, the Socialist Party candidate for president.[34]

The most concise statement of this political position came in the fourteenth proposition, which talked about creating "a free and universal society" and "a shared life in a shared world."[35] The humanists understood socialism as an effort to create a democratic order that could rein in the excesses of capitalism. Though heartened by the Roosevelt administration's early New Deal activism, the humanists believed that the president had not gone far enough. Like many liberal intellectuals of the 1920s, they criticized the "acquisitive profit-motivated society" that deadened men to the higher spiritual values of life.[36] It was a left-wing ideology that Sellars and the Unitarian humanist preachers had long espoused, a view they held in common with the Social Gospelers and the progressive reformers. Many of them had been personally active in social movements to alleviate poverty, and some, like Curtis

Reese, who by then had become dean of the Abraham Lincoln Center settlement house in Chicago, remained at the forefront of such movements.

In 1933, during the Great Depression's darkest hour, the socialist program was particularly appealing. The manifesto was written during the heady first one hundred days of Roosevelt's presidency. It urged "a radical change in methods, controls, and motives" in order to establish a "socialized and cooperative economic order [and] the equitable distribution of the means of life." As partisans of the political left, the humanists believed that true freedom was impossible without financial freedom, and this would only emerge from careful, rational management of the economy. Humanist faith in the scientific method extended to a faith in rational control over the economy because they believed that such management would support a freer and more equitable society.

The Humanist Movement in the 1930s

Thirty-four men had signed the manifesto. The names of the signatories alone highlight the fact that the movement was no longer just a Unitarian effort. Although Unitarian ministers spearheaded the movement and continued to sustain it for the next decade, institutionally its scope was much broader. In addition to Unitarians, there was a Universalist minister, a rabbi, and two leaders of the Ethical Culture movement, all members of liberal groups outside of the Christian mainstream. Universalism, like Unitarianism, followed a path that led from its roots in the colonial era as an unorthodox Christian sect that preached universal salvation to a period of radicalization in the nineteenth century and finally, by the time of the manifesto, to a group that straddled Christian and non-Christian liberal religion. It would eventually merge with Unitarianism to form the Unitarian Universalist Association (referred to today simply as UUA) in the early sixties.[37] The presence of a rabbi among the humanists is telling. Jews, who were religious outsiders in a predominantly Christian nation, found many ways to accommodate themselves to American society. Many liberal, urban, and intellectual Jews did so by minimizing the role of religion in their lives, and many of them drifted away from theism altogether while still retaining a cultural identity as Jews.[38] The Ethical Culture movement, five decades old at this point, was a prestigious and financially secure liberal sect that in many ways presaged humanism. Its founder, the ex-rabbi Felix Adler, had, like many Protestant seminary students in the nineteenth century, gone to Germany to study theology and returned radicalized. His movement avoided "God talk" and eschewed the term *religion* but retained many of the qualities of a religious organization. "Deed not creed" was its byword, and several of its leaders worked closely with Unitarian humanists, frequently identifying Ethical Culture as a variety of humanism.[39]

During the 1930s, local humanist groups thrived, and humanist leaders in their different settings—Unitarian, Universalist, Jewish, Ethical Culture, and even some independent churches—continued to minister to congregations of various sizes. At the same time, scholars, writers, and other intellectuals began to promote a humanistic outlook outside of the churches as a secular philosophy. It was the blending of these two groups that turned out to be so invigorating for the movement, especially when scholars became major nodes of influence in the humanist network. The writing of the manifesto by Sellars was only one of many ways in which nonclergy had a hand in the development of humanism. The religious leaders were able to maintain these networks despite the institutional changes that occurred, and those institutions provided the spaces where ideas and activities could be shared, bridging the humanist community, clergy and nonclergy alike. During this time, a large body of literature was published by humanists from these different groups.

The intellectual powerhouses behind the movement tended to be the scholars, and there were two geographical nodes where this scholarly community was primarily situated: Chicago and New York City. In the former city, there was the University of Chicago with the adjacent Meadville Seminary, and in the latter, New York's Columbia University just up the street from the Ethical Culture meetinghouse and across town from Potter's independent humanist church. At both Chicago/Meadville and Columbia, professors, especially philosophers and religious studies scholars, drove its intellectual development. These two networks, which intertwined the religious and academic communities, offered different and unique support to the young movement as it advanced both secular and religious agendas. The work of these highly educated and well-placed professionals would give humanism an outsized influence within American culture. In the next chapter, we will look at the nature of that academic network more closely.

CHAPTER FOUR

Philosophers in the Pulpit

Were the naturalistic foundations and bearings of religion grasped, the religious element in life would emerge from the throes of the crisis in religion. Religion would then be found to have its natural place in every aspect of human experience.

John Dewey (1934)

Columbia University philosopher John Dewey was the most famous signatory of the Humanist Manifesto. An international scholar and a major public intellectual, Dewey was influential across American culture and was recognized for his contributions to pedagogy, philosophy, and political thought. When, at age seventy-five, he wrote a book about religion, entitled *A Common Faith*, one might have expected it to be a late-life memoir of an aged celebrity—but it was not. It was a serious argument for the importance of a religious sensibility in the modern world.

Dewey's book was not an outlier; many other philosophers made both scholarly and popular arguments defending religious humanism. They wrote books and articles, gave lectures and sermons, and appeared frequently in humanistic venues—churches, liberal synagogues, and meetinghouses. In other words, humanism was not incidental to their lives. Nor was it incidental to their work as philosophers. Defining and defending the humanistic worldview was as much a professional interest as it was a personal one, and that intersection between the academy and the church proved to be especially fruitful for both parties involved. It allowed the philosophers to spearhead an area of thought that they believed to be critically relevant to life in the modern world, and it abetted an agenda that opposed a hyperprofessionalization developing in American academia. The humanist movement also gained a lot by this affiliation. The philosophers bolstered humanism by expanding its intellectual reach beyond the seminaries and churches and gave it a measure of intellectual status that the ministers alone did not have.

Early religious humanism was such a close collaboration between ministers and philosophers that one might go so far as to say that it was as much a
philosophers' campaign as it was a ministers'. It is quite striking, in fact, both
how many philosophers were involved, given the small size of the group in
its early days, and how strongly they were committed to the movement. No
other academic discipline was as active, and none had the dedication that the
philosophers had. This chapter sets out to answer the question of why this
was so: what caused this tight network of American philosophers to be so
closely involved with the religious humanist movement?

The Professionalization of American Philosophy

One part of the answer seems to be precisely the mutual benefit that arose
in the collaboration: humanism offered a public role for scholars in return for
intellectual status. The philosophers were able to participate in humanist venues in ways that they could not so easily do in the academy. The kind of work
that these philosophers did as humanists was becoming increasingly difficult
in the professionalized academy, even in philosophy. This was a time in which
changes in the expectations and roles of university-based scholars were making it harder for them to apply their scholarship to the kind of public activism
that had drawn them to philosophy in the first place. Progressive politics was
elevating scientific expertise in many fields, but it was ignoring the humanities. The social sciences, for instance, were benefiting from a widespread interest in building a managerial society that would apply scientific knowledge
to the nation's endemic social problems.[1] But what were philosophers to do?

The predicament of American philosophy at the turn of the twentieth century can be better seen by comparing the field then with how it was many
years earlier, before the Civil War and before the academy had become so
professionalized. During that period, a young man who was able to attend
college would have gone to a small denominational school, and there he would
have encountered philosophy as the linchpin of the curriculum: a course of
study that supported the religious mission of the school. These small colleges
were teaching institutions organized around a liberal arts curriculum, and
their faculty were not research scholars.[2] The philosophy courses were taught
by men trained in theology, and the concepts that were taught harmonized
with and supported the school's religious tenets, so much so that the pinnacle
of the curriculum was usually a course in moral philosophy—almost invari-
ably taught by the president of the college himself, who was usually a minister. Because it was designed to prepare the next generation of men for leader-

ship by instilling in them an ethical character, this course of study was planned so that by the time the young men graduated they understood themselves as moral creatures living in a moral universe.[3] Since their training was essentially moral in nature, philosophy was seen as both an essential component of this system and a partner of traditional faith. The men who did philosophy were religious men and moralists, not trailblazing scholars.

By the early twentieth century, this relationship that had subordinated philosophy to religion had disappeared. The American college had become a place to prepare for a career and was no longer considered a finishing school. The teachers—at the large universities, at least—were now expected to do research, and an entirely new professional ethos emerged that centered on being a good scientist. Disciplinary boundaries emerged as scholars specialized and their research became ever narrower. Academic researchers were enjoined to engage in the disinterested pursuit of knowledge, to separate their scholarship from their personal prejudices, including, especially, one's religious faith.[4] The effect of these changes on the discipline of philosophy was significant: the content of the discipline turned toward logic and epistemology and away from ethics, theology, and questions of meaning and value. As a consequence, philosophers spoke less on moral issues and less to the public. The historian Daniel Wilson, who looked at Harvard's graduate program in philosophy at this time, found that this professionalization created a tension within these young men. The students who had entered the discipline expecting to address big-picture questions of personal and spiritual relevance found themselves instead being trained to write about narrow technical problems. Many were disheartened to find that they would not be trained to be the public intellectuals of the sort they had imagined when going in.[5]

The University of Columbia Philosophy Department

The aggressive professionalization at Harvard was not as evident in Columbia University's philosophy department. This was due in part to the close ties between the philosophy department there (its faculty and graduate students) and a vibrant community of religious radicals of different sorts in the surrounding city. New York City at that time was already a bastion of liberal and radical religion, and many of those men ended up at Columbia's philosophy department. As a result, the school became a magnet for scholars seeking to engage in a publicly oriented philosophy, one that looked back (consciously or not) to the older religious agenda of philosophy but was determined to modernize it. The presence of two scholars, in particular, created an environment that was especially conducive to the nurturing of humanism there: Felix Adler and John Dewey.

Columbia had grown by this time from being a small and unexceptional college to one of America's preeminent educational institutions. Located in Morningside Heights, on the Upper West Side of Manhattan, it was home to an elite group of scholars who were well known throughout the country. It had never been a denominational school like Harvard or Princeton, but in the twentieth century it came to be a place where a new spirit of public philosophy, one with strong religious overtones, flourished.[6]

The older religious mission of philosophy had a strong supporter at Columbia even at the turn of the twentieth century. The philosopher Frederick J. E. Woodbridge joined Columbia's faculty in 1902 after having taken a degree from neighboring Union Theological Seminary, one of the country's preeminent independent theological schools, which was known, like the Chicago Divinity School, for its liberalism. Woodbridge soon became head of the department and later dean of the graduate faculties. He occasionally spoke of the philosophy department in ecclesiastical terms and took pains to identify it with spiritual values. This is especially clear in some of his statements about the department: comparing professors' offices to "churches," he said that they "are not idle during the hours in which they may be unoccupied. . . . [Men] without sacred places, readily lose their sense of relative values. But shrines are a warning even to the profane."[7] Under Woodbridge's direction, a religious sensibility permeated the department of philosophy at this secular school. In a fashion not unlike that found in the earlier denominational colleges, Woodbridge believed that philosophy could give people a sense of security in the world by providing understanding and guidance to the inquiring mind.[8]

Woodbridge may have set the tone for the department, but it was religious reformer Felix Adler who was the true conduit at the philosophy department to nontheistic religion. Adler taught at Columbia only part-time because he was primarily concerned with the management of the large and growing Ethical Culture movement. Adler had established Ethical Culture in the late nineteenth century after he had had a falling out with the synagogue that he was to inherit from his father. The son of a rabbi, Adler had been destined to become the next leader of New York City's most prominent Reform temple, but after returning from Germany, where he had studied biblical higher criticism, it became quickly evident to both him and his father's congregation that the radical academic view of Judaism he now held was out of synch with Reform theology. He was no longer a suitable rabbi for a group that continued to hold traditional ideas about God and the world. So, Adler set out to establish a new organization that would be more in keeping with his philosophical outlook, convinced—just as the humanist ministers would be a few decades later—that fellowship of the sort that churches provided remained essential even for people who had come to disbelieve in God. The end result of his efforts was a new religious form that combined three elements: the ethical spirit of Judaism, the naturalism of modern scholarship, and the activist impulse of the Social Gospel. It was a universal faith that would "embrace in one great moral state the whole family of man."[9]

The resulting Ethical Culture movement sponsored numerous projects, ranging from settlement houses to schools for children of lower-class working parents. Regular meetings of the group took place on Sundays, when Adler spoke to its members. During his lectures, he de-emphasized traditional religious themes and spoke on politics or other secular issues in their stead. "Deed, not creed," he proclaimed, was the primary goal of the movement.[10] The New York Society for Ethical Culture thrived; other societies were also formed under Adler's supervision, so that by the end of the nineteenth century New York, Philadelphia, Chicago, and St. Louis all had active congregations. The East Coast congregations tended to be supported by wealthy Jewish families, but the movement as a whole quickly became much broader, with gentiles assuming leadership in many other locations.[11]

By the 1920s, the New York Society for Ethical Culture had become a fixture of the City's West Side, and it was richly endowed by its wealthy members. The Ethical Culture School, initially organized for the children of workingmen, began to attract middle- and upper-class families; it eventually became known as one of the city's premier educational institutions, a school that provided a superior educational environment that was grounded firmly in the moral development of the child.

Adler envisioned himself as a philosopher and intellectual as well as a religious reformer, and to that end he worked his way into the academic community. His position at Columbia enabled him to teach social and political ethics there on Friday afternoons. He was far more than an adjunct, however; he was widely regarded as an important philosopher by the academic community outside of Columbia. He served a term as president of the Eastern Division of the American Philosophical Society and, just as significantly, spearheaded a formal philosophy discussion group in New York City that included some of the most influential philosophers in the Northeast.[12]

Adler had a charismatic personality, authoritarian and witty, that attracted a number of Columbia philosophy students to his Ethical Culture Society. Three of these men—John Herman Randall Jr., Horace Friess, and James Gutmann—later joined the Columbia faculty, further strengthening the Adlerian influence on the department.[13] Friess even married Adler's daughter. As an ex-student once said about the place, "Most of the members of the Department are either sons of ministers, son-in-laws of ministers, or ex-would-be ministers."[14] These men had more than a superficial connection to religion; they all lectured occasionally at churches and religious societies, and they frequently wrote about moral and ethical issues as public philosophers. They did not confine themselves to just publishing academic monographs. All of this contributed to the "churchly" air of the department that Woodbridge had remarked on—despite the fact that these men were unbelievers.[15]

Given Adler's background, one might think that he would have been an active supporter of humanism, but he was not. He was suspicious of it because he felt that it overemphasized rationalism, empiricism, and science. Also, he was philosophically disinclined to embrace this movement because he was a Platonist by conviction. All of this meant that as long as he was alive, there were limits to the kind of collaborative work that could take place between humanism and Ethical Culture.[16]

John Dewey was the other crucially influential figure in the department, one whose work ultimately opened up room within academic philosophy to accommodate spiritual and religious thinking of precisely the sort that humanists promoted. Dewey arrived at Columbia in 1904, a forty-five-year-old, well-established scholar who was already highly regarded for his accomplishments in education. Dewey's work in philosophy during the next forty-eight years of his long life, which he spent at Columbia, was not primarily concerned with religion or religiosity in the way that the other humanist philosophers were, but his approach to modern thought and its relationship to the social world had an impact on the way he and his followers conceptualized religion. He and other pragmatists built on his particular form of naturalism and formulated a coherent and appealing defense of humanistic religion. Much of Dewey's own work was squarely within the realm of public philosophy because of its wide-ranging character and social concern.[17]

Dewey is primarily known for his role in developing and extending the reach of the philosophical school of pragmatism and in spearheading a radically new form of pedagogy. As a philosopher, he was one of three or four men who developed what some consider to have been the first quintessentially American philosophical school. Dewey came to pragmatism somewhat later than did Charles S. Peirce and William James, the two other scholars recognized as founders of this school, but his efforts to extend its application to nearly all areas of life and experience brought him fame and notoriety around the world. Peirce had developed pragmatism in an effort to grapple with a theory of meaning based on notions of experiment, experience, and observation in the context of late nineteenth-century ideas about the nature of science. In Dewey's hands, pragmatism's applications were expanded to include a reasoned understanding of how men and women in a democracy ought to act and behave, and he explored questions that extended into such far-flung areas of human experience as art and religion. As an educator, he advocated creating classrooms that would offer a more experiential form of learning, one in direct contrast to the more didactic and rote methods traditionally used. In both philosophy and education, his ideas were formed out of a critical and dynamic evolutionary perspective that saw human action as an experimental response to one's conditions. Some have called him a democratic populist in terms of the way that he interpreted pragmatism. All in all, his pragmatism merged an experimental approach to life with an egalitarian and democratic ideology.[18]

Understanding Dewey's own religious development helps to put his views on that topic in perspective. He had grown up in Burlington, Vermont, and had been brought up in a strict Calvinist household ruled by a stern mother who was a Congregationalist and an active Social Gospeler. Religion was a strong presence in his home, and as the years passed Dewey became increasingly resentful of his upbringing. He remained bitter about it even into old age.[19] Although he gradually gave up his Christian faith, he remained an active—if increasingly liberal—church member even into his thirties, when he started his career as a professor at the University of Michigan. There he attended church, taught courses on the Bible and the history of Christianity, wrote extensively on religious topics, and spoke at gatherings of the Students' Christian Association. For a while he espoused mainstream religious modernism, identifying God as the evolutionary and transformative force of the cosmos and the Kingdom of God as an egalitarian democracy. However, as he radicalized politically, he drifted away from religion and turned his attention to labor issues, settlement houses, and progressive education, and eventually he left the church altogether.[20]

From his early years as a pious schoolteacher who fervently led an opening prayer each day in class to his elder years as a humanist, Dewey recognized the force of inner religious commitment. He described having had occasional transcendent mystical experiences throughout his life and claimed to have come to a deep and abiding feeling of satisfaction in living. The most significant of these experiences occurred during an extremely lonely and troubled period after college while he was teaching high school in Oil City, Pennsylvania. He reported a brief feeling of unity with the universe, which he described in undramatic, but typically Deweyan, language: "Everything that's here is here, and you can just lie back on it." This event left a deep impression on him, a sense of calm and contentment.[21] He did not see it as anything supernatural; indeed, quite the opposite: he considered it to be a feeling of the materiality and solidness of the world around him and, ultimately, a feeling of security as part of it. At one point, quoting the Apostle Paul, Dewey even claimed, "I enjoy the peace which passes understanding."[22]

His religious background informed his philosophy in different ways. One might even consider Dewey's philosophy as a response to the need for a secular alternative to religion. His philosophical vision was both sweeping in its ability to engage in questions about man's place in the universe and practical insofar as it could help provide ways of thinking about moral, political, and social issues. Starting with a Darwinian view of mankind—not the teleological version of the Protestant modernists, but one that was contingent and open-ended—Dewey built a philosophy based on strictly scientific and naturalistic principles as he understood them.

Dewey gave great importance to experiential knowledge, and he believed that the empirical focus of natural science should be central to philosophy. Too much of modern philosophy had taken a wrong turn, he said: it was far too abstract and intellectualized. He saw endless speculation about the nature of underlying reality that had little relevance to the way that people actually lived their lives. What was needed instead was a philosophy that attempted to grapple with real concerns and that dealt with lived experience.[23]

He acknowledged that "philosophy inherited the realm with which religion had been concerned" and believed that it was essential that philosophers learn to fulfill its newly inherited obligations. First of all, philosophy must be scientific, and Dewey was deeply critical of the ways in which some philosophers who retained an orthodox religious worldview took on problems that no longer had any bearing on modern knowledge.[24] It was essential that philosophers start with modern science and organize that knowledge in humanly significant ways: "When philosophy shall have co-operated with the course of events and made clear and coherent the meaning of the daily detail, science and emotion will interpenetrate, practice and imagination will embrace. Poetry and religious feeling will be the unforced flowers of life."[25]

His most mature reflection on religion came out in his small book *A Common Faith*, published in the year following the Humanist Manifesto. Dewey had already had a long and noteworthy career and was at that point in his life when he sought to reflect on deeper and more personal matters. His book was a philosophical exploration of how religion ought to function in the modern world. The work presented Dewey's view of religion as an integral part of the human experience. The argument was by this time familiar to humanists: even nontheists could live their lives with a religious intensity like that found in traditional God-focused religions. Religious experience did not have to contradict a scientific—which for Dewey meant an entirely naturalistic and nonteleological—understanding of the world.[26]

All in all, Dewey and Adler held very different and often incompatible views, and they each stood somewhat apart from the humanist movement proper; and yet, despite this, they provided a scaffolding within the Columbia philosophy department that supported humanism. This is not something that one might have expected. Adler, as we have seen, never called himself a humanist and even distrusted the movement's overly rationalistic approach. The fact, however, that he was such a strong force for Ethical Culture and could demonstrate the vitality and richness of a nontheistic community proved critical. The Columbia department would have been a very different place without his presence and the ties to Ethical Culture that he brought with him. Dewey, by contrast, signaled his steadfast support of humanism by signing the manifesto, but he did so only in the fashion of an occasional cheerleader. He was not a rank-and-file member of the group, a strong activist, or a participant. Rather, he was a fashioner of ideas, and it was those ideas in their breadth and variety that proved inspirational for philosophers and other academics. His vision of a new kind of philosophy—one that coincided closely with the mode of thought that the religious humanists were developing in their churches—provided both intellectual and spiritual nourishment for the school's students and attracted them to the humanist movement.

We can better see how Adler's and Dewey's competing visions created an environment conducive to the flourishing of humanism at Columbia by looking at one more faculty member, John Herman Randall Jr., who arrived at Columbia as an undergraduate and retired about fifty years later as one of its longest-serving professors. Randall was inspired by both Adler and Dewey, but the humanism that he espoused had its own unique flavor, a result of both personal and professional influences. He was a prodigy by any standard; by the age of twenty-three he had already finished his doctorate at the department, and within four years he had written a textbook on the history of modern philosophy. This book, *The Making of the Modern Mind* (1926), was enormously successful and was assigned in college courses across the country for decades. Like many of the philosophy students at Columbia, upon graduation he joined the department that had trained him and remained there until his retirement in the late 1960s.[27] He was a fixture at Columbia throughout much of the twentieth century.

Randall, like Adler, was the son of a religious leader. His father, John Herman Randall Sr., was a lapsed Baptist preacher who had radicalized. He and another ultraliberal Unitarian, John Haynes Holmes, established an independent nondenominational church in New York City. Over the years, the older Randall adopted an ever more eclectic and pluralistic religiosity, mixing Buddhism, Hinduism, and the Baháʼí faith with his liberal ex-Protestant ideas. In the late twenties, the two Randalls, father and son, worked together on a book, *Religion and the Modern World* (1929), that merged a humanistic faith with religious pluralism and a critique of modern materialistic culture.

This book presented a more conservative approach to religion than the typical humanist account because the Randalls held a generally sympathetic and tolerant view of the great variety of human religious experiences around the world. Their point of view was ethnological and literary, open to non-rational religion in a way that most humanists were not. The authors also resisted the tendency of some scholars to try to locate a common thread throughout all religious experience. Each religion ought to be seen on its own terms since each one interpreted human experience differently. The authors discussed religion in terms of semiotic (or at least symbol-based) and ethnological interpretations of religious experience, and in this way they foreshadowed a postmodern outlook that would come into vogue decades later: "To see all religious beliefs as metaphors of discourse, as the symbolic renderings of deep human experiences, frees one from the vain attempt to find a core of religious truth common to every historic faith." There was no "common prose for all this imaginative symbol."[28] It gave credibility to premodern religion even as it denied that that religion had any factual or scientific authority.

Likewise, the Randalls were much more critical of modernism than one might expect, calling the present era one of "moral opportunism." The modern world presented an existential dilemma to many people, offering an unsatisfying culture that left ordinary people longing for more. Science and technology were part of the problem: "Our bodies," Randall Jr. stated, "live in a world where machines are triumphant and where science is their servant. But men do not yet feel emotionally at home in the bare halls of the factory or the laboratory. . . . They inhabit the new civilization but they are not really living its life."[29] Modern culture was the embodiment of the machine age—a world without God, cold and unfeeling—and because of this, masses of people retreated into the sensual materialism of the Jazz Age and became "cunning beasts." This was one reason why we should not discard the old forms of religion altogether, which is precisely what most liberals and most humanists taught. The Randalls believed that, on the contrary, religions were "symbolic and poetic expressions of man's deepest experiences in past centuries."[30] As such, the old forms should be preserved and used.

The younger Randall's historical work led him to many of these same conclusions, and he remained much more tolerant of orthodox Catholicism than most liberals. The medieval world was a fascinating place for him, and he found elements of it laudable and good. This was in sharp contrast to humanists, who on the whole disparaged medievalism as an unmitigated impediment to knowledge. Randall went even further, arguing that some aspects of medieval thought, such as Aristotelian scholasticism, actually paved the way for modern scientific thinking.[31]

Yet despite these differences with mainstream humanism, Randall Jr. considered himself a humanist and signed the manifesto. When push came to shove, science was essential, our measuring rod for seeing our world clearly. When new knowledge contradicted literal, orthodox beliefs, we were liberated from the latter. History, anthropology, and psychology allowed us to study religion and helped us to unshackle ourselves from dogmatic faith. Modern religion required that people adopt "new loyalties and habits of mind" that were more akin to science than to what people usually thought of as religious.[32] We needed to think in provisional and experimental terms, not dogmatic and fixed ones.

On strictly philosophical grounds, Randall was still a naturalist, which made him at home among other humanists, and his eclectic temperament made him at ease within the meeting halls of Ethical Culture. Indeed, he spoke and participated in both venues, giving talks with titles like "The Experimental Attitude in Morals," "The Ethical Life and the Humanist Temper," and "The Ethical Challenge of Pluralistic Society." In one notable address, he defended the vitality of modern religion in a Columbia University chapel service in

1932, which "about 100 persons, including virtually the entire Philosophy Department, attended."[33] In this respect, Randall was not different from the other younger members of the department who found religious conversation appealing and who were likewise active proselytizers of humanism.[34]

Roy Wood Sellars: The Architect of Evolutionary Naturalism

The same kinds of church-academy interactions took place outside of the Columbia circle, and the same goals motivated these other humanist philosophers: to maintain a place for public philosophy while addressing man's spiritual concerns. This was certainly true for the University of Michigan professor Roy Wood Sellars. We have already noted how important he was to the history of the humanist movement by his drafting the Humanist Manifesto. That document helped to make philosophical naturalism a foundational principle of religious humanism.

Sellars was fifty-three when he wrote the manifesto and was well known among his peers. He had twice been elected to leadership positions in the professional community, once serving as vice president of the Eastern Philosophical Association, at another time president of the Western association. In many respects, Sellars was the model of a career academic. His father had moved to Michigan from Canada to study and practice medicine, and Sellars grew up there and spent most of his adult life in Ann Arbor connected with the University of Michigan. He attended the university for both his undergraduate and graduate degrees, and then, like so many men of that period, he stayed on as a faculty member.[35]

In religion, he was an active Unitarian, and his religious beliefs seem to have been guided more by his intellectual convictions than by emotional or psychological experiences.[36] This primarily intellectual focus on religion can be seen in the direction his studies took when, at one point, he explored religion at the Hartford Theological Seminary a short time after his undergraduate work: he studied comparative culture—not theology, as one might expect at a seminary. It was the diversity of religions that seems to have most engaged him. Religions were cultural forms, something he had come to believe as a youth growing up in rural Michigan surrounded by Anglo-Canadian, French Canadian, and Norwegian families, all of whom adhered to different faiths.[37]

Sellars was also strongly influenced by the modernist writings of the Chicago Divinity School faculty, which brought him directly into liberal Unitarian circles, where he became an active participant in the humanist-theist debates of the late 1910s and early 1920s. It was during this time that he published his pathbreaking defense of humanism in *The Next Step in Religion*.[38] Like so many of the other professors we have looked at, Sellars was frequently welcomed as a speaker by liberal congregations. His lectures had titles like "Is the Universe Friendly?," "Has Man a Cosmic Companion?," and "Has Human Life Intrinsic Meaning?" Not surprisingly, with all of these connections, Sellars was a frequent lecturer and participant at the Humanist Fellowship at the University of Chicago.[39]

As we have seen, he was both a political and a religious radical. He was an outspoken socialist, and his radical politics went hand in hand with his religious views that were developed during the height of the Social Gospel movement, when liberal religion and reformist politics worked together. In Sellars's view, religion was a holistic integration of the social and the personal. He talked about humanism as "a free man's religion, a religion for an adult and aspiring democracy."[40] Like his fellow humanists, he emphasized not just courage and self-reliance but also social integration and community.

The topics of Sellars's books point to his efforts to reach the public and to keep philosophy relevant. We see this not only in his books of humanist apologetics, *The Next Step in Religion* and *Religion Coming of Age*, but also in his book on socialism, *The Next Step in Democracy*. For Sellars, philosophy helped people better understand the world and their place in it. This is why Sellars was so interested in religion as a practice and as a language to talk about the human experience. Religion spoke the language of human aspirations and loyalties that Sellars wanted to salvage from technical philosophy.[41] "It is out of man that religion arises, out of man's passionate struggle with life," wrote Sellars. "Religion is beyond all things an *expression* of human life; and, as man's spirit deepens and his imagination quickens, and his knowledge broadens, the form which religion takes is bound to alter."[42]

In 1922, Sellars published a rigorous book-length defense of naturalism. Naturalism, as we have seen, rose in prominence in the nineteenth century as a challenge to the prevailing supernaturalist worldviews. Sellars explained that naturalism was nondualistic and nonteleological. It entailed a number of ideas: first, that the universe is self-contained and everything in it can be understood in terms of matter, forces that operate on matter, laws that govern the behavior of matter, and effects produced by matter; second, that the physical world can be understood only through rigorous scientific study, and nothing outside of nature is needed to explain how it arose or why it is the way it is; and finally, that there are no minds or forces outside of nature and thus no supernatural entities.[43]

The evolutionary and nonreductive character of the world was especially important to Sellars's philosophy. His rejection of reductionism was based on the recently developed notion of levels of organization, that is, that complex objects in the universe such as biological organisms and, especially, mankind cannot be understood simply by reducing them to their individual components. In the universe, matter had evolved into ever more complex organizations, and each level of complexity had its own properties, each requiring its own explanatory principles. The different sciences demonstrated this: "Chemistry, biology and psychology have become autonomous, concrete and profoundly expressive of evolutionary ideas." This meant that naturalism was not simply a materialistic, reductionistic philosophy. "It is no longer possible for a *fair* critic to identify naturalism with the mechanical view of the world. Scientists are tentatively reaching out for more flexible and less dead-level ways of approach. Evolutionary naturalism is not a reductive naturalism."[44]

Crucially for the understanding of human beings, Sellars argued that we cannot reduce human behaviors, actions, and thinking to properties of matter alone. He is especially critical of materialism in this regard. He decried earlier versions of naturalism that attempted to explain human life reductively in terms of physical, chemical, or low-level biological principles. Herbert Spencer, T. H. Huxley, and Ernst Haeckel all had proposed such simplistic and naive accounts. Although Sellars agreed with these men about the material basis of the universe and the nonexistence of supernatural forces, he found their mechanistic views to be inadequate to explain the richness and diversity of human life.[45] Sellars's evolutionary naturalism was "no narrow naturalism limited to the physical sciences. The *whole* of man must be included in nature, and nature so conceived that his inclusion is possible."[46] His form of naturalism fit well with the humanist outlook because it depicted the human being in holistic terms—the human person was the category of analysis, and it was essential that we did not attempt to study people by reducing them to chemical or biological structures. "The evolutionary naturalist desires to combine the total information he can gather about the human organism. He knows himself to be an evolved substance of tremendous complexity whose material is that which he consumes each day in food and drink. He thinks, acts, adjusts himself to complex situations, minsters to his body, experiences fatigue, has sicknesses, etc. *That is the kind of thing he is.* He is all of these cooperative functions at one and the same time."[47]

This line of argument was comfortable for humanists. Central to the humanistic outlook was a value system founded on the notions of human freedom and the worth of the individual, ideas that were unsustainable in a strictly deterministic account of human nature. The holistic view of the person that Sellars argued for ensured the importance of these notions. Whereas materialists had gotten carried away by the grand scope of "cosmic emotion" and had lost sight of the human scale, Sellars asserted that his evolutionary naturalism did not do this. Though he eliminated God and the soul from his philosophy, he asserted that the individual person had inherent value: people could think, make decisions, and behave morally, and in this way he made room for the spiritual life of man.

Looking at the larger picture, we can see how Sellars, throughout his career as a professional philosopher, also worked as a public intellectual, and how humanism was central to his work. He showed how science and religion could be integrated under a broad philosophical tent. Religious humanism was the vehicle that made this possible. As part of that religious community, he could talk about human aspirations and loyalties and how they were linked to the cosmos as we now understood it.[48]

Max Otto: Advocate for a Personal Pragmatism

The University of Wisconsin professor of philosophy Max Otto was active in the humanist movement from early on. Like the other humanist philosophers, he had an abiding interest in religious questions and the use of science and philosophy to address them. Indeed, he became a philosopher as a result of his own personal religious quest. He was born in Germany in 1876 to a strict Lutheran family. When he was five years old, the family moved to the United States. Unable to endure his father's violent rages, he ran away from home at the age of fourteen and ended up in Chicago, where he became a social worker at the YMCA. In that capacity, he preached the Gospel in the seedier parts of the city by night and worked as an errand boy by day. In unpublished autobiographical manuscripts, he states that he became almost an ascetic, driven by the principle of Christian charity and constantly fearful of Satan's temptations. In an effort to untangle his psychological and religious torments, he turned to philosophy. Through philosophy, he hoped to gain a clearer understanding of God's will.[49] He enrolled as a student at the University of Wisconsin but was quickly disappointed by his classes. His professors seemed to treat philosophy as little more than "a game of intellectual chess."[50] Nonetheless, he stayed on, completed his PhD, and was soon offered a permanent position on the faculty.

By the time Otto had graduated, his Christian faith had ebbed. However, since he still valued spiritual fellowship, he became a member of the First Unitarian Church in Madison, where he would occasionally give sermons. This Unitarian connection brought him into the humanist network in Chicago, which was only a few hours down the road. As they did with Sellars, the students in the Humanist Fellowship elected him an honorary member of their group.

Over the years, he revised and explained his views in three books: *Things and Ideals, Natural Laws and Human Hopes,* and *The Human Enterprise.* In these works, Otto turned the tools of philosophy to topics of concern for the average man. In particular, he grappled with the existential problem of how to live in a universe without either a God or a cosmic direction.[51] In each of these works he spelled out a philosophical outlook that he had derived from his close association with John Dewey. The two men knew each other well, and Otto considered the older philosopher a mentor. In one letter to Dewey, Otto explained how deeply he was indebted to him by describing his impression of a recently sculpted bust of Dewey by the noted artist Jacob Epstein: "The vision is there; the travail of soul is there; . . . As I look at the bust I get a sense of a vast and complex striving of a people coming to some sort of focus in you. I am urged up and on to give it such concrete existence in my own life and in my local situation as I may be able to."[52] Dewey's ideas and work dominated the philosophical thinking of all of the early humanist philosophers, but perhaps none more than Otto.

Otto's own humanistic philosophy was built on three principles. The first was antideterminism: human social relations had their roots in biological evolution but were not determined by human biology; laws, rules, and social behaviors developed and evolved as people encountered new problems and responded to them. The second was public philosophy: like Dewey, Otto believed that philosophy must give practical solutions to real-world problems. The third was human spirituality: he gave great weight to the problems that engaged people's innermost concerns, their needs and desires, fears and hopes. These were as important as social and political issues.[53]

The first pages of Otto's book *The Human Enterprise* illustrate how he tried to make his philosophy engaging and relevant in ways that would meet the needs of a public philosophy. Using a first-person narrative voice, he introduces the reader to a vivid scene in which he, the writer and philosopher, inhabits the world. He and his context are affirmed at the outset. His thinking and philosophizing arise out of an actual experience: "It was Sunday afternoon and I had climbed a hill to get a view of the surrounding country. From the base of the hill stretched a strip of oak wood, the billowy tree tops motionless in the mild September sun. . . . Seated on this hill, it was easy to indulge a languid half attention and to mistake this dreamy state of mind for profound intuition, the intuition of an all-comprising unity behind the landscape's palpable diversity."[54] Not only was the philosopher present in the landscape; he was also feeling and reacting to it. In a later passage, Otto's account resonates with Dewey's description of his own mystical experience with the world: "It was not, however, the mystic's sense of absorption. . . . It was a plain man's perception of a unity within and through the rolling expanse spread out from hill to horizon and beyond. . . . As I looked from this hilltop, responsive to the quiet yet animated scene, it appeared obvious that in direct experiences of this kind man touches metaphysical bottom."[55] Philosophical argument thus goes hand in hand with the physical reality of existence. Otto tried to make his philosophy concrete and tactile: these ideas were things you could grasp with your hand as well as with your mind. His philosophy was meant to be a discussion of lived human experience, as well as thoughts about how to live as a person inhabiting our cosmos.

One might not have expected such a strong imagistic and experiential element in the work of a philosopher, but this is precisely what Otto believed was necessary in order to move philosophy out of the professional academic conversation and into the public realm. This rhetoric sharply differentiated him from the professional philosophers whom he thought of as only playing intellectual games. For Otto, philosophy had to be both practical and useful. It also had to encompass human values and deal with ethics and questions of social and political justice incorporating human hopes and aspirations in a way that was accessible to the common man.

Summing up the complex interactions between religious humanism and the various philosophers discussed in this chapter, we find that the philosophical agenda for all of these men had a common thrust, namely, to transcend professional limitations and to put scholarly work to use in meaningful ways for the layperson. Joining the humanist movement and participating in public outreach from that dimension gave the humanist philosophers an especially broad canvas at a time in which the professionalizing forces of philosophy were narrowing the discipline's focus and audience. Their tie to humanism was in part a reaction against that limiting and deadening of the philosophical mission. For them, philosophy was not simply a career: it was, at its best, a way of understanding life in the broad sweep of existence, a place to go to grapple with questions of meaning and value, and a resource for ordinary men and women. That orientation would shape the way that humanists responded to outside criticism in the coming years.

Science informed their thinking, and through their philosophical work they were able to unite science with wider social values. But the way they thought about science and nature, especially in their opposition to reductionism and determinism, put early humanism on a collision course with other science-based philosophies. Eventually the philosophical commitments of these early humanists would create challenges when philosophers and scientists with a very different outlook joined the movement. In addition, these men's philosophical mission—which highlighted the holistic and, in their view, *religious* elements of human experience—would find less traction in the movement in subsequent years. Liberal religion in America began to lose its hegemony, and when that happened, humanists confronted ever more aggressive attacks from the religious right, and the rhetoric of early religious humanism began to lose its effectiveness. That rightward shift in the culture and its effect on humanism are where we turn next.

INSTITUTIONS OF RELIGIOUS HUMANISM. *Top*, the Meadville Theological School, as shown in this May 1925 photo of faculty, students, and visitors, was critical to the long-term survival of religious humanism. The young Edwin H. Wilson, who would become a pivotal figure as editor and director of humanist institutions over most of his life, can be seen just above the two women at the bottom of the photograph. *Bottom*, Felix Adler, the founder of Ethical Culture, presides over the laying of the cornerstone for the well-funded Ethical Culture School in New York City in 1903. Meadville Theological School, 1925. Meadville Lombard Theological School, Archives and Special Collections. Felix Adler laying the cornerstone of the Ethical Culture School. The American Ethical Union.

FOUNDERS OF RELIGIOUS HUMANISM. *Left*, popular minister John H. Dietrich, one of the founders of religious humanism in the 1910s, fostered interest in humanism in the Midwest through his lively sermons in the First Unitarian Society of Minneapolis. *Right*, University of Michigan philosopher Roy Wood Sellars, one of the top American philosophers of the early twentieth century, advocated for religious humanism through books like *The Next Step in Religion*. Sellars wrote the first draft of the Humanist Manifesto of 1933. John H. Dietrich. The First Unitarian Society Minneapolis. Roy Wood Sellars. Portrait of Roy Wood Sellars, Roy Wood Sellars Papers, Bentley Historical Library, University of Michigan.

WORK FOR PEOPLE WHO LIVE AND SUFFER

THE ACADEMY AND THE PULPIT. *Top*, scholars formed the intellectual backbone of the religious humanist movement in its early days. In this news story in the *Chicago Daily News*, we see University of Chicago's A. Eustace Haydon and Max Otto of the University of Wisconsin speaking at a Humanist Press Association event in 1938 on the third anniversary of the group's founding. *Bottom*, this photo, probably taken in the 1940s, shows the esteemed philosopher John Dewey shaking hands with Jerome Nathanson, head of the New York Society for Ethical Culture. A. Eustace Haydon and Max Otto in the *Chicago Daily News*, ca. 1938. Special Collections Research Center, University of Chicago Library. John Dewey and Jerome Nathanson. The American Ethical Union.

BUILDING A MORE DIVERSE MOVEMENT. *Left*, when writer Priscilla Robertson took over as the second editor of the *Humanist*, she brought with her a feminist perspective on science, ethics, and human social interactions, as well as an interest in the nascent movement of humanistic psychology. *Right*, Lewis McGee, Unitarian minister and one of the staff at the AHA during Wilson's editorship, was something of an outlier as one of relatively few African Americans in the movement in the early 1940s. Priscilla Robertson. American Humanist Association, with the permission of Henry P. S. Robertson. Lewis McGee. Meadville Lombard Theological School, Archives and Special Collections.

HUMANIST HOUSE. *Top*, with strong financial support of a few wealthy donors such as Cyrus Eaton—sponsor of the Pugwash Conferences on Science and World Affairs—and Corliss Lamont, the AHA was able to erect a new modernist-style building in Yellow Springs, Ohio, to support the growing activities of the group around the country. *Bottom*, it was there, in the early 1960s, that a portrait of John Dewey was dedicated and unveiled, marking Dewey's importance in early humanist thought. Humanist House, Yellow Springs, Ohio. Photo courtesy of the American Humanist Association. Portrait of John Dewey hanging in Humanist House. Photo courtesy of the American Humanist Association.

COLLABORATIONS. *Top*, the movement's many agendas were represented by leaders with quite different backgrounds. In this detail from a 1962 photo taken at the annual conference, we see religious humanist minister Edwin Wilson, science writer and editor Gerald Wendt, and secular activist Vashti McCollum. *Bottom*, local chapters flourished around the country during this time; this image shows a chapter meeting in Minneapolis, Minnesota, with future president Bette Chambers sitting on the far left. Edwin H. Wilson, Gerald Wendt, and Vashti McCollum. Photo courtesy of the American Humanist Association. AHA Chapter meeting in Minneapolis, Minnesota. Photo courtesy of the American Humanist Association.

CHAPTER FIVE

Humanists at War

Suppose a group of well-intentioned democrats were to accept an invitation to attend a conference ostensibly convened to further the democratic way of life and to stem the tide of totalitarianism. While there, instead of hearing proposals for the defense and extension of the democratic faith . . . they find their own philosophical position abusively denounced as responsible . . . for Hitlerism.

Sidney Hook (1942)

The rise of fascism and the advent of the Second World War deeply worried humanists, who believed that it threatened not only the nation's future but also the course of Western civilization and the overall Enlightenment project. The ideological struggle was especially worrisome because it endangered democracy itself. Humanists were not alone in this view. American intellectuals of all stripes reflected on the ideological underpinnings of the war, but they held sharply different ideas about the crisis. The traditional opponents of humanism, religious conservatives, laid the blame for the war in large part on ideas that humanists championed, charging that atheism and technocratic rationalism created the conditions that set the stage for war.[1]

The New York philosopher and public intellectual Sidney Hook, quoted in the epigraph above, was furious with conservatives who attacked humanistic ideas in this vein. Of course, in response, the humanists themselves were just as willing to condemn their conservative enemies as both antidemocratic and authoritarian. Religious orthodoxy, in the humanist view, was simply authoritarianism in another guise, and they feared that this reactionary sentiment was now in a position to gain strength by cloaking itself in patriotism. Over the decade of the 1930s, humanists had watched uncomfortably as Catholics and Protestants around the country propounded ideas antithetical to the liberal vision of a democratic world, and by the time that actual war broke out in Europe, the ideological battle had come to a head. Humanist leaders around the country followed Hook in their defense of humanism. Speaking for the humanists, Jerome Nathanson, leader of New York's Ethical Culture Society, wrote in 1945, "Above the drone of the fighter planes, the shriek of the bombs, and the roar of the heavy shells, the hoarse voice of reaction rises over the earth. The war itself is only symptomatic; it is evidence of the world's sickness, not the sickness itself."[2]

This chapter documents a series of episodes involving intellectuals with radically different agendas, who saw the fight abroad as "symptomatic" of ideological battles that were fundamental to the character and future of the nation. These debates pitted competing worldviews against each other and highlighted the ways that beliefs about modernism and science were tied to moral and political affairs. Humanists promoted the scientific spirit, seeking to expand its influence across all areas of life. Their conservative critics, by contrast, wanted to limit science's influence in order to prevent American culture from collapsing into an amoral technocracy. These two views entailed two utterly incompatible conceptions of science. The two sides did agree, however, on one point: the ideological struggle at home was as important as the military struggle abroad, and the resolution of that struggle would determine the future of the country after the guns had gone quiet and the bombs had stopped falling.[3]

The humanist movement was in transition during the decade in which these events took place. The Depression had taken the wind out of the movement's sails, but the Unitarian minister Edwin Wilson remained at the helm as editor of the *New Humanist*, and two years after the 1933 Humanist Manifesto a core group of mostly Unitarian ministers incorporated a new organization to advance humanism, the Humanist Press Association. Their goal was to "popularize and expand" humanism and to "study and extend educationally principles and ideals concerning human progress, values and welfare." Wilson started a newsletter to supplement the work of the journal and, later, a more polished bulletin. These efforts ultimately proved too much to sustain in the last years of the decade, and the journal itself folded, leaving only the thin bulletin to keep the members of the movement in contact with each other.[4]

The advent of the war and the economic upturn that it brought about signaled new possibilities, however, and the nonprofit American Humanist Association (AHA) was founded, replacing the short-lived Humanist Press Association. The journal was revived under the shorter and simpler title the *Humanist,* and its first issue was published in March of 1941, with Wilson (now ministering in the small upstate New York town of Schenectady) as its editor. The same desire to make humanist ideas more accessible to the mainstream culture underlay the goals of this new organization and the new journal. It was in this context that the movement addressed the wartime challenges to their ideology.

The Conference on Science, Philosophy and Religion

One of the flash points in what we might call a battle over America's soul took place in September 1940 with the first meeting of a high-profile conference series held in New York City at the Jewish Theological Seminary. These conferences riveted humanists and would play a pivotal role in the rebuilding of the movement after the Depression. They became a unifying event, not because they exemplified humanist ideals, but rather the reverse. They highlighted reactionary and antihumanistic voices and, as a result, became a rallying point for humanists and an opportunity for rebuttal. To see why the humanists reacted as they did, we must better understand the conference series itself.

These conferences have come to be seen by many historians as important in their own right, a signal moment in midcentury American intellectual history that brought together an extraordinary roster of elite scholars and thinkers and continued for an astonishing thirty years. The founders had the broad ambition of building a coherent moral foundation for America in the modern world.[5] With the sweeping title "The Conference on Science, Philosophy and Religion," the first meeting was held a year into the European war and only days since the president had announced the first peacetime draft in American history. The fact that this conference was initiated during the dark early days of the war is critical to its understanding—the wartime situation shaped the motives, ideas, and rhetoric of both participants and observers alike—and it tells us much about the humanist response.

The head of the conference was conservative Jewish theologian Louis Finkelstein. He and his planning committee brought eighty men and women together from widely different fields. The speakers included physicists, astronomers, social scientists, and psychologists, all highly regarded in their disciplines, and many of them well known outside the academy. A wide spectrum of faith traditions was invited, represented by such internationally renowned figures as Harry Emerson Fosdick (a Protestant Modernist and bête noire of American Fundamentalists), Paul Tillich (a German-born existentialist Protestant theologian), and Jacques Maritain (a French Catholic neo-Thomist philosopher).

Although the agenda was ostensibly to find a common framework that would unite men and women of all faiths and ideologies, it did not actually do so. Indeed, looking back on the assumptions behind the conference over half a century later, it is clear that Finkelstein and his co-planners set themselves an impossible task. Disunity was unavoidable. For humanists especially it was antagonistic. It had a religious tone from the outset that alienated them with a lot of theistic language, which was perhaps not surprising given the large number of theologians on the planning committee. The religious thrust was ecumenical—extending across Jewish, Protestant, and Catholic traditions— but also conservative and traditional. An early press release, for example, claimed that "the democratic American way of life . . . must be based ultimately on the religious principle of the Fatherhood of God and the worth and dignity of Man when regarded as a child of God."[6] Even Finkelstein's own agenda troubled humanists by claiming that the convention should seek to create a grand synthesis of knowledge similar to what existed in the medieval period when thinkers like Maimonides and Aquinas presented a unified world picture within a religious framework: "We need . . . a School which shall continually present to the religious world the facts of Science and interpret them at once in terms of religious and ethical values."[7] These ideas were in stark contrast to the naturalistic sentiments that underlay so much of humanist thinking.

Finkelstein's opening remarks give a clear picture of his views: he saw the European military conflict as a problem of "intellectual disunity," a lack of coherence in the culture that bred social instability. This in turn created weakness in the face of the existential threat posed by fascism and Communism. A significant part of this incoherence, he thought, arose from a widening gap between the realms of knowledge and morality: scientific research had become separated from ethical reflection, and in his view, it was the job of religion to rebuild that bridge.[8]

As we have seen, this idea of integrating science, religion, and ethics already existed within humanism, and it was a compelling idea. The difference was that humanists wanted to remake religion in a scientific and modern mold, while Finkelstein and other traditionalists believed that the task was to bring the wisdom of conventional religious thought to bear on science. As another speaker at the conference put it, "our way of life" was threatened with "the passing of ancient sanctions and the collapse of traditional loyalties." If the path forward meant that we must return to ancient principles and values, then humanism had nothing to offer. It was modern and naturalistic and utterly opposed to transcendence. The prominence given science was misguided: it should not lead; rather, it needed to be led. The conservatives sought harmony and agreement—"a unity of thought and effort"—but only within America's common Judeo-Christian heritage, and humanists and other nontheists seemed to be excluded from this.[9]

If this weren't bad enough, the acerbic University of Chicago professor of philosophy Mortimer Adler presented a now-infamous speech on the first day of the conference that only reinforced the humanists' predisposition to see the entire conference as a conservative, theistic, and antimodernist project. Adler entitled his speech "God and the Professors." In it, he extolled medieval scholastic philosophy and Aristotelianism in particular. Adler expressed a consciously backward-looking point of view that resonated with some of the themes already laid out by other speakers. Though Adler himself was Jewish, he sympathized with the recent efforts of Roman Catholic thinkers to resurrect their most famous theologian, Thomas Aquinas, who had managed in his day to transform pagan Aristotelian philosophy and put it to use to empower Christian theology.

Adler agreed with Finkelstein that there was a spiritual vacuum in the modern world—a yawning chasm that separated values and knowledge. He went further than Finkelstein in arguing that this was not just a problem for some scientists who overemphasized the power and role of the scientific method but endemic to the entire professoriate. Moreover, Adler was combative, and his extreme uncompromising assertions rang louder than those of most other speakers at the conference.

The crux of the problem, as he saw it, was logical positivism, a philosophical outlook that he believed had utterly overtaken the academy: "The most serious threat to Democracy is the positivism of the professors, which dominates every aspect of modern education and is the central corruption of modern culture."[10] Logical positivism (today often referred to as *logical empiricism*) was relatively new to American philosophy at the time. It had originated in Vienna in the early decades of the century and been enthusiastically promoted by prominent philosophers in Britain and America. Recent scholarship on the positivist movement shows it to have been much more complex and sophisticated than Adler admitted, but many contemporaries, supporters and opponents alike, also understood positivism in similarly simplistic ways.[11] In seeking a way to differentiate meaningful statements from meaningless ones, positivists asserted a strict separation of value statements from statements of verifiable fact and sought to differentiate all knowledge along these lines. Adler denounced what he saw as a takeover of philosophy by scientific method, diminishing philosophy's relevance and eviscerating the basis for moral reasoning.

Adler's angry attack on the academy gained additional rhetorical force because it was delivered to an anxious audience fearful of the dangers of war overseas. Perhaps the most quoted passage of his speech was his following claim: "Democracy has much more to fear from the mentality of its teachers than from the nihilism of Hitler. It is the same nihilism in both cases, but Hitler's is more honest and consistent, less blurred by subtleties and queasy qualifications, and hence less dangerous."[12] He decried the weakness of the philosophical position of the intelligentsia, calling it a form of idolatry, "a false religion." Explicitly contradicting the notion of democracy as proposed by John Dewey and his followers, Adler believed that true democracy was orderly, hierarchical, based on authority, and founded on the notion of natural rights. In a final swipe against his opponents, he stated that "science contributes nothing whatsoever to the understanding of Democracy."[13]

Adler's talk was followed by a blistering critique from New York University philosopher Sidney Hook, who gave a point-by-point refutation. Hook's response highlighted the erroneous assumptions made by Adler regarding his understanding of science, democracy, and religion. In a later reminiscence, Hook claimed that his own speech bothered Finkelstein so much that the theologian stamped hard on Hook's foot under the table to get him to stop talking. The exchange—Adler's over-the-top reaction to the entire American professoriate and then Hook's sharply incisive response—was a defining moment for the conference and has been the focus of historical evaluations ever since.[14] (It is always tempting to compare Hook and Adler because the two men represented such different factions of American thought, and they both had outsized influence on the country's intellectual life. One might be forgiven for mixing up their biographical details: born eight days and only a few miles apart, these two New York Jews obtained their doctorates in philosophy at Columbia in 1927 and 1928, respectively.)

The historian Fred Beuttler has cautioned that the Adler-Hook exchange was a one-off event that dramatically oversimplifies what happened at the conference series, which was a much more multifaceted and less divisive project than these two papers show.[15] Few papers were as reactionary as Adler's. Moreover, many liberals and scientists who had no religious agenda and even some humanists participated in this and future conferences. Indeed, Beuttler, who has studied this conference in detail, has argued that the primary motivation of the planners was liberal and pluralistic, an effort meant to bridge America's diverse religions and ideologies. Beuttler also reminds us that Adler left after the first conference and refused to participate later.[16] Those caveats are relevant to the wider understanding of this conference, but it remains a fact that the framing of the conference did antagonize humanists, and it did provide a stage for many speakers to either dismissively or antagonistically denounce the Deweyan pragmatic tradition and, by extension, naturalism and the humanist outlook.

Historical Shifts in American Theology

There was good reason why the humanists felt threatened by the language of the participants. For the past decade, the country had been shifting rightward on religious matters, and these shifts stung the humanists, who had been used to seeing themselves on the cutting edge of religious change. Having built their movement on top of liberal Protestant modernism, they envisioned a somewhat linear path toward a humanistic America—and even talked about humanism as the "religion of the future." America did not move in that direction, however; during the Depression, the population turned toward a more sober religiosity that incorporated ideas like suffering and sin, vocabulary seldom used by liberals, let alone humanists. Some historians have attributed this change to the changing economic and political conditions—depression and war—that gave the lie to the modernist promise of progress, peace, and prosperity. This trend toward traditionalism and conservatism took place in all three of America's core faith traditions: Judaism, Catholicism, and Protestantism. Further, it changed the way that science was perceived and discussed. Modernism's optimistic faith in scientific thinking faded in the wake of a traditionalist portrayal of science that was much narrower and frequently negative. For traditionalists, it was folly to imagine that the instrumental and utilitarian methods of science would be able to overcome the tragic aspects of our humanity. Scientifically based reason needed moral guidance.

The conservative shift was explicit in the preachings of the popular theologian Reinhold Niebuhr, who had himself taken an intellectual journey from a Marxist-inspired Social Gospel theology to a much more conservative outlook that he called neo-orthodoxy. Niebuhr's book *Moral Man and Immoral Society* (1932) expressed his own disillusionment with modernism and left-wing politics. In it, he attacked the philosophy of naturalism and sharply criticized John Dewey as the main architect of the failed modernist experiment.[17]

The liberals, said Niebuhr, overemphasized science and misunderstood mankind, and this led to an inability to create a coherent moral order. These ideas, as we have seen, were echoed by many of the conservative speakers in the Finkelstein conferences. Niebuhr believed that scientific naturalism was unable to produce an effective moral foundation for society because it lacked transcendence. Naturalism alone left people to the whims of their passions. Any action could be justified. Convenience, not ethical precepts, was the only possible guide in a world without a supreme lawmaker, he thought. In addition, and perhaps more importantly, Niebuhr rejected what he saw as modernism's naive faith in mankind: people were not innately good or inherently rational, assumptions that modernists erroneously insisted on. Everything that Niebuhr observed about human nature led him to see humanity as weak and easily corrupted by greed, envy, and desire. Since modernists had an insufficient grasp of these aspects of humanity, they failed to predict or to understand the contemporary world crises. This more tragic point of view made it possible to revive the notion of sin, which the modernists had thrown out. In sum, Niebuhr's outlook was a dramatic shift away from the ideas that had dominated the liberal seminaries and pulpits of the first part of the century.

Conservative Catholics also promoted a remarkably antimodern outlook, despite the fact that many Catholic writers were finding ways to reconcile such ideas as evolution and big bang cosmology with their faith. Conservatives, however, embraced the official Catholic theology based on Thomism, a philosophical framework developed by the Church's leading scholastic, the thirteenth-century philosopher Thomas Aquinas. Aquinas had been the touchstone of orthodoxy during the Counter-Reformation of the sixteenth century, so the twentieth-century revival of Thomism seemed to be nothing short of a reaction against modern philosophy.[18] The Thomist critique of modernity was similar to that of Niebuhr and other neo-orthodox Protestants insofar as it criticized liberal theology for overestimating the power of human effort alone: without God's support, they believed, there could be no just order. The Thomist view of human nature, like Niebuhr's neo-orthodoxy, contradicted many assumptions of humanists and naturalists. The American proponents of Thomism found ways to harmonize it with the preeminently American concern for democracy, and they did so in quite conservative terms, drawing on their philosophically complex theology and arguing that it could only be supported with a theistic faith.[19] This meant, of course, that the kind of science-based pragmatic conception of democracy found in Dewey and others was rejected. Catholics of this mind demanded that science be kept in its place.[20]

The Protestant and Catholic critics of liberalism and scientific naturalism were part of a new kind of defense of traditional religion that was no longer strongly sectarian. This broader ecumenical movement recharacterized America's religious heritage in a more pluralistic way. The Western intellectual heritage, it was claimed, was not simply Protestant, nor even exclusively Christian, but rather the product of a richer, more inclusive "Judeo-Christian tradition." The notion of such a tradition made it possible to bring most American theistic faiths into a single large family, a religious counterpart to the "melting pot" ideal of America.[21] The Conferences of Science, Philosophy and Religion were among the first places that this notion of a Judeo-Christian tradition was expressed.

The language of a common Judeo-Christian family was adopted by orthodox theologians and came to be used as another way to bludgeon secular philosophy, naturalism, and humanism. Whereas orthodoxies of the past were generally sectarian and exclusivist in nature, the Judeo-Christian framework reconfigured conservatism to make it more accommodating in a pluralistic America. But this was not a liberalizing move. When we look at the way that it was employed by Finkelstein and others, this becomes clear. Religious modernism, with its deference to scientific knowledge, is nowhere to be seen. In its place, we find talk about the foundations of American thought in ancient and medieval times. By turning back to ages past, when the three religious traditions purportedly shared common origins, the spotlight shifts from modernity and science to metaphysics and theology. In doing so, it abetted conservatives like Finkelstein and Adler who wanted to ground an American ethical system in orthodoxy. The secular Enlightenment as the humanists understood it was not part of the Judeo-Christian notion of America.[22]

Humanist Response

For all the reasons just described, humanists responded to the Conferences of Science, Philosophy and Religion with passion and intensity. Three factors help to explain the strong reaction: historical shifts in theology, wartime rhetoric, and organizational necessity. The shifts in theology diminished the authority of liberal modernism, and the wartime rhetoric heightened the differences among different groups. In terms of organizational necessity, the fact that humanism was still a relatively young movement and undergoing a revival meant that the conferences served as a useful foil that could help the humanists popularize their own agenda. This organizational factor should not be underestimated. What emerged from this period was a somewhat changed humanism, a movement with agendas different from those of the 1933 Humanist Manifesto generation.

Edwin Wilson, editing the newly revived *Humanist*, became a central figure in coordinating a response to the conference. The first issue of the *Humanist* came out in the spring of 1941, half a year after the first of Finkelstein's conferences. Over the next several issues, Wilson published five articles by different critics reviewing various papers from the conference series. One of those articles, a rousing piece by Sidney Hook in the autumn of 1942 entitled "Theological Tom-Tom and Metaphysical Bagpipe," became a lightning rod that spawned a fifteen-page symposium, pro and con, on Hook's contentions. Humanist Max Otto was so impressed that he wrote Hook a personal note: "If *The Humanist* printed such direct, incisive, able articles more often we'd get on with it and its objectives more conspicuously than we do."[23] In that article, Hook called for a counter-conference that would provide a competing message, and this suggestion became the nucleus for a new project that Wilson immediately jumped on.

Where Wilson provided the organizational muscle, Hook was the one who lent an intellectual punch and tenacity to the humanist response. It is hard to imagine that the humanist reaction would have materialized without Hook's presence. His involvement helped reshape the humanist movement in several respects.

"If the life of the mind were a street fight, Sidney Hook would be without peer among modern American intellectuals." So opens the article on Hook in the 1995 *Companion to American Thought*. That statement pretty much summarizes Hook's intellectual temperament.[24] He was a perfect sparring partner for Mortimer Adler on the conference stage on that first day of the Finkelstein conference. Moreover, Hook was a staunch defender of John Dewey, under whom he had obtained his degree in philosophy at Columbia, and to whom he gave his loyal support throughout his life. Hook's willingness to speak out in articulate and impassioned prose on controversial political and social matters gained him notoriety as one of America's most important public intellectuals of the twentieth century.[25]

He was also a Marxist, and the most significant accomplishment of his early scholarship was the refashioning of Marxism into a form that was compatible with American pragmatism and sharply opposed to much European-style authoritarian socialist thought. With zero interest in religious practice, he affiliated with the humanist community solely as a result of his Columbia connection, most especially through his close relationship with Dewey. In this respect, Hook was unlike nearly all of the philosophers in the humanist movement who preceded him—including Dewey himself—who, in one form or another, understood humanism in religious terms. A Jew by birth, Hook was not a member of the Ethical Culture movement, nor was he a Unitarian. Religion was, if anything, anathema to him, and while he may never have attacked liberal religion as such, when he spoke about religion, he derided it.[26] This is nowhere more evident than in the attacks he leveled on his targets in the Conference on Science, Philosophy and Religion.

Hook excoriated the conference in print. Just over a month after he faced Mortimer Adler on stage at the Finkelstein conference, Hook's speech was published in the *New Republic*, warning of "the new medievalism" that the conference represented.[27] Two years later, on the occasion of the publication of the second set of conference papers, Hook wrote "Theological Tom-Tom and Metaphysical Bagpipe," where he followed up his criticisms with the claim that the theologians had taken over. The starting assumption of the conference, he said, was that "a belief in supernatural religion is the necessary, if not sufficient, condition for genuine democracy. Those who do not share this belief are regarded as philosophical fifth columnists—'atheistic saboteurs' a leading speaker calls them at the concluding session—who are doing Hitler's work." The conference papers demonstrated to him that humanism and scientific method were under attack and that democracy was doomed unless this authoritarian attitude was rebuffed.[28]

Hook followed his "Theological Tom-Tom" article with another piece in the influential *Partisan Review*. In that article, he developed a wide-ranging historical argument about the dangers of supernatural religion to the stability of civilization. He announced that a "new failure of nerve" was at hand, a civilizational crisis similar to the loss of confidence that spread throughout the Roman world at the outset of the Middle Ages. Hook's analysis drew on the ideas of an earlier scholar, Gilbert Murray, who used this psychological explanation to describe the retreat of Greco-Roman culture during the decline of the Roman Empire, a loss of confidence in human creativity and the fearful turn to supernatural certitudes.[29] We moderns, Hook said, faced a similar peril unless we regained our courage: "A survey of the cultural tendencies of our own times shows many signs pointing to a new failure of nerve in Western civilization. . . . [A]t bottom it betrays . . . the same flight from responsibility, both on the plane of action and on the plane of belief, that drove the ancient world into the shelters of pagan and Christian supernaturalism."[30]

There were ample signs that our civilization was already in retreat: more and more people had begun to believe in the original depravity of human nature, leaders were being deified, there was much posturing about spiritual purity, and people discussed "mysteries" rather than "problems."[31] In all, Hook turned the tables on the orthodox writers who had attacked uncertainty and naturalism as the basis for fascism. To the contrary, not the lack of certainty but the search for absolutisms was the threat. Nazism exemplified what went wrong when people embraced a misguided lust for the absolute, when militarism and nationalism overwhelmed a society's foundation in scientific thinking and democratic behavior.[32]

Other humanists writing for Wilson also published denunciations of misleading claims and skewered many of the conference papers. They emphasized Adler's extremism and focused on the theistic rhetoric of Finkelstein and others, depicting the entire conference series as deeply reactionary. In an article entitled "Take It Away!" Max Otto complained of the supernaturalist and the Judeo-Christian emphasis. He especially attacked a paper written by a group of Princeton scholars arguing for an explicitly Judeo-Christian foundation for democracy. Otto objected to the way that the authors discussed science and the fact-value distinction. Like Adler, they claimed that it was impossible to justify democratic values through any strictly scientific thinking because science had no ability to assess moral claims. Their conclusion? Supernatural faith was essential to democracy. Otto, not surprisingly, found this idea outrageous; it was dualistic and neglected science's usefulness in supporting democratic values. (Otto, as you may recall from the previous chapter, saw values and ideals emerging from the human experience, which was in his view little different from the scientific experience.)[33]

Another *Humanist* author objected when the religious principle of *agape*, or selfless love, was proposed as the basis for social development. "Scientific living," he declared, not agape, should be our goal. He went on to argue that by adopting science as our mode of thought, we would embrace a "*tentative loyalty*" to hypotheses that would be constantly tested. To be scientific, to be modern, one needed to leave absolutism behind.[34]

The Scientific Spirit Conferences

In "Theological Tom-Tom," Hook proposed that "humanists and genuine democrats" boycott the conferences and hold meetings of their own. Wilson thought that an excellent idea and began to organize a planning committee in New York early in 1943.[35] The committee consisted of university professors and humanistic religious leaders. Columbia historian J. H. Randall Jr. was elected chairman, and John Dewey was elected honorary chairman. Ethical Culture leader Jerome Nathanson took over the day-to-day organizational and planning duties. Sponsors included humanist ministers John Dietrich, Charles Potter, and Curtis Reese; philosophers Corliss Lamont, Max Otto, and Roy Wood Sellars; and religious scholar A. Eustace Haydon. All in all, the composition of the group included most of the first-generation humanists who had been involved in the manifesto work a decade before. The presence of Hook and a few other new voices, however, was notable. The "Conference on the Scientific Spirit and Democratic Faith" convened in May of 1943 in the Ethical Culture meetinghouse. It was attended by between four hundred and five hundred people.[36]

Like Finkelstein, the humanists were instilled with a sense of urgency by the ongoing war. Participants of the Scientific Spirit conferences saw the war as an illness infecting the body of modern civilization. Like their orthodox opponents, the humanists frequently talked about these dangers as psychological and moral, not just political. Moreover, they also saw this as a battle in which the enemies at home were perhaps as dangerous as the enemies abroad.

The printed program of the first Scientific Spirit conference put the scientific focus of the meeting front and center with this epigraph by Thomas Jefferson: "Reason and free inquiry are the only effectual agents against error." The conference members made much of the relationship between scientific and political values. The underlying assumption of the conference was that science and democracy went hand in hand—that democratic habits of mind came about through the scientific spirit. Indeed, science was everywhere in the papers given by the conferees, and the key to understanding what the humanists were trying to do at this point lies in how they talked about science and its role in society. The characterizations of science echo the words of leading humanist intellectuals, such as John Dewey's praise of tentativeness and John Dietrich's talk of courage.[37]

One thing that both the humanists and the orthodox writers seemed to agree on was that science and religion needed to be distinguished carefully. Moreover, both groups agreed that in science only tentative knowledge was possible, whereas most religious positions sought or claimed to have attained certain knowledge. For humanists, however, tentativeness was a principal value, and the claim to certain knowledge was dangerous. Unlike the orthodox religionists, the humanists believed that this tentativeness brought with it habits of mind that extended beyond the laboratory and could be embraced by all people. For the humanists, the scientific method taught values that arose from an acceptance of uncertainty, in particular, tolerance and flexibility. Religion, by contrast—in this case, orthodox religion—taught an "allegiance to fixed principles, inflexible rules of morality, and unquestioned acceptance of a supernatural interpretation of human experience."[38] The latter qualities were unsuited to the modern world. They were dangerous and bred authoritarianism and a dogmatic temperament. Science, not orthodox religion, contained the values that best served modern life.

The first humanist conference on the scientific spirit was in many ways a recap of long-developed humanist ideas. It explained why the scientific mind was of crucial importance to democracy. The second conference, one year later, focused on "the authoritarian assault on education" by Catholics and other traditionalists. The topic was by no means arbitrary. It would not be an exaggeration to say that the entire debate between liberal humanists and proponents of traditionalism—starting with Adler's "God and the Professors" address—was fundamentally about the nature and future of education in America. It was a fight that was already raging at the University of Chicago and had everything to do with the influence of John Dewey.

Dewey was a towering figure in American education. He was the figurehead for the progressive education movement even though many educators departed quite dramatically from what he wrote. Several traditionalist critics of Dewey took him to task for his view that the *process* of learning should be the focus of education rather than its content. The conservatives' argument against progressive education mirrored their antagonism to Dewey's naturalism: essentially, they charged that he left no room for moral and religious instruction, which ought to be the centerpiece of early childhood learning. Roman Catholic neo-Thomist thinkers especially believed that there was a hierarchy of knowledge and that this hierarchy was contradicted directly by Dewey's dynamic evolutionary worldview. Most significantly, Dewey was a humanist and a naturalist, so the educational system he proposed was based on scientific method and viewed religious dogma as irrelevant and dangerous.[39]

Adler's "God and the Professors" talk extended this line of argument to higher education. Adler and Robert Maynard Hutchins, president of the University of Chicago, had been building a case against the Deweyan view of education since the early 1930s. Hutchins's extremely popular 1936 book *The Higher Learning in America* argued for the complete restructuring of both common school curriculum and the academy based on a "great books" approach to learning, one that demanded that all students study a core curriculum of the most important books of Western civilization. Hutchins lambasted process-based education with its practical and technical focus, and he promoted in its place a return to morality-driven learning that would serve as the foundation for all thought.[40]

In rebutting the conservative educational agenda, Edwin Wilson explained the purpose of the second Scientific Spirit conference in a donation request that he mailed to colleagues and contacts in the days leading up to the meeting: "Reactionary religious groups in the past year have been especially aggressive in the effort to capture our schools for orthodoxy. The aim of [our] Conference is to counteract the antiscientific and antidemocratic reaction in contemporary thought with a positive application to present problems."[41]

At the conference, several speakers attacked the great books curriculum because it was reactionary and antiscientific and enshrined "the ghosts of departed wisdom."[42] Participants included many men already in the humanist movement, as well as some who would soon come to play more prominent roles. There were philosophers, educational theorists, scientists, and editors.[43] Randall and Haydon participated again, as did Hook, and even the aging Dewey presented a paper at this conference. In the concluding session, there were two men with close scientific ties: A. J. Carlson, president of the American Association for the Advancement of Science and signer of the Humanist Manifesto, and Gerald Wendt, science editor at *Time* magazine. Carlson would later be named Humanist of the Year, and Wendt would much later assume editorship of the *Humanist*.[44]

Yet a third conference met the following year on the theme "Science for Democracy," which examined the interconnections between the scientific and political spheres.[45] Together, these humanist-sponsored conferences offered proposals for integrating science more fully into American culture, especially in the realm of political and personal values. Some of the papers focused on educational reform, some on the popularization of science, and some on the need for scientists to get more involved in world affairs. All of them promoted the view that a culture with a strong scientific focus could stand up to authoritarian politics and reactionary religion.

The Problem of Religion

Humanism's status as a religion became a much-debated topic at this time of transition in the group, and the new organizational structures provided the forums where these debates took place. It was at this time, in 1941, that a core group of ministers established the AHA, whose main focus was to publish the quarterly journal the *Humanist*. The journal published a wide diversity of material, including commentary on issues of the day. Significantly, it was no longer directed toward humanist ministers and their congregations. The association was conceived as a means to extend beyond the churchgoing humanists and reach a broad public. The question now was, What was humanism? Was it still a religion? The bylaws of the new group characterized the AHA as an "educational" project of "religious and other humanists," not "a new sect or cult."[46]

The case for religion was propounded by most of the old guard and was made again and again in various ways and in various forums. Those who saw humanism in a religious context were content to discriminate between liberal religion and conservative religion. As humanists had long known, liberal religion was chiefly characterized by its accommodation to science and liberal social views. One humanist, Harry Overstreet, made the point that science and religion could be reconciled if religion were redefined and purged of its historical institutions, creeds, and priests.[47] E. Burdette Backus, a manifesto signatory and long-time Unitarian minister, went even further at the first Scientific Spirit conference when he said that this new accommodation represented the birth of a new religion: "It was clear, from the deliberations of this conference, that a new religion is in the process of development, a religion in which the spirit of science with its passion for the truth is joined with the spirit of democracy with its passion for mankind. . . . [The conference] was in itself a splendid manifestation of the power and resources of this religion and a validation of our hope that it may indeed prove to be the religion of the future."[48]

The *Humanist* magazine featured many articles defending the idea that humanism was a religion. The banner issue in the spring of 1941, for example, opened with article-length definitions by many of the old guard who had fought early battles with modernists to get humanism accepted as a religion. Bryn Mawr psychologist James H. Leuba emphasized the biological nature of *Homo sapiens*'s urge toward "spirituality." This was followed by an article entitled "Humanism as a Religion" by Roy Wood Sellars. Humanism, he contended, was a kind of religion for "adult humanity." Later issues continued the trend. Archie J. Bahm, a professor of philosophy and sociology at Texas Technical Institute, suggested that although individual humanists belonged to separate denominations, humanism, broadly conceived, was a religion. It was unlike other religions, however, because it embraced the entire brotherhood of man. As such, he thought that humanism was attempting something unique: where other religions excluded people on the basis of belief, humanism accepted every person. This supported the common humanist belief that all religious traditions would eventually metamorphose into varieties of humanism.[49]

Similar support for the pro-religious position appeared the following year. Bahm and Leuba wrote again, in addition to articles by a philosopher, a poet, a publisher, and a Unitarian minister. Leuba argued that the words *religion* and *religious* could be used by humanists because they did not have to carry supernaturalistic implications. Those words were too valuable to be left to the supernaturalists, he said, because they gave unique expression to "the manifestations of devotion to whatever is conceived as the highest aim of humanity."[50]

But what of those "other humanists" who were mentioned in the AHA bylaws? One of the loudest voices against the religious definition of humanism came from the wealthy New York Marxist Corliss Lamont. Lamont asserted that it was wrong to characterize humanism as religious. Reacting against the language of the Humanist Manifesto of 1933, he contended that it contained "probably the worst and loosest definition of religion that I ever saw." "Such definitions," he continued, "add nothing but evasion and confusion, confounding their creators, their users and the public at large."[51] While Lamont was teaching philosophy at Columbia, he grew tired of the kind of humanism espoused by other Columbia professors: it was far too polite. He was disappointed in the faculty's cowardly and hypocritical stance on religion: while they embraced atheistic naturalism, they refused to attack religion, but instead redefined God, immortality, and other religious terms so as not to hurt anyone's feelings. It was simply "verbal hocus-pocus."[52]

Lamont complained that "most of the members of the Department are either sons of ministers, son-in-laws of ministers, or ex-would-be ministers. Only, I suppose, when one is no more than the *grand*son of a minister, like myself, is it possible to view religion as objectively as any other phenomenon." When the professors remained too closely attached to their Protestant roots, they allowed their sentiments to get in the way of their scholarly judgment. Questions about religion and death, he felt, were far too important to be obscured by vague and sentimental language. According to Lamont, the term *religion* should follow popular usage, which linked it to belief in a supernatural order.[53]

The irreligious stance of some members generated friction within the group. The aggressive secularism of Sidney Hook, for instance, angered John Herman Randall Jr., who resigned as chairman of the Scientific Spirit conference planning committee as a protest against the "anti-Catholic bias" of some of its members, one of whom was most certainly Hook.[54] As an intellectual historian who was interested in the contributions of the Catholic Church (which he believed had nurtured the Western intellectual heritage), he could hardly concur with Hook's openly prejudicial attitude.

But even Edwin Wilson was cautious about the religious label. A Unitarian minister, Wilson remained wedded to the notion of a religious humanism, but as editor of the *Humanist*, he was careful to avoid excluding those who did not see humanism in this way. He earnestly wanted humanism to be a big tent, and he was as eager to print articles critiquing the religious label as he was the reverse. Moreover, even as a Unitarian, he had reasons for minimizing the explicitly religious elements of humanism in the AHA. In particular, he did not want the AHA to be viewed as a separate, competing church. Humanism, in his view, was a movement that had a religious expression within the liberal churches, but it also had a secular expression in the pages of the *Humanist*. For the next two decades Wilson worked hard to keep the AHA and the humanist-oriented Unitarian churches from clashing. In his words, the AHA should "supplement, not supplant," humanistic Unitarian congregations.[55]

While old-line humanists still considered humanism a religious movement, Hook's and Lamont's antagonism to the very concept of religion points to a new sensibility creeping in. To some degree, this was because religion itself had come to look a lot more conservative than it had in the 1910s and 1920s. Hook and Lamont did not want to identify themselves with traditions that they considered essentially reactionary. But there were other reasons as well. Humanism was a movement of intellectuals, and it attracted academics from all over who sympathized with its progressive, scientific, and modernist sensibilities. There was a lot of diversity in that group. Among the new recruits were people who had very different ideas from those of the religious old guard. Among them were a host of ideologically motivated scientists who came to play a major role in the movement going forward, men who brought with them not only a passion for science quite different from that of the ministers and philosophers but also a strong internationalist attitude.

CHAPTER SIX

Scientists on the World Stage

The alternative offered as a third way out of the present crisis of civilisation is humanism, which is not a new sect, but the outcome of a long tradition that has inspired many of the world's thinkers and creative artists and given rise to science itself.

Amsterdam Declaration of the International Humanist
and Ethical Union (1952)

In the immediate postwar years, American internationalism reached a peak. These were the years when the United Nations arose and American influence was framed in terms of cooperation and aid to nations around the world. Of course, it was also a time of fierce anti-Communism. In fact, much of that internationalism was part of the ideological warfare that the country was waging against Russia.[1] Humanists entered this heady moment decidedly on the side of internationalism. Although some of its members were staunch anti-Communists, even they were cosmopolitan and internationalist in sentiment. The humanists embraced the new global order, and in so doing, they came to see humanism as a global movement.

The American Humanist Association supported internationalism in various ways. The organization voted on and passed a wide variety of resolutions promoting international causes: disarmament, population control, universal calendar reform, and even Esperanto. More than once, they resolved that the United Nations ought to be more powerful and influential. In addition to the resolutions, the organization was able to get accredited as a nongovernmental organization, something that allowed them to send representatives to UN briefing sessions.[2]

The composition of the movement began to change during this time. One of the most surprising things about early humanism is that for nearly three decades hardly any humanist leaders were scientists. Given how important science was to their ideology, it is odd that the movement grew and flourished throughout the first part of the century with relatively few practicing scientists in its ranks. As we have seen, the early movement was dominated by ministers and philosophers, with the latter being the main force behind its ideology, bringing philosophical rigor to the religious vision of the ministers. The philosophers were the ones who defined science and characterized its role vis-à-vis religion and society. And this meant that the humanists' idea of the scientific spirit—a concept central to the movement's identity—was an outgrowth of the views of its largely Deweyan cohort of philosophers (with Sellars's own brand of evolutionary naturalism integrating easily into that framework).

Not until the postwar period do we find natural and social scientists assuming prominent roles in the movement, and with their arrival, humanism changed. This was in part because the scientists' overall understanding of what science was, as well as what its role should be, differed from that of the philosophers and ministers. And it was in part due to the scientists' detachment from religion: although some joined the Unitarian church, and some talked about their emotional and psychological connection to nature, they seldom discussed the topic of *religion*. In other words, they spoke a different language than the founders of humanism, one that was more secular and more utilitarian. At the same time, the scientists also opened new doors as activists. Their broadly cosmopolitan outlook prepared them to participate in the newly arising international organizations, and many of them assumed positions of leadership within the intergovernmental groups and agencies—the United Nations Educational, Scientific, and Cultural Organization (UNESCO) and the World Health Organization (WHO), in particular. Some of them even played central roles in the international antinuclear movement by spearheading one of the most significant antiwar institutions of the period, the Pugwash Conferences.

The International Humanist and Ethical Union

One of these men, Julian Huxley, was a key figure in promoting humanism as an international movement. This British biologist was the grandson of T. H. Huxley ("Darwin's bulldog"). In the years following the conclusion of the war, he and Edwin Wilson, director of the AHA, were writing to each other. Wilson reported in the humanist's newsletter, for instance, that "in January Dr. Julian Huxley wrote us that 'some form of Humanism will be the world's next important religion.'"[3] Huxley, one of the few scientists in the movement who used religious language to talk about humanism, was keenly aware of the limitations of the new intergovernmental and UN organizations, having just completed a disastrous short tenure as the first director general of UNESCO. He said that these groups, whose work was essentially humanist, "can't adopt an explicitly Humanist philosophy because they're official organizations" made up of governments of different philosophies and religions.[4] As a result, he was anxious to put together an international coalition of humanist groups to lead what he saw as a coming worldwide revolution in thought.

This new international humanist coalition considered merging with some preexisting freethought groups but ultimately determined that this would be too narrowly focused on anticlericalism to be the home for a broad movement like humanism.[5] The result of this work was that in 1952 the AHA joined with other humanist and Ethical Culture groups from around the world to found the International Humanist and Ethical Union (IHEU). Huxley was its first president. The seven groups composing the union were the AHA, the American Ethical Union, the British Ethical Union, the Vienna Ethical Society, the Dutch Humanist League, the Humanist League of Belgium, and the Radical Humanist Movement of India. Several common threads united them. They espoused democracy, promoted the creative uses of science, recognized the dignity and freedom of man, rejected all creeds and all methods of indoctrination, emphasized personal liberty and responsibility, and advocated a way of life that resulted in maximum fulfillment for each individual. Huxley's language to the contrary, the group did not consider itself a religious organization.[6]

Its principles differed little from many earlier versions of humanism. Apart from a vigorous debate over whether or not the word *ethical* should appear in its name to recognize the participation of the Ethical Culture societies, the participants were in agreement on most issues. The recent world war and the rise of a very chilling Cold War imbued the 1952 meeting with a cautious outlook on the present. Yet even with these acknowledged global dangers, an air of exhilaration permeated the meeting. The participants felt that they were standing on the edge of a new era when national boundaries would virtually disappear and a thriving global culture would arise. Their main concern was how the humanist movement could best help catalyze this leap into the future.

At the first congress of the IHEU, the conferees agreed on the centrality of science to humanism. The first session included several papers that dealt with "the meaning of science and democracy in human progress."[7] The talks in this session highlighted the tensions within the humanist movement between the rationalistic humanists, who had complete faith in science, and other humanists who took a more cautious attitude toward scientific or technical solutions to human problems. All were clear that humanism represented the future and frequently disparaged traditional religion, or at least their caricature of it.

In one of these papers, the Dutch schoolteacher Libbe van der Wal spoke on the need for scientific ethics. Scientists had too long ignored ethical concerns, he thought. Ethics should be based on the scientific method rather than religious tradition, and it was incumbent on humanists and scientists to apply the intellectual rigor of science to such areas of human conduct as morality, education, justice, and politics. Similarly, Curtis Reese, now the dean of a Chicago settlement house, continued to press for a "scientific religion," one that, unlike traditional religions obsessed with metaphysics, would address the mundane problems of daily life. Internationalism was central to the way Reese understood this project, as the new religion would go "beyond parochial and provincial biases, national and ethnic prejudices, and [would be] capable of universal application." Despite his modern ideas, he used old-fashioned religious language from the manifesto days to talk about scientific religion: it "can infuse laboratories and factories with the spirit of holiness. It can throw the mantle of sacredness over the common affairs of man, and it can make of human economy a divine adventure." Barbara Wootton, a British social scientist and magistrate, focused on politics rather than religion or ethics. She emphasized the need to transform democratic institutions in light of science. She argued for a technocratic, managerial system that employed psychologists to identify tyrannical personalities. She wanted to replace elections with opinion polls and to promote eugenic programs to raise the intelligence and public-spiritedness of the population.[8]

A few people voiced opinions that opposed this strong scientistic approach. The Columbia philosopher Horace Friess, an Ethical Culturist, cautioned against the overzealous idolizing of science. A disciple of Felix Adler, Friess's views reflected the warier attitude toward science typical of Ethical Culture members. In his view, both science and democracy had to be understood as imperfect, human creations. Neither should be oversold as complete solutions to humanity's problems. Friess's main concern was the promotion of "spiritual diversity" and toleration in a pluralistic society. Not everyone will gain "the fullest meaning and the deepest peace" by adopting "the same attitude and belief about the universe" that humanists did.[9]

Though not all members of the international coalition championed this more focused emphasis on scientific utility—what I am calling here *scientific humanism*—the overall ethos of the movement with regard to science was quite different from that of American humanism in earlier years. These men and women placed more weight on the utilitarian power of science to address human problems and encouraged technocratic solutions. Most importantly, they argued that scientists themselves should be given more power and influence within society. The overall outlook was mixed, but the more evocative language of "scientific spirit" that had been developed by the first-generation humanists disappeared to a large extent (with a few notable exceptions, as we will see).

The Scientists of the Chicago Circle

Back in America, the AHA was changing too. The group became more secular, with fewer clergy in leadership positions. When the AHA was established in 1941, nearly all of the board members were practicing ministers. Five years later, still not much had changed: the president, two vice presidents, and the treasurer were all ministers. After a decade, however, the board looked very different. The president of the AHA in 1951 was a businessman, Lloyd Morain, and the rest of the board included two scientists and a lawyer. Only one minister remained.[10]

The decline of ministers in the movement was accompanied by an increase in the number of scientists. The paths that took these scientists into humanism varied widely, as did the nature of their involvement. Some of them were simply given honors by the group but otherwise did not actively promote themselves as humanists. Some wrote articles in the *Humanist* magazine, where they were able to speak out on the relation of science to humanity. And a few of them became enthusiastic supporters and assumed influential positions within the organization.

Starting in 1953, the AHA began offering an award for Humanist of the Year, a person who, "through the application of humanist values, has made a significant contribution to the improvement of the human condition."[11] Of the first seven awards in the 1950s, six went to a practicing scientist or doctor: the physiologist Anton J. Carlson, the behavioral scientist A. F. Bentley, the medical doctor James P. Warbasse, the neuroscientist C. Judson Herrick, the biologist Oscar Riddle, and the psychiatrist Brock Chisholm. Even the 1957 award to the birth control activist Margaret Sanger could be considered a vote for science, dealing as it did with women's health and the technology of contraception. The list of scientists continues into the 1960s, with awards going to the physicist Leo Szilard; the chemist Linus Pauling; the biologists Julian

Huxley and Hermann Muller; the psychologists Carl Rogers, Hudson Hoagland, Erich Fromm, and Abraham Maslow; the doctor Benjamin Spock; and the engineer Buckminster Fuller. All of these people either employed their scientific or technical expertise to solve major social problems or rose to prominence based on their scientific or medical credentials.[12]

The path into humanism for these scientists varied. Four of the earliest scientists involved in the movement belong to the first generation of religious humanists and came to it as a result of the religious humanist community in and around Chicago. Some biographical information about these four men provides insight into the connections that bound them to humanism. Bernard Fantus, a Hungarian immigrant, established himself in Chicago after the turn of the century and soon became a professor of pharmacology and therapeutics at the University of Illinois. His signature achievement was the establishment of the nation's first refrigerated blood bank in 1937—indeed, even the term *blood bank* was his coinage. The idea was relatively simple, but it involved rethinking the mission and capabilities of hospitals, which were not accustomed to working in a larger regional infrastructure with such technological sophistication. The collection, storage, and distribution of human blood required complex coordination among disparate organizations, and the blood bank system was a prime example of how modern, twentieth-century technology could transform institutions.[13] Fantus was a member of the Unitarian Church where Edwin Wilson ministered at the time, and in his obituary Wilson praised Fantus for both his scientific achievements and his involvement in the movement. As a manifesto signatory, Fantus had given much thought to the document, and in the years after its publication he began drafting a revision.[14]

Anton J. Carlson was a Swedish immigrant who arrived in the United States as a young man. He took his medical degree at Stanford and moved to the University of Chicago, where he eventually became the chair of the department of physiology. Carlson had originally planned to go into the ministry but abandoned the idea when he ran into trouble at seminary. He caused an uproar when he suggested a way to scientifically test the efficacy of prayer with an experiment: ministers would ask their congregations to pray for rain, and then, later, the prayer days could be statistically compared with weather bureau data.[15] After switching to science, he made a name for himself as a physiologist doing research in cardiology and digestion, and he even landed on the cover of *Time* magazine. His presence on the faculty of the University of Chicago brought him into the humanist network there. Having signed the manifesto, Carlson also went on to write articles for the journal. One piece, entitled "Biology and the Future of Man," focused on how our understanding of biology, as it related to reproduction, food supply, pollution, housing, and health, would affect our ability to create a better future for ourselves.[16]

Maurice Visscher wrote a column called "Science for Humanity" for the *Humanist* in the forties and fifties. He took his doctorate in physiology from the University of Minnesota and studied under Carlson at Chicago. Although he was raised in the Reformed Church, he ultimately abandoned that denomination's conservative Calvinist theology. A year after the Humanist Manifesto was published, Visscher became a member of the Third Unitarian Church of Chicago; like Fantus, he was one of Wilson's congregants. Upon his return to Minneapolis to head the University of Minnesota's physiology department, he joined John Dietrich's church, maintaining his close Unitarian humanist connections. Visscher published widely on medical topics and was eventually elected to America's prestigious National Academy of Sciences.[17]

In his "Science for Humanity" column, Visscher discussed science and politics. In particular, he worried about the resistance that political groups had toward science and scientific solutions to social problems. On the national front, he lambasted the US Congress for underfunding the National Science Foundation, for preventing it from supporting research in the social sciences, and for its paranoid restrictions against leftist scientists. On the international front, he reported frequently on the efforts of the United Nations, UNESCO, and nongovernmental organizations, favorably describing how they used scientific knowledge and techniques to address pressing social problems. Echoing the manifesto, he declared that "the spirit of scientific enlightenment is being breathed into the world through these international efforts."[18]

The neurologist C. Judson Herrick grew up in a religious family. His father was a Free Will Baptist minister, and at one time he also considered the ministry. His religious ambitions faded, however, and he turned to science. The focus of Herrick's research was in "psychobiology," which sought to explain human nature as a product of our biological and neurological makeup.[19] The same year that he received a Humanist of the Year award, he published *The Evolution of Human Nature*, in which he argued that in order to solve our social problems, we must understand our physical nature. "The human body," he explained at the outset of his book, "is the most complicated mechanism . . . in our known universe."[20] We must understand this mechanism if we are to save ourselves from destruction. Herrick decried the slow pace of cultural evolution that brought us to the brink of nuclear annihilation. Science could help if we put it to work to find "the biological principles with which all human conduct must conform if civilization is to survive and prosper."[21]

Humanism gave these four scientists—Fantus, Carlson, Visscher, and Herrick—a way to integrate their scientific materialist outlook and their professional calling with a pragmatic social agenda. For Visscher and Fantus, the church remained important as a social center that gave organizational impetus to their humanism. Despite their religious upbringings, they all seemed less interested in the religious aspects of humanism and more interested in its ability to promote science as an effective tool for solving human social problems.

The term *scientific humanism* began showing up at about this time as a way to describe this type of humanism. It was an outlook in sharp contrast to the highly philosophical concerns of the earlier religious humanists. The involvement of these scientists changed the nature of the movement. It was no longer wedded to the larger philosophical understanding that the Deweyan pragmatists espoused or to the liberal religious ideas of the ministers. While the scientists here did nothing that explicitly contradicted these other ways of explaining humanism, and in many ways provided a very compatible alternative, the outlook came with different assumptions and different goals. Scientific humanism had a new texture.

The UNESCO Affair

Scientific humanists brought with them a cosmopolitan perspective that harmonized well with the postwar enthusiasm for international collaboration and world governance. The discovery and use of nuclear weapons played a dramatic role in this. The bombing of Hiroshima and Nagasaki terrified the world and created a dark shadow over science and its potential for destruction. At the same time, it brought enormous prestige and stature to scientists themselves; the sheer power of scientific knowledge caused many people to turn to scientists for answers to all kinds of questions. In this new environment, scientists were thrust into the limelight and into positions of political responsibility. The scientists who readily embraced this new role often found themselves cheered on by the humanists.[22]

One of the most enthusiastic defenders of scientists as leaders in the new world order was Sir Richard Gregory, president of the British Association for the Advancement of Science. Gregory advocated his technocratic vision in a 1943 paper for the Conference on the Scientific Spirit and Democratic Faith, discussed in the previous chapter. Gregory's paper, entitled "Science as International Ethics," explained how important science was to world affairs. Contrasting science and politics, Gregory said that the former was objective and empirical; the latter, prejudiced and irrational. Many of mankind's problems were due to the fact that political decisions were being made without proper scientific grounding. The two activities remained separate, yet the future of humanity depended on our ability to make them work together. Gregory argued that science could make politics more rational and evidence based. If people could apply "the principles of independent inquiry and impartial judgment" to political affairs, they would be able to make better decisions. All of this, he believed, was compatible with democratic governance.[23]

Gregory went further, arguing that scientists should replace the politicians. Their objective, problem-solving mentality, he thought, was particularly well suited to positions of political power. Moreover, he believed that their work obliged them to assume a responsibility for the proper application of their discoveries. "The powers [scientists] have released have not been rightly used in the services of mankind as a whole," he said, but if scientists had the ability to help guide the way that those powers were used, the world would be better off. Indeed, scientists have a moral obligation, he thought, to enter the political sphere and "assist in the establishment of a rational and harmonious social order."[24] This style of governance put scientists and experts in the seat of power. This was not a new idea. John Dewey and most early twentieth-century progressives supported technocracy through rational planning and the consultation of scientists and experts. Unlike these thinkers, though, Gregory expected the scientists themselves to wield power.[25]

Two men, in particular, were lauded by humanists for their contributions in the new international order: Brock Chisholm and Julian Huxley. Both were given directorships in the international governing bodies affiliated with the United Nations, Chisholm at the WHO and Huxley at UNESCO. Chisholm and Huxley were exemplars of Gregory's scientist-politician.

Chisholm, voted Humanist of the Year in 1959, was a Canadian psychiatrist. He had started his career by studying ways to apply psychology to the training of soldiers. From there he moved directly into the Canadian bureaucracy as deputy minister of health. Chisholm championed technocracy of the type advocated by Gregory: unless scientists who studied and understood human behavior were allowed to lead society, we would never achieve world peace.[26]

Chisholm made no bones about his antipathy to religion. He was a staunch anti-Catholic and spoke out on a number of issues that brought him widespread notoriety in his country. At one point, for example, he suggested that children not be encouraged to believe in Santa Claus, a fiction that he considered "one of the worst offenders against clear thinking, and so an offence against peace." It was a statement that led one reporter to call him "Canada's most famously articulate angry man."[27] Even more controversial was his assertion that religious concepts like sin were the cause of "much of the social maladjustment and unhappiness in the world." Even more extreme, he contended that adults should avoid teaching children ethics: "freedom from moralities means freedom to observe, to think and behave sensibly." Children, he said, should be allowed the independence to come to their own ethical beliefs. Only when children had developed their own intellectual integrity would they have attained an honest moral education.[28]

Chisholm's public statements on religion and international issues brought him to the attention of Edwin Wilson, who published an article of his in the *Humanist*. This was in 1948, the year Chisholm assumed directorship of the WHO. In the article, Chisholm spelled out his ideas on "the future of psychiatry and the human race," in which he argued that scientific research in psychiatry should be at the forefront of the WHO agenda. The survival of the species was dependent on finding ways to "prevent personal and social maladjustment." Psychologists and psychiatrists needed to become "leaders in the planned development of new kinds of human beings."[29]

Of all the scientists who became active in the American humanist movement at this time, Julian Huxley was arguably the most famous. He was the grandson of T. H. Huxley, the man who coined the term *agnostic*. Following in his grandfather's footsteps, Julian became an evolutionary biologist, and also like his grandfather, he held a strikingly unconventional faith stance. His 1927 book *Religion without Revelation* was an apology for religious humanism; its inspiration came from a fellow British statesman who asserted that "the next great task of science will be to create a religion for humanity."[30] The book promoted an agnosticism reminiscent of T. H. Huxley's but blended it with his own brand of personal mysticism.[31] Huxley's sensitivity to the emotional side of religion came out in surprising ways, and he spoke about being "mystically united with nature." He related one cosmic experience in which he felt as if he "could *see* right down into the center of the earth, and embrace the whole contents and its animal and plant inhabitants." He exclaimed further, "For a moment I became, in some transcendental way, the universe."[32] He used the words *mystical* and *transcendent* but did so in ways that did not admit of any supernatural existence. There was no divine spirit, no God, no transcendent realm. For Huxley, the sacred was to be found in nature. His use of religious language and his discussion of his mystical experience are especially distinctive. These are not the things one finds among most of the other scientific humanists of this period, but in Huxley's hands they were deftly integrated into his much more utilitarian ideas about how science could be put to use in a way that was typical of scientific humanism.

By the time Huxley was appointed director general of UNESCO, it had been twenty years since the publication of *Religion without Revelation*. UNESCO was founded at the end of the war with an internationalist and pacifist agenda. The justification for the organization was clearly summarized in a single phrase: "Since wars begin in the minds of men, it is in the minds of men that the defenses of peace must be constructed." UNESCO's mandate, set out by the ministers of culture of the allied nations, was to undo the propaganda that the Germans had imposed on schoolchildren in their conquered territories. Its broader goal was to promote cultural and scientific exchange among the nations of the world. The preamble of UNESCO's constitution announced that peace required not just political and economic cooperation but "the intellectual and moral solidarity of mankind."[33]

Huxley turned out to be too much of an ideologue to be an effective director of the organization. His views were out of step with many of the people he was trying to enlist in the project, and his ideas ran into strong opposition. Although he wrote a booklet entitled *UNESCO: Its Purpose and Its Philosophy* in 1947, he used it to promote his own ideas about scientific humanism. Because it was so strongly humanistic in outlook, it aggravated his relationship with people in the project who held more conventional religious viewpoints. Huxley's general promotion of science was probably not especially controversial. "UNESCO," he said, "must see that its activities and ideas are not opposed to . . . established scientific doctrine," and "it must encourage the use of the scientific method wherever it is applicable."[34] Where he got into hot water was when it came to his intolerance for supernaturalistic religion: "[UNESCO] cannot and must not tolerate the blocking of research or the hampering of its application by superstition or theological prejudice," he inveighed. "It must disregard or, if necessary, oppose unscientific or anti-scientific movements, such as anti-vivisectionism, fundamentalism, belief in miracles, crude spiritualism, etc."[35] Moreover, he declared that the organization should adopt a humanistic outlook "on a truly monistic unitary philosophic basis."[36]

Evolution was a key element of Huxley's worldview, and he believed that it was essential that all human beings understand its implications. If everyone around the world saw themselves as part of the same animal species, *Homo sapiens*, united in our biology, we would be able to transcend our national and cultural differences. Minimizing those differences would benefit everyone. Huxley's insistence on the transcendence of national and cultural differences was somewhat at odds with his mandate as head of UNESCO, which had ostensibly been created to protect cultural diversity.[37] But Huxley, like Chisholm and Gregory, was a committed internationalist who believed that national and racial divisions should give way to a united world federation. In Huxley's view, this was the inevitable outcome of evolutionary progress.[38]

Huxley's proposal elicited strong objections from the representatives of the United States, France, the Netherlands, and Greece. The antinational and, by most standards, irreligious ideology set off alarm bells, and member states saw his agenda as a threat to their political autonomy. His book was never even put up for a general discussion.[39] Huxley's politically naive radicalism caused him to vastly underestimate the opposition. He apparently did not realize that his denunciation of supernaturalistic religions would be so vehemently rejected as intolerant and narrow-minded.

The Pugwash Movement

Huxley's UNESCO debacle demonstrated the difficulties that humanists faced when working within intergovernmental agencies. The Pugwash Conferences illustrate a more successful and influential form of humanist internationalism. This series of meetings brought together scientists from both sides of the Iron Curtain and made real progress in reducing the likelihood of nuclear war. While not an explicitly humanist undertaking, it was inspired and funded by a number of self-described humanists from America, Canada, and Great Britain.

The Nobel laureate Hermann Muller, who served as the AHA's president from 1956 to 1958, would be a key player in the Pugwash Conferences, which began during the years of his tenure. Muller was born in 1890 in New York City and completed his PhD at Columbia. There, he studied genetics in the *Drosophila* lab of T. H. Morgan. After leaving the lab, Muller pursued work on genetic mutations and received the Nobel Prize in physiology and medicine for his discovery that X-rays could cause mutations in genetic material.[40]

Muller, like so many American intellectuals in the first third of the century, became a convinced Marxist as a young man but moved rapidly away from that ideology later in life. While working as a professor at the University of Texas in the early 1930s, he gave secretive support to a Communist student group that advocated such radical ideas as free tuition, uncensored speech, and the elimination of racism and ethnic discrimination. The outrageous, rampant anti-Communist activism in the United States during those years drove him out of the country and to the Soviet Union, where he took up a laboratory post. Expecting to find a progressive political climate there, he instead found himself on the wrong side of the powerful Russian biologist T. D. Lysenko, who enforced his ideas on genetics through political might rather than scientific evidence. Witnessing firsthand the brutal and authoritarian dictatorship of Stalin, Muller escaped back to the United States, and he was staunchly anti-Soviet thereafter.[41]

Muller's political experiences merely reinforced his beliefs that science ought to play a more prominent role in human society. In particular, he thought, scientists must not shrink from policy discussions that touched on their knowledge and expertise. So Muller assumed a more activist role after the war and became part of a coterie of top American scientists (many of whom had worked on the atomic bomb) who lobbied for international control over atomic energy and atomic weapons. As Cold War tensions increased and the dangers of world destruction became ever greater, Muller and several other scientists began organizing a meeting of experts to discuss nuclear energy concerns in closed-door meetings. Amazingly, their proposals were approved by both the Soviet and Western governments. The result was a series of conferences that began in 1957 devoted to "Science and World Affairs." They soon came to be called the Pugwash Conferences, after the town in Nova Scotia where the first couple of meetings were held.[42]

The original idea for the Pugwash Conferences came from the famous pacifist and British humanist Bertrand Russell. His seminal article of 1903, "A Free Man's Worship," had become a classic expression of humanism that explained how he had learned to live without God or any kind of supernatural support system. The article discussed the moral fortitude that human beings must adopt in order to survive. Russell's popular tracts included such titles as *Why I Am Not a Christian* (1927), *The Scientific Outlook* (1931), and *Religion and Science* (1935). In all of these books, he expressed his disdain for traditional religion and his praise of the scientific method. For several years, Russell was president of the Rationalist Press Association, the British counterpart of the AHA.

In early 1955 Russell drafted a manifesto calling for an international scientific meeting "to appraise the perils . . . of weapons of mass destruction, and to discuss a resolution." People must set aside their political prejudices and consider themselves "only as members of a biological species." His faith that an international assembly of scientists could find solutions to the nuclear dilemma was based on the assumption that scientists were more objective and analytical and were therefore well equipped to cut through the political difficulties. Albert Einstein and eleven other scientists, mostly Nobelists like Hermann Muller, signed the document, which came to be known as the "Russell-Einstein Manifesto."[43] Two years later, the first of a long series of international scientific conferences was held at Pugwash. At that meeting, twenty-two scientists came together to discuss the hazards of atomic energy, the control of nuclear weapons, and the social responsibilities of scientists.

The humanist connection to the Pugwash Conferences was quite close. The men behind the initial proposal—Muller, Russell, and Einstein—were all affiliated in some way with humanism: Muller and Russell were presidents of their respective countries' main humanist organizations, while Einstein was more marginally connected to the movement as a nominal member of Charles Potter's Humanist Society in New York.[44] AHA member and future Humanist of the Year Brock Chisholm was also invited to the meeting. But the most important connection was probably the wealthy Canadian industrialist Cyrus Eaton. Eaton was the main financial backer of the Pugwash Conferences and the one who arranged for the meetings to be held in Nova Scotia in the first place; Pugwash was Eaton's hometown. Eaton was also a major contributor to the AHA, at one point donating $5,000 annually, quite a hefty sum in those days. This made him one of the group's most important supporters. In a long interview in the *Humanist* in 1956, Eaton explained his views on religion: "There is nothing that the people of the United States need more than an education in world religions and an emancipation from the superstition and ignorance that prevail among us concerning man's place in the universe."[45] An ardent internationalist, he also called for understanding and cooperation "in a world of alien and opposed philosophies."[46]

One other connection is worth noting. Pugwash organizer Eugene Rabinowitch, a chemist and editor of the *Bulletin of the Atomic Scientists*, gave the opening address at the 1958 AHA annual meeting. The address, entitled "Atoms and Man," called attention to the continuing dilemma that nuclear power posed to mankind.[47] Later that year, the *Humanist* published an article by Rabinowitch in which he lamented the tendency of politicians to either "ignore [scientific advice], or pick out, among dissenting scientific opinions, the ones which best fit their political plans, and not the ones which carry the best scientific support." Most importantly, he felt, it was necessary to humanize science, by making scientists aware of their responsibilities toward "the fate of mankind in the scientific age."[48]

Given the large number of humanists at Pugwash, the meetings allowed for humanist business to be conducted on the side. At least in one case, Wilson used it to lobby Eaton for more financial support for the AHA. Knowing that both Muller and Chisholm would be there, Wilson asked Muller to encourage Eaton in this regard, telling him that the Pugwash goal of a stable world peace was at base a humanist effort "even though it may not go by that name."[49]

The Pugwash movement illustrated the value of the scientific enterprise as a vehicle to promote international cooperation. The universal qualities of natural science, its oft-touted objectivity, and its categorization of human beings as a single species were central elements of the humanists' internationalist agenda. Pugwash also highlighted the socially responsible scientist, the person who humanized science by applying it to political problems, transcending the narrow technical boundaries of the discipline. The Pugwash Conferences illustrated how scientific values could help promote international cooperation and do so within a humanist mode.

This chapter has shown that scientists found their place in humanism in large part because their worldview, shaped by their scientific practice, reinforced the outlook that humanists already held. Their scientific research as well as their professional associations and ethical commitments brought their ideals in line with those of the major thinkers in the humanist movement up to that time. Science, however, is not a unitary enterprise, and the ideas and moral outlooks vary considerably. The next chapter turns to ways in which a cohort of scientists brought new and controversial ideas to the humanist movement. The notions of biological determinism and positivism, in particular, created challenges that would reverberate throughout the rest of the century.

CHAPTER SEVEN

Eugenics and the Question of Race

[What man] will know and be able to do along biological lines only a few generations hence would seem like a science fiction dream to most of us of
today. . . . I am convinced that, unless he shortsightedly destroys himself, as by
means of radiation, he will remake himself.

H. J. Muller (1957)

When the *Humanist* published Hermann Muller's two-part article "Man's
Place in Living Nature" in the late 1950s, Muller was president of the American Humanist Association and a Nobel laureate. As a geneticist, having made
his name doing pathbreaking work on radiation-induced genetic mutations,
he had a very biocentric view of human beings. His *Humanist* essay explored
the way that human biotechnology would one day be able to transform our
species by simply changing our genetic makeup. Muller's presidency and importance in the AHA during these years mark the beginning of a change in
humanism. The humanism that he expressed in his "science fiction dream"
in the epigraph above arose out of philosophical presuppositions foreign to
the religious humanism of the movement's first few decades. Muller's understanding of human beings was shaped by his genetics work, giving him an
outlook utterly different from that of the early ministers and philosophers.
Muller was one of a growing number of scientific humanists who adopted a
reductionistic perspective, based on hard science and what they considered
to be an unflinching dedication to scientific facts.

All humanists—scientific humanists like Muller as well as earlier religious
humanists—were evolutionists and recognized the extreme importance of
biology to our human nature. But the two groups drew widely different messages from evolution and biology, and their depictions of humanity were
often in conflict. The crucial differences had to do with the degree to which
one could accept a definition of mankind that was sharply constrained by our
biology or other internal mechanisms that shaped our behavior. As we have

seen, the early religious humanists found it impossible to say meaningful things about humanity by focusing on the mechanistic aspects of our makeup. Sure, the human body was a material thing that was constrained by the biological building blocks that made it up, but the human being as a whole was not a determined or mechanical object. It was in this way that the early humanists adopted a holistic vision of mankind, resting on, but not wholly dependent on, its material foundations. The social world and the environmental context in which we lived were crucial in understanding and explaining humankind. John Dewey and his followers were especially clear in this regard. They believed that experience was the means by which an organism interacted with the world. The complex psychology of the evolved human organism mediated our lived experience, and so we transcended our biological roots. Likewise, Roy Wood Sellars spoke of emergent properties, precluding any kind of simple reductionism.[1] Even more important was the political framework of humanist ideology: its progressive core enshrined human freedom and self-determination, which mitigated all forms of outright determinism.

Early humanists not only recognized but also highlighted the significance of our evolutionary origins. In fact, a Darwinian conception of nature stood behind nearly everything they wrote. Still, for them, evolution was not strictly a biological process; it was a general law of nature that permeated all change in the universe. Everything evolved in some form or fashion, and the humanists were less concerned with the details of how it happened than with the fact that it did and what it meant. This is especially clear in the work of Sellars, who explained the evolutionary universe in terms of levels of organization. Evolution made the world more complex and differentiated, and at each level of complexity, new possibilities arose. *Homo sapiens* were, of course, material and biological beings, but we had crossed a threshold at some point in our evolution that set us apart from our biological limitations. We were not determined by our cells, our genes, or the chemicals in our brains. Modern humans were first and foremost cultural animals. And in embracing the cultural aspect of humanity, the early humanists adopted a holistic outlook that was deeply suspicious of biological reductionism.[2]

The four biomedical scientists in the Chicago circle discussed in the previous chapter, Anton Carlson, Bernard Fantus, Maurice Visscher, and Judson Herrick, also resisted the kind of determinism and mechanistic outlook associated with reductionism. They studied human problems that revolved around the control or manipulation of human bodies and tissues but did so in ways that sought to enhance human freedom at the personal and social levels. They all propounded a holistic view of the human person. Fantus always used the religious humanist rhetoric that inevitably shed any trace of mechanism, determinism, or reductionism. In his work on a revision to the Humanist Manifesto, he stated, "Man is the greatest single evolutionary factor on this earth today and, to that extent is divine. By an intelligent appreciation of his powers and responsibilities he will make the world much more fit to live in."[3] Herrick, too, explicitly rejected biological determinism, even though he acknowledged that the biological body was a mechanism. He founded a school of psychobiology that was strongly influenced by John Dewey's ideas, and his beliefs about humanity all contradicted a strictly mechanistic analysis: organisms were to be understood in terms of their interaction with the environment, not simply their biological makeup; education was the primary mover of human change in the world; and our cultural heritage—not our biological inheritance—was what defined us.[4] Carlson similarly objected to simplistic techno-utopian ideas, as when he reviewed Hermann Muller's 1935 book *Out of the Night*: "The author . . . proposes that we remove the fertilized ova from the uterus of women and implant them for development in the uterus of another species," he explained, adding that "when a noted zoologist talks that way, we can only pause and wonder."[5]

The fact that a number of scientists and philosophers after the 1950s came to the defense of a more mechanistic view of mankind, often embracing one or another form of biological determinism, demonstrates a marked shift in humanist philosophy. The dominant antireductionist view was not overturned, but it was strongly challenged. These new scientific humanists brought in ideas that would have been largely unacceptable earlier: strong eugenic beliefs, as well as ideas about the heritability of intelligence. The espousal of biological determinism within the humanist movement created serious tensions. This chapter looks at those tensions by pointing to three different episodes: the humanist embrace of the population and birth control movements, the advocacy of progressive eugenics, and the controversy over race and IQ. It concludes with a discussion of the political ramifications of this mechanistic form of scientific humanism in terms of how it affected race relations in the movement.

Birth Control and the Population Problem

The significance of mankind's biological nature was brought to the fore in a very specific way within humanist circles in the postwar period insofar as articles about population growth appeared regularly in the pages of the *Human-*

ist. Seldom was the problem of population discussed in terms of ecological necessity—that is, about the earth's carrying capacity—rather, the pressing concerns at the time were social, related to the economic well-being of poor families and their access to birth control. Much of the discussion revolved around Margaret Sanger, the woman who had spearheaded the birth control movement since the early years of the century. One *Humanist* article, entitled "Birth Control and World Peace," championed Sanger's ideas about how uncontrolled population growth increased the likelihood of war. Another issue featured an interview of her by *Humanist* editor Edwin Wilson that highlighted her activism, and in it Wilson pointed out how Sanger's work fit well into the humanists' agenda. Not only was she tackling a global problem head-on, but she was also battling the Roman Catholic Church, a fact that endeared her to many humanists. This theme was repeated in an article entitled "Planned Parenthood and the Modern Inquisition" that focused on the Catholic Church's fierce opposition to birth control despite the clear social benefits of contraception.[6]

It was the AHA's recognition of Sanger as Humanist of the Year in 1957, however, that best broadcast the importance of Sanger's work in the humanist movement. She was the fifth person to be given this award and the only woman to be so honored in the prize's first two decades. She was also the first—and for many years, the only—honoree who had no academic credentials. The rationale for the award centered on her courageousness, determination, and independence, not her scientific expertise. One must remember that when she began her advocacy of birth control in the early 1900s, it was still considered obscene to talk about it in public and illegal to send materials mentioning it through the mail. She encountered tremendous opposition to her work and even spent time in jail.[7]

Sanger, like many humanists, was antipathetic to religion (though she often downplayed her views when she was looking for allies in the early years of the birth control movement). Sanger's mother had been Roman Catholic, but her father was a freethinker, and his unbelief ultimately won her over. At mealtimes when she gave thanks in prayer, he would needle her: "Why are you talking to your bread?" he would ask; "Is God a baker?" In time, she came to appreciate this constant prodding, saying that her father had taught her to think. Her irreligion and her activism blended together at times, as when she gave her first birth control journal the motto "No Gods, No Masters."[8] By the 1950s, she had become a member of the AHA.

As much as anyone, Sanger was responsible for identifying Catholicism as the main enemy of birth control. In fact, she built her movement, in part, by playing on anti-Catholic prejudices, which endeared her to many humanists. By the 1950s, Catholicism was becoming more acceptable to the Protestant majority, and humanists began to openly denounce what they saw as reactionary Catholic influence in American democratic institutions. The author Paul Blanshard, in his regular *Humanist* column "The Sectarian Battlefront," frequently sounded the alarm on Catholic issues, and his book *American Freedom and Catholic Power*, first published in 1949, was an immediate best seller. For Blanshard and many other humanists, Sanger's movement represented a valiant fight against the nefarious Catholic Church with its medieval ideas of humanity and culture.[9]

Throughout this period, the humanists spoke very little about the importance of birth control for the lives of women, however. The language was nearly always universalist and implied that men and women together would benefit from the ability to have small families. Sanger's own work in the cause of women's welfare was not ignored—Wilson acknowledged that her humanism "centers around her concern for woman's place in society"—but articles about her in the *Humanist* gave this less importance. Further, her citation for the Humanist of the Year award spoke only of how her work would "better man's future destiny and the welfare and happiness of individuals." The one major exception was an article entitled "Women, Democracy, and Birth Control," which began with the declaration, "Women are sacrificed on many altars—and have been for ages. They are victims of institutions which promote . . . cruelty and which prey on their ignorance. They are asked to risk their lives without question to build a greater population for 'the fatherland,' to increase the number of the faithful, or to prove the masculinity of their mates. When the woman is treated as an animal, . . . she, her children, and the society in which she lives all suffer."[10]

Between 1957 and 1962, the AHA issued five resolutions related to birth control and population. The overall thrust of these resolutions was to encourage discussion, research, and individual action on measures that would help deal with the problem of population increase and the closely related problems of poverty, war, and political instability. For the most part, these resolutions focused on birth control as a way to lift nations and families out of economic distress.[11] In several cases, the resolutions emphasized the importance of individual autonomy and democratic process in promoting and implementing social policies. In 1960, for example, the resolution that endorsed the control of population growth and supported the international Planned Parenthood

Federation concluded by insisting "upon the observance of principles that are in harmony with democratic and humanist ideals. Most particularly, we insist upon recognition of the worth and dignity of the individual."[12]

The Question of Eugenics

In America and around the world, the topic of eugenics was never far from the discussion of population control. Early on, Sanger joined the eugenics movement and began to speak of birth control as a means of eugenic planning. In particular, she represented birth control as one technology that could help prevent undesirable members of the population from having children and thereby weakening the gene pool. Some historians have suggested that Sanger's advocacy of eugenics was simply pragmatic: by getting the powerful eugenics community on board, she gained valuable and powerful allies. Other historians are more skeptical and find her support of eugenics to be genuine. In either case, there can be no doubt that even late in her life, when the high tide of eugenic activism had passed, she did not mind linking birth control to eugenics.[13]

At the 1927 World Population Conference in Geneva, Switzerland, Sanger joined forces with mainline eugenicists around the world. Here and elsewhere she espoused quite reactionary eugenic ideas on the use of birth control to advance class- and race-based social agendas. When Sanger was named Humanist of the Year, the humanists conveniently ignored her views, even though her acceptance speech included a reference to the founder of eugenics, Francis Galton, as a "long-neglected genius."[14]

The question of eugenics is especially relevant to the question of scientific humanism in this period because it represents the kind of biological determinism that was accepted among some of the most prominent members in the movement. In addition to Sanger, Hermann Muller and Julian Huxley both adopted some form of eugenics in their thinking. Both men believed that our status as evolved biological animals was critically important to how we thought about our present and future problems here on earth. They argued that in order to achieve human betterment over the long term, it would eventually become necessary to take control of human reproduction. Only in this way could we ensure continued progress in the centuries to come.

Unlike Sanger, Huxley, and Muller, though, the first-generation humanists were not eugenicists. Humanist philosophers by and large discounted it. Dewey himself hardly ever mentioned eugenics, and when he did so once later in life, he endorsed a dissertation by one of his graduate students that attacked it for being a failure as a social reform movement.[15] The elevation of Sanger, Muller, and Huxley within humanism therefore bears some discussion. Both Muller and Huxley were politically liberal and cosmopolitan and claimed that their eugenic policies were compatible with the ideals of democracy, freedom, and individual self-worth. Their eugenics was meant to be progressive and nonauthoritarian. To a large extent, Muller's agenda did conform to this goal. He was adamant that until the social conditions of society improved and it became possible to practice eugenics in a fair and unbiased manner, any kind of eugenic policy would be a moral disaster that would only benefit the elites.

Muller's progressive eugenics was very different from what the Nazis had enacted. He opposed discrimination on the basis of race or genetics. When he wrote his book-length eugenic tract *Out of the Night*, he was living in Russia, and his left-wing politics deeply influenced his views, which set him apart from mainstream eugenicists in America and Germany. In contrast to the Nazis and to the American nativists, Muller advocated using eugenic technologies in ways that would give men and women greater freedom. Individualism and equality were advanced in radically new ways in his thinking. First of all, he rejected efforts to "purify" the gene pool through sterilization, forced marriage, and other policies often advocated by mainstream eugenicists.[16] He also asserted unequivocally that in order for human beings to advance themselves through new biological technologies, they must first adopt "a wisdom that can be gained only by genuinely free inquiry . . . and backed by the broadest most unbiased good will." In other words, there must first be an egalitarian social revolution before any kind of eugenic policies were attempted; otherwise, grave injustices would occur.[17]

That said, he envisioned a future in which genetic technologies would be freely available for all people of any class and race. Artificial insemination would give couples freedom by separating reproduction from love: men and women could marry for love but use new reproductive technologies to give birth to genetically superior children. Other technologies like "ectogenesis" (that is, growing fetuses outside of the womb) could give women more social and political freedom by alleviating the burdens of childbearing, and this would increase their equality relative to men. In other words, the eugenic technology that Muller outlined here could be used—under the proper social conditions—in ways that would not merely perpetuate existing social inequalities, which was a clear danger in mainline eugenics.[18] Of course, behind all this was the assumption that manipulation of the human gene pool would create a new and transformed human.

The distant future, in his account, held out the possibility of even more extreme genetic manipulation and even greater freedom. With advanced reproductive technologies, the human race could eventually gain so much control of its own biological destiny that it could remake itself. There was almost no limit to what could be done. We could engineer new kinds of people by inventing "new characteristics, organs, and biological systems." Even our mental abilities could be modified. The only thing stopping all of this was "social inertia and popular ignorance."[19] Muller's ideas may have been far-fetched, but they were not unique. There was already a literature about future technologies of this sort, some enthusiastic and some admonitory, the most famous of which was Aldous Huxley's sharply critical novel *Brave New World* (1932).

Muller's colleague and friend Julian Huxley (Aldous Huxley's brother) shared a similar vision of the future, but his eugenic ideas were more reactionary than Muller's. This may have been due in part to the fact that Huxley was more wedded to mainline eugenic ideas. He had attended the 1927 World Population Conference in Geneva, Switzerland, one of the early efforts of Sanger to internationalize the birth control movement, a key event in the history of eugenics when Sanger herself embraced the reactionary political views of mainstream eugenics. Unlike Muller, Huxley did not talk about waiting to implement eugenic policies. "The ideologically most important fact about evolution," Huxley wrote, is "the fact that the human species is now the spearhead of the evolutionary process." Mankind had the "duty and privilege" to take the reins and begin to control evolutionary change.[20] Huxley, moreover, was less careful than Muller in discussing the design and enactment of eugenic programs, which meant that some of his views were more in line with the reactionary eugenicists.

One of Huxley's goals at UNESCO, in fact, was his attempt to ground the program's mission on evolutionary and eugenic principles. This is evident in his self-published vision statement *UNESCO: Its Purpose and Its Philosophy*, discussed in the previous chapter. The book is extraordinary in its attempt to place the mundane policy work of a Cold War–era intergovernmental agency into a sweeping vision of man's place in the cosmos. "From the evolutionary point of view," he wrote, "the destiny of man may be summed up very simply: it is to realize the maximum progress in the minimum time. That is why the philosophy of Unesco must have an evolutionary background, and why the concept of progress cannot but occupy a central position in that philosophy."[21] Our biology, he said, posed a difficult problem, namely, how to reconcile "our principle of human equality with the biological fact of human inequality." The only way to deal with this, he believed, was to employ eugenic science to help us. He didn't shy away from using reactionary language to discuss how this might be done. He talked about the need to deal with "weaklings, fools, and moral deficients"; about the "dysgenic" effects of modern civilization; about "the dead weight of genetic stupidity, physical weakness, mental instability, and disease-proneness"; and about "the quality of the human raw material."[22] Thus, in both his language and his ideas, he betrayed an allegiance to conventional eugenic doctrines. Similarly, he asserted that some people were genetically incapable of getting advanced degrees: "Only a certain fraction of any human population is equipped by heredity to be able to take full or even reasonable advantage of a full higher or professional education." Intelligence, just like physical strength, was a genetic endowment, he thought, and as a result, UNESCO needed to closely study the distribution of intelligence in population groups around the world.[23]

All in all, Huxley and Muller closely linked their eugenic ideas to their vision of humanism, but they were outliers during their time. Apart from Sanger, it is hard to find any other humanists who espoused similar notions in the pages of the magazine. That situation changed over the next decades, however, and it was given a push by a young philosopher eager to promote controversial ideas in the pages of the *Humanist*.

The Jensen IQ Controversy

Jumping forward to the late 1960s, we can see how the evolutionary view of mankind and its eugenic implications influenced the movement in a period in which the older Deweyan pragmatist thinking was on the decline. It was also a period in which scientific humanism with its more positivist outlook was rising, the decade in which both Muller and Huxley were named Humanist of the Year. This flirtation of the scientific humanists with biological determinism and eugenics had significant implications for the humanist ethos. By that time, the magazine was being edited by the philosopher Paul Kurtz at the University of Buffalo, a protégé of Sidney Hook. Kurtz was becoming a powerhouse in the humanist movement, and his confrontational style frequently provoked dissention and division within the movement. Though he had roots in the Deweyan tradition, he came to epitomize the more positivist and iconoclastic strain of thought embraced by the scientific humanists. The social fallout of some of the more incendiary ideas proposed by this school of thought highlights the pitfalls it had for a socially progressive movement like humanism.

One especially challenging moment arose when Kurtz opened the pages of the *Humanist* to a reactionary and racist idea based on eugenic principles. The controversy arose after he published an autobiographical statement by noted British psychologist H. J. Eysenck, who sympathized with the view that American Negroes were innately inferior to people of European ancestry. Kurtz printed this piece in the midst of one of the most intense periods of civil rights activism of the decade, and campuses were already on fire about an article published by an American researcher, Arthur Jensen, that defended this racist notion. Both Eysenck and Jensen claimed that the cumulative data collected through IQ testing over the previous decades provided conclusive evidence of innate and hereditary Negro inferiority.[24] Jensen, a Berkeley psychologist, had ignited the protests in 1968 after his 123-page article was published in the *Harvard Educational Review*. In it, he asserted that blacks were genetically inferior to whites and then went on to argue that this fact ought to be taken into account in formulating social policy toward Negroes. It became a nasty flashpoint in the charged atmosphere of civil rights–era America. "Jensenism" came to be synonymous with "racism" among leftists and students of color on college campuses around the country, and Berkeley even had to hire bodyguards for Jensen because he was getting death threats.[25]

The autobiographical Eysenck piece was introduced as a statement by a "leading humanist," and indeed, Eysenck would later sign the 1973 Humanist Manifesto II. In his article, he explained that humanists are rationalists who should follow the scientific evidence wherever it leads and must combat all kinds of superstitions that are standing in the way of human progress. After attacking religious beliefs and patriotic political agendas, he explained that "liberals and progressive thinkers, too, have their superstitions and shibboleths. Thus, for instance, a belief in the intellectual equality of the Negro is widely held and accepted as axiomatic: yet the evidence of hundreds of scientific studies ... almost uniformly supports a belief in the innate inferiority of at least the American Negro." The rest of the article goes on to explain Eysenck's strongly deterministic view of human beings, in all areas of psychology.[26]

Kurtz followed this piece six months later with a series of short articles focused on Eysenck's claims about the inferiority of black Americans. Of the five response pieces, the NYU research scientist John R. Dill's comments were the most critical and incisive. "As a black man, a psychologist, and a humanist," he began, "I find Eysenck's contentions about the nature of man replete with anti-humanistic doctrine." Eysenck's effort to connect his views with humanism, Dill thought, was outrageous: if we used Eysenck's own definition, then anything accomplished in a rational and scientific manner, even Hitler's extermination of the Jews, could be called humanism. The Negro's performance on intelligence tests, Dill claimed, must be evaluated in relation to the historical and contemporary oppression of black Americans.[27]

Two years later, Kurtz published an "Ethical Forum" on the topic of IQ and race. The opening piece was written by Jensen. The other seven articles included both supporters and detractors of Jensen's work, producing a pro-and-con intellectual debate that had become typical of Kurtz's editorial style. The articles presenting "The Case against I.Q. Tests" were all written by Harvard professors of human development or psychology who objected to various aspects of the IQ research itself. These were followed by a few articles defending "The Case for I.Q. Tests," in which Jensen and Eysenck both wrote pieces arguing that intelligence was inherited. What really focused the debate, however, was an extraordinary article defending IQ testing written by Nobel Prize–winning physicist William Shockley, the inventor of the transistor. Shockley argued that society should encourage people with low IQs to submit to sterilization, offering a person $1,000 for every IQ point that they scored below 100 if they agreed. To link his ideas to the humanist tradition and show his humanist bona fides, Shockley quoted a statement by Muller that had appeared in the *Humanist* years earlier. Though his ideas were fundamentally different from Muller's, Shockley was not wrong to recall that scientific humanists had already tested these waters.[28]

The reader fallout was significant; letters to the editor came in for the next several months. Many were short positive statements congratulating the *Humanist* on such a daring and intellectually stimulating forum, but other readers were outraged by what they called racist and pseudoscientific ideas. A few letters pointed out that the definition of the concept of race itself was never addressed—a lacuna, they thought, that damned the entire discussion. One of these letters, by the radical leftist group Students for a Democratic Society, noted that no one in the forum, not even the writers in opposition, made any attempt to rebut the notion of black inferiority.[29] Likewise, Joseph P. DeMarco of the Tuskegee Institute reacted to the lack of any content (even among letter writers) on the meaning of the crucial term *race*. He pointed to the 1950 and 1951 UNESCO statements that provided scientific arguments undermining the entire basis on which Jensen's claims rested.[30]

The debate gives us some idea of the priorities of some scientific humanists in this period. The individuals defending IQ testing argued that even if the ideas turned out to be wrong, it was essential that scientific research investigating the question not be stopped. This was, in fact, the rationale that Kurtz himself gave. In a telling editorial statement, Kurtz framed the controversy as a debate over two basic principles of humanism: "an open society" and "a free mind." It was an implicit defense of the IQ researchers against anyone who wanted to stop science simply because he or she didn't like the results. Claiming the moral high ground, Kurtz reprimanded the critics, saying, "Perhaps those most committed to helping the black man should be in the forefront of free and unrestricted research into the social and biological determinants of human behavior."[31] Eysenck also voiced outrage at the closed-minded radicals on the left who were so viciously attacking his friend Jensen. We must reject superstition, he said, and in the current climate, the liberals were behaving like religious zealots, irrationally attacking ideas and evidence that disagreed with their social dogma.[32]

But the most extended argument in this vein came from none other than Sidney Hook, who had drifted over the years from Marxism to a strikingly reactionary conservatism. The New Left, with its strident student activism, was at this point one of Hook's bête noires, which helps to explain his visceral response: "Whatever the outcome of further inquiry into the genetic components of intelligence, the intellectual intolerance of those who would taboo such inquiries should be exposed and the disruptions of the lectures, classrooms, and laboratories of the scientists engaged in the study of these problems should be firmly repudiated and punished. There are more racists among the intolerant dogmatists who would ban attempts to discover the facts about race than among those they currently accuse of racism."[33]

The fact that Hook is commenting here on the IQ testing debate, in which he has no scientific expertise, is particularly revealing and sheds light on Kurtz's motivations in promoting this debate. In Hook's view, the debate was less about race and more about defending science against intolerant critics. Hook, Kurtz, Jensen, and Eysenck all saw this as a replay of the Galileo trial, in which left-wing activists assumed the role of the church: the scientists themselves were the courageous fighters for truth, attacking superstition and ignorance. The past injustices and abuses that made eugenics, IQ testing, and race such fraught topics for radical activists simply did not register with these men. This is something particularly surprising in the case of Hook and Kurtz, who were Jews. None of this was destined to endear the humanist movement to African Americans.

Race in the Humanist Movement

The history of unbelief in the African American community is a relatively long one, going back at least to Frederick Douglass in the nineteenth century. Today, one can find several specifically Black and African American humanist and atheist groups, some directly associated with the humanist movement. This was relatively slow in coming. In fact, early humanism had little support from the black community overall, even though the movement espoused a very progressive line on race, apart from the discussion presented above.

At the founding of the AHA in 1941, there was outreach to the African American community, but it was not particularly effective. The black Unitarian minister Lewis A. McGee was a member of the AHA board of directors during its first decade of existence. Not long after this, he joined the AHA staff as a full-time administrative assistant, working as a liaison with humanist groups and chapters around the country. McGee was slightly older than Wilson but a relative newcomer to humanism. Ordained as a pastor of the African Methodist Episcopal Church, McGee had served as a US Army chaplain during both world wars, after which he returned to seminary and received his doctoral degree at Meadville. He was one of the first African Americans to minister a Unitarian congregation. He and his wife, Marcella Walker McGee, founded the Free Religious Fellowship in south Chicago, which was the city's first interracial and noncreedal faith group.[34]

McGee's early presence in the AHA was an indication of the group's strong concern about race issues and civil justice. In the organization's member newsletter, *Free Mind*, one finds occasional articles in support of racial equality and reports of on-the-ground civil rights activism by humanists.[35] Racial concerns also arose periodically in the pages of the *Humanist*; McGee had a column of his own that dealt with legal and political rights in America—and he frequently discussed the concerns of African Americans. There were also other articles that affirmed the importance of the contemporaneous civil rights movement. In addition, the AHA passed several resolutions affirming its overall support for African Americans. These statements tended to be either quite vague or excessively narrow. A 1948 resolution, for example, enjoined "all members of the AHA . . . to devote themselves to the improvement of race relations as a major project of the AHA." Two years later, the group wrote a resolution opposing segregated housing in the District of Columbia. Over the next decade they issued two more resolutions on civil rights: one on desegregation and one on "open occupancy" housing.[36] Of the roughly seventy resolutions issued between 1946 and 1961, eight dealt with race in one form or another.

In the mid- to late 1960s, the group continued to tackle racial issues under the passionate leadership of its executive director Tolbert McCarroll. McCarroll wanted to make humanism less intellectual and more engaged in the problems of the lived experience. Unlike Kurtz's idea-focused humanism, McCarroll promoted activism. He listened to young people, civil rights activists, and counterculture communities. He expanded the organization's program to train and certify counselors for their local communities, and some of them began working in prisons and addressed the rising problem of drug addiction, issues that tended to hit African Americans harder than others. Also, as part of his agenda to find meaningful applications for projects being developed by humanistic psychologists, he hosted the nation's first black-white encounter group.[37]

During the same years as the Jensen controversy, the *Humanist* published articles that dealt very sympathetically with the difficulties of black Americans (so one cannot consider Kurtz to be entirely blind to the problems of African Americans). One such article on "Inner-City Education" discussed a small pilot project that focused on the problem of integration between the races. It explained how this unique program provided education and counseling to a small, integrated group of boys, and how its apparent success encouraged the authors to use it as a model for larger programs elsewhere around the country.[38] Around the same time, the AHA gave its first Humanist of the Year award to an African American. A. Philip Randolph was a labor and civil rights activist, whose work as a union organizer had thrust him to prominence. Since the early years of the century Randolph had been a secularist and a labor activist, and in his view the two went together. In an article in a socialist journal, he and a coauthor wrote that "prayer is not one of our remedies; it depends on what one is praying for. We consider prayer as nothing more than a fervent wish; consequently the merit and worth of a prayer depend upon what the fervent wish is."[39]

All in all, the AHA frequently engaged with the problems of race in American culture and was often at the forefront of many issues promoting racial equality. The group, however, never achieved anything remotely like a racially diverse membership. Humanism was mostly a well-to-do white person's movement during most of its history, and this clearly affected the nature of the intellectual debates.

So what does this say about humanism as a whole, and especially scientific humanism in the postwar period? For humanists, biology provided a crucial framework for understanding humanity, but the way that they used biology—and especially evolution—changed over their history. The unanimity that existed in the early years gave way to division when scientific humanists who were also biological determinists entered the movement, challenging the previously dominant antimechanistic view of humankind. Of course, individually, humanists differed greatly and did not always fall into a simple dichotomy. Sanger, despite her desire to make life better for women and the poor, held reactionary eugenic notions on race and mental disability. Huxley both was a champion of liberal Enlightenment values and held an elitist—though probably not racist—view of biologically based intelligence. Eysenck, Kurtz, and Hook gave credence to the scientific evidence for black intellectual inferiority but, in other respects, were strong advocates for freedom, democracy, and individualism.

The historical ties between Jewish intellectuals and humanism make the positions of Kurtz and Hook especially surprising. Only two decades since the American discovery of the Nazi concentration camps and the terrible revelations of the Holocaust, one might have expected most humanists, especially Jewish intellectuals, to be skeptical of scientific claims about race. But this was not the case. The rightward shift among many Jews in America during the Cold War was mirrored in the humanist movement as well. Kurtz and Hook were naturally suspicious of left-wing politics and considered the opposition to Jensen and Eysenck to be politically motivated and therefore a threat to freedom of inquiry. In their view, science was a neutral ground on which these debates could be resolved; one just had to allow open, disinterested study to proceed. Surprisingly, neither man reconsidered this position at the time, even when it ran counter to long-standing humanist values.

The intellectual shift that we have traced in this chapter was indicative of a more substantial transformation in the philosophical underpinnings of the movement, and these would have far-reaching implications for the nature of humanism in subsequent years. It is to those other related changes that we turn in the next several chapters.

CHAPTER EIGHT

Inside the Humanist Counterculture

I have been trying to show that love, which has always been one of the theological virtues, is also a scientific virtue—that love is needed for science even to exist, and that it is at the same time a product of the scientific spirit. . . . Science is not detached from human feeling—it can give us more and richer values, not less and fewer.

Priscilla Robertson (1952)

The writer and historian Priscilla Robertson was the second editor of the *Humanist*. Working from her family's tobacco and beef farm in Kentucky, she held only a short, three-year tenure as editor from 1956 to 1959 before sad circumstances cut short her affiliation with the group. It has therefore been easy to overlook her presence in the organization. She came and left too quickly to have a lasting influence. For our purposes, however, she is especially interesting because of her strikingly original ideas about twentieth-century science and its power to reshape modern social relations. Moreover—and perhaps more importantly—her ideas foreshadowed a radically new kind of science-based worldview that would emerge in the decade to come, one founded on new advances in psychology. This new worldview directly challenged the reductionistic form of scientific humanism and also departed from the older philosophically based religious humanism.

Robertson, who had studied the history of women and family life at Vassar, was keenly interested in the different roles that mothers and women played in human societies. In fact, her writing represented an explicitly feminist rethinking of science. Although these feminist notions did not survive her departure, other aspects of her work did. From her first months as editor, she highlighted recent discoveries in the human sciences, showing how they could help us understand ethics. She also published and commented on essays by several social science theorists, especially a group of maverick psychologists who embraced ideas similar to her own. These ideas would usher in a countercultural revolution within the humanist movement itself.[1]

Science and Human Love

At the time Robertson took over, Edwin Wilson had been editor for fifteen years. The divestment of this responsibility would allow Wilson, then fifty-eight years old, to devote more time to the American Humanist Association's "many-faceted program," as he described it. The original mission of the organization was to publish a magazine, but now it took on a much more activist agenda: Wilson spent time lobbying to protect the separation of church and state, he threw great effort into building the international humanist movement, there were lectures and educational materials to organize, and there were local chapters to support.[2] As a result, Wilson simply could not give enough attention to the journal. A new editor was needed.

Unlike Wilson, whose professional identity was tied to institutionalized religion, Robertson grew up in a secular home. Her grandfather had been a Presbyterian minister but was convicted of heresy for acknowledging that there were historical errors in the Bible. In the early 1900s, he had taught at the Meadville Unitarian seminary.[3] Her father abandoned even that liberal heritage, and as a young man he gave up all religious convictions. A Vassar professor of history, he provided his daughter an entirely secular education. Robertson credited her father for raising her with humanistic values: he taught that "the happiest living comes from loving your family and working at something you think will benefit humanity," and God was simply not discussed at home.[4]

Robertson's enthusiasm for science was evident in the editorial she wrote for the first issue under her management, where she highlighted the importance of "the scientific revolution of the twentieth century." This editorial and two other articles written for the *Humanist* that same year outlined her unique perspective on science. In these articles, she pointed not to science per se but to specific developments that constituted this new scientific revolution—she drew heavily on work in neurology, psychology, anthropology, and sexology. Her thinking was shaped by a close reading of recent books by the neurologist Kurt Goldstein, psychoanalyst Roger Money-Kyrle, and anthropologist Weston La Barre. In these articles, she made two distinct arguments.[5]

First, she claimed that science is absolutely *not* value neutral. As illustrated by the epigraph of this chapter, from an article published a few years earlier in the popular magazine *Harper's*, Robertson considered science permeated with values and emotions, so much so that she considered even the human emotion of love, which seems most distant from rational thought and empirical discovery, essential to the scientific discipline: "The ability to see things straight is more emotional than intellectual, in my view, and the kind of emotion you need in order to perceive truth is love."[6] Taken out of context, this statement sounds poetic and figurative, an evocative phrase that one might find in a minister's sermon, but that's not how she intended it. She was being precisely literal.

This brings us to Robertson's second claim: that human society is rooted in our biology, though in ways quite different from the views of most scientific humanists, as we saw in the previous chapter. She based her arguments on psychological and anthropological findings concerning sex and motherhood, two functions fundamental to human reproduction. According to Robertson, these two aspects of our biology represented distinctly different forms of love. The maternal bond, she claimed, was the one that lay the foundation for scientific thought. This might seem an odd notion, but in her view the respect for truth is learned through the mother-child relationship. Respect was the critical notion here; science was an emotional connection to the world around us, not merely an intellectual one. This striking idea reflected her experiences as a mother and wife as well as a scholar and professional editor. One has the feeling that her life on the farm also played a role in shaping her views—a rural perspective that reflected a familial world that tied nature and economy together in a way quite different from the urban social experience out of which so much humanist thinking emerged.[7]

During her tenure as editor, which coincided precisely with Hermann Muller's presidency, Robertson worked with and published the works of many different authors. In particular, she discovered several American psychologists who in a few years' time would found a new school of psychological thought. Robertson published an article by pioneering clinical psychologist Carl Rogers, in which he explored ideas about freedom and creativity. A year later, she picked up the same theme with an entire issue devoted to psychology and human values. The lead article was written by Abraham Maslow, a maverick in the field and the man at the center of a network of psychologists who opposed the prevailing positivistic bias of the discipline. These articles grappled with the role of authority and the importance of individual uniqueness in a culture that demanded conformity. In assessing this symposium,

Robertson explained that "psychology is a critical test of scientific method, because the measuring and abstracting techniques of science can do violence to human personalities."[8]

Robertson was sacked abruptly in 1959 in a dispute over editorial freedom, specifically her unwillingness to run a series of articles by one of the group's main donors, Corliss Lamont.[9] For several years after Robertson's editorship, the movement followed a relatively unadventurous path. Then, in 1965, the lawyer Tolbert McCarroll (who had already succeeded Wilson as executive director) picked up the thread when he was given editorial control of the magazine. The years of his tenure would constitute one of the most tumultuous periods in the organization's history.[10] They were a time of transformation throughout American culture, and the historical forces that drove it refigured humanism too. Historians acknowledge the sixties as a crucial turning point in American culture. It was a decade in which young people revolted against an establishment they saw as dehumanizing, in which civil rights activists fought defiantly against deep-seated racial prejudices, in which a nationwide pacifist movement denounced the war in Vietnam, and in which nontraditional ideas of gender, sexuality, and drug use flourished.[11] Those same forces transformed humanism as well. Under McCarroll, the humanist project listened to these many voices of change, experimented with new forms of activism and intellectual engagement, and embraced those aspects that seemed to fulfill the fundamental agenda of making the world better for humanity.

Toward a Countercultural Humanism

When subscribers received their issue of the *Humanist* in January of 1965, it must have been something of a shock. In place of the familiar white cover with its sober Lydian typeface, readers found a dizzying array of type styles and a mosaic of words, photographs, and drawings. A profile of Albert Camus was accompanied by a still from the Beatles' *A Hard Day's Night*. The titles of the articles—such as "The philosophy of a criminal" and "With other methods of birth control available today, why sterilization?"—were similarly provocative. Inside the cover of the magazine, an editorial proclaimed, "*This magazine must issue challenges and calls for action. . . . We are rebels, but rebels with a concern for man.*"[12] The *Humanist* was, in a single makeover, transformed from a stolid philosophical-literary journal into a punchy monthly review. It still contained many of the topics humanists had always followed—church-state issues, international affairs, humanistic ethics—but it now featured explicit discussions of sex, psychedelic drugs, the youth movement, and race relations.

These issues had seldom been visited by the previous editors, and on the rare occasions when they had, it was a high-altitude flyby.

The transformation of the *Humanist* reflected a major identity crisis within the movement. When the AHA handed McCarroll the keys to both the executive office and the journal, he refocused the priorities of the movement and redefined what it meant to be a humanist—or, perhaps it is more accurate to say that he opened up new possibilities. His version of humanism departed from both the religious and scientific visions of the early movement. To begin with, he challenged the positivistic and deterministic scientific humanism that had arisen in recent years. In its place, he focused on the individual person and the existential aspects of lived experience. He still considered humanism a religious movement, but he de-emphasized reason and focused on personal experience.

And it was a good time for humanism. By the early 1960s, the movement was thriving, having grown rapidly in the postwar years, rising to a membership of nearly five thousand (up from about one thousand a decade before). The AHA board of directors at this time was a diverse lot: scientists, professors, lawyers, doctors, and businessmen. There were still some ministers, but many fewer than before. Some of the board members were Unitarians and attended church, but the leadership was no longer dominated by clergymen. With McCarroll at the helm, the movement stood at a turning point.[13]

Portland, Oregon, lawyer Tolbert McCarroll stood over six feet tall and had a friendly, engaging personality. One acquaintance claimed that he was so charismatic that he "created an altered reality" when he entered the room.[14] When McCarroll came to the AHA, he had already worked as an organizer for progressive causes, ranging from the simplified burial movement to Planned Parenthood. He had also helped establish the Portland Ethical Study Society, which consisted of people from various religious traditions who wanted to "humanize" their spiritual experience. This organization welcomed people of all beliefs, not just atheists. In fact, the very question of belief versus unbelief was considered irrelevant to the pursuit of the good.[15] In this sense, he was more closely aligned with the Ethical Culture Societies than with the humanists, yet, in the end, McCarroll decided to make his Portland group an AHA chapter.

Unlike Wilson, McCarroll didn't mind competing with established religious groups. The Portland society, for example, offered weekly meetings, held a Sunday school, and offered other services traditionally associated with regular churches. Wilson, by contrast, had been keen to assert that humanism was not a church; it was there to supplement the liberal churches, not "supplant" them. Wilson feared that if the AHA were to promote local chapters, Unitarian ministers would see that as a threat that might drain off members of their own congregations. Since McCarroll had never been a Unitarian minister, he did not feel any compunction in this matter and pushed forward to strengthen AHA chapters around the country.[16]

Like Robertson, McCarroll recognized the singular importance of human sexuality for both society and the individual. Humanists, he believed, needed to grapple with this taboo subject. He realized that a truly humanistic approach to sex had to go much deeper than the social agendas of Planned Parenthood. Thus, as editor of the *Humanist*, he devoted a special issue to sexuality in its many manifestations, drawing on scientific sexological studies to help frame a new humanist outlook. The cover of the issue was dominated by a copy of Picasso's *The Love of Jupiter and Semele*, a line drawing of two nudes, the reclining Semele in the god's arms. Above the picture, a quotation by Havelock Ellis read, "Sex lies at the root of life, and we can never learn to reverence life until we know how to understand sex." The issue was guest edited by Lester A. Kirkendall, professor of family life at Oregon State University and one of the nation's foremost sexologists. Kirkendall had recently published the book *Premarital Intercourse and Interpersonal Relationships*.[17]

The articles in the sex issue ranged from the obvious "Sex: Is There a Humanist View?" and "The Woman Rebel" (another portrait of Margaret Sanger) to the more provocative "When Is Sex a Crime?" The latter article, written by McCarroll, argued for less criminalization of human sexual relations and a "more enlightened and creative sexual understanding." Kirkendall offered a strong defense of sex education for adults and provided frank discussions of many previously taboo subjects, such as homosexuality, masturbation, and abortion.[18] In an issue not long after this, McCarroll ran an article on Hugh Hefner and the *Playboy* attitude toward sex, accompanied by a photomontage that included several pictures from *Playboy* itself. The article studied the contradictions in the *Playboy* ethic and debated its merits.

People took notice. There were continued requests for extra copies of the sex issue and so many letters to the editor that McCarroll continued printing them for months afterward. Yet as radical as all of this was for the *Humanist*, many readers still found the articles too conservative: "I still think," wrote one reader, "that we are only skimming the surface of the significance of sex in this kind of approach."[19]

McCarroll's tenure as executive director was also marked by the establishment of humanist youth groups around the country. The energy and enthu-siasm of the youth culture could be drawn into humanism, he thought, if hu-manists were willing to shed their stodgy ministerial and academic robes and address problems relevant to young people.[20] With the influx of young people into the organization, the AHA established a Humanist Youth Fund to en-courage students to attend humanist conferences. McCarroll's efforts to in-crease student participation in the AHA paid off handsomely: according to a 1965 AHA conference report, students in the affiliated Humanist Student Union of North America composed fully 20 percent of the AHA membership.[21]

It was to the burgeoning counterculture that McCarroll turned in his effort to make humanism more relevant to the young. He printed a feature article on the Free Speech Movement at Berkeley, which had riveted the country the previous fall. The article, written by Ward Tabler, a Berkeley faculty member, decried the arrogance and ineptitude of the administration and praised the students for their activism.[22] Other articles sympathetic to the youth culture continued to be published under McCarroll's editorship.[23]

Not all humanists were sympathetic to the student protests, however. "The Movement" had split the faculties of many universities, and it was creating a sharp divide among some humanists as well. A letter to the editor by septua-genarian humanist Harry Elmer Barnes, for example, railed against the "long-haired beatnik, Beatle, rock'n roll, and allied quasi-degeneracy" of the youth culture.[24] Opposition to student protests increased as their tactics became more militant, and by 1968 students were breaking into buildings, disrupting classes, and striking. Faculty members who had long ties to the humanist movement—Sidney Hook and J. H. Randall Jr., especially—decried these tac-tics. Even McCarroll himself worried about some of the youth behavior. In an article entitled "The Undead," he attacked the narcissism of some parts of the cultural revolution.[25] Unlike the conservative critics, however, McCarroll was quick to differentiate the dangerous trends from the progressive ones, separating the self-centered and destructive forces from those that were cre-ative and socially productive.[26]

Humanism and the Third Force

McCarroll also worked to strengthen the humanists' counselor program. This was a training program for people who wanted to minister to those needing emotional and spiritual support. Humanist counselors were not professional psychological counselors but volunteers trained by professionals. They held office hours, helped local family counseling bureaus, went to people's homes, visited prisons and hospitals, and led group discussions.[27] In this respect, the counselors were secular counterparts to clergymen, so much so that they even performed marriages and funeral services. It was this counselor program, as much as anything, that characterized McCarroll's ecclesiastical style of humanism and began to turn it into a direct competitor with the churches. The counseling focus also had philosophical ramifications that put it on a collision course with other forms of humanism.[28]

Outside of McCarroll, the person most responsible for building the counseling program was clinical psychologist Rudolf Dreikurs. An Austrian by birth, Dreikurs immigrated to the United States, and by the end of the Second World War he was living and working in Chicago. Dreikurs's clinical emphasis was family relationships, and he authored several popular books on marriage and parenthood. Dreikurs had joined the humanist movement in the 1940s and served a term as vice president of the AHA. He wrote for the magazine on a wide variety of topics, including daily living, sexuality, the youth movement, democracy, and international humanism. By the mid-1960s, he was supervising the training of new counselors and was closely involved in the management of the whole program.[29] Dreikurs's influence helped McCarroll elevate the science of psychology to a singularly important position in the movement, and one particular theoretical outlook came to dominate.

American psychology by this time was divided into three main schools of thought: psychoanalysis, behaviorism, and humanistic psychology (often called *third force*). McCarroll and Dreikurs aligned themselves with the last school. The term *third force*, with all its bombastic connotations, was invented to distinguish it from its two main rivals. Humanistic psychology did not have a particularly unified theoretical framework; it was more of a collection of theorists and clinicians who employed a variety of methods, technologies, and theories. What brought them together was their isolation within the academy due to their rejection of many of the basic assumptions of the other two schools: behaviorism, they thought, reduced people to machines, while classical psychoanalysis had become a dogmatic and strictly medical intervention that focused on the dark side of the human psyche. Humanistic psychologists, by contrast, were characterized by their holistic and integrated approach to the human experience. They celebrated the wide variety of emotional, sexual, and spiritual experiences that made up a rich and full human life. These scientists believed that their work was ethically more respectful of the individual and his or her autonomy. Many of the theories held by this group of psychologists resonated with popular ideas inside the counterculture: the development of individuality and consciousness, the ethical concern for autonomy and freedom, and the desire to help people achieve their full potential.[30]

Many of the ideas central to third force psychology had already begun to show up in the *Humanist* during Robertson's editorship. By the early 1960s, articles by leading humanistic psychologists regularly appeared, as did reviews of their books. Third force became so central to the humanist perspective that within the space of four years, three of the most well known of these psychologists were named Humanist of the Year: Carl Rogers in 1964, Erich Fromm in 1966, and Abraham Maslow in 1967—dates that correspond closely to the period of McCarroll's greatest influence. Fromm was noted for having transcended Freud's deterministic ideas and for applying psychoanalytic approaches to existential questions about modern society. It was his commitment to freedom, however, that held particular appeal for the humanists of this period. The work of Rogers and Maslow connected most closely to the community-centered humanism promoted by McCarroll and Dreikurs.[31]

Carl Rogers's work on "client-centered" therapy had revolutionized the way that clinicians approached their work, but it was his attack on behaviorism that became a defining feature of his involvement with the humanists. In Rogers's acceptance speech published in the *Humanist*, he denounced the morally impoverished ideas of behavioristic psychologist B. F. Skinner and the mechanistic model of human behavior Skinner popularized. It was not a new argument; Rogers had been attacking behaviorism for years. In fact, a 1956 showdown between the two men at the American Psychological Association in Chicago continues to be recognized as a major moment in the history of twentieth-century psychology.[32] In his view, the behaviorist, by describing people as conditioned machines, removed those aspects of humanity that most distinguished us. Rogers argued that the human experiences of freedom, commitment, love, and spirituality were fundamental to human life, but to the behaviorist they were mere fictions of the mind because they could not be measured in the laboratory. The positivistic, quantitative scientific method of behaviorism, therefore, was in fact incapable of dealing with the complexity of human existence.[33]

In contrast to the machine-man model proposed by Skinner, Rogers talked about an inner subjectivity that could only be studied outside of the laboratory. Moreover, the notion of clinical work went beyond simply healing mental illness; Rogers saw its potential to help people discover more enjoyment and meaning in their lives. At the crux of his thinking was the idea that "Man is subjectively free; his personal choice and responsibility account for the shape of his life; he is in fact the architect of himself. A truly crucial part of his existence is the discovery of his own meaningful commitment to life with all of his being."[34]

Maslow attracted the attention of humanists for similar reasons. He had studied under the arch-behaviorist Harry Harlow, whose emotional deprivation experiments on monkeys became notorious for their cruelty among animal rights activists. Maslow rejected Harlow's behaviorism for many of the same reasons Rogers rejected Skinner's work: it was reductionistic, it portrayed animals (and by extension, men) as machines, it was amoral, and it ignored some of the most basic and important features of human experience.[35] Maslow focused his research on human motivation and complex human needs such as love and esteem, and he set out to understand the psychological factors that contributed to full and successful lives and allowed people to achieve their greatest potential as "self-actualizing individuals."

One of Maslow's most noteworthy ideas was the "peak-experience," a mental state of clarity, exuberance, and serenity. Such experiences, he claimed, were like the intense spiritual encounters described by religious men and women of every age. Maslow was a passionate man but certainly not religious. Since childhood, he had rebelled against what he saw as his mother's superstitious outbursts, so much so that he claimed she had taught him to despise everything about religion. Even in discussions of peak-experiences, his anticlericalism never lay far beneath the surface: "I want to demonstrate that spiritual values have naturalistic meaning, that they are not the exclusive possession of organized churches, that they do not need supernatural concepts to validate them, that they are well within the jurisdiction of a suitably enlarged science, and that, therefore, they are the general responsibility of *all* mankind."[36] Maslow claimed that the same feelings of oneness, mystery, inexplicable joy, and fulfillment that religious mystics had described in past ages could be goals of modern men and women. "The sacred," he proclaimed, "is *in* the ordinary, . . . it is to be found in one's daily life, in one's neighbors, friends, and family, in one's back yard."[37] In the end, Maslow believed that psychology could provide people with "a positive, naturalistic faith."[38]

Peak-experiences offered humanists like McCarroll a way to discuss the emotional side of human life without adopting a supernatural belief system. By claiming that even typical religious experiences could be both natural and fulfilling, Maslow helped humanists argue for the sufficiency of naturalism while acknowledging the reality and significance of numinous experiences.

Supernatural agencies were not needed to explain or provide meaning to these events. He also helped these humanists develop a new formulation of religious humanism, one that was personal and spiritual rather than social and intellectual, one that did not depend on religious or ethical organizations.

Humanists like John Dewey and Julian Huxley had already talked about their own transcendent experiences and noted their singular importance. The experience of themselves in relation to the entire cosmos had strengthened, not weakened, their naturalism by providing an emotional understanding of nature that supplemented their intellectual one. Maslow even cited John Dewey's *Common Faith* as an example of a nontheistic religious outlook and considered the AHA to be a humanistic sect that might one day become a "religion surrogate."[39] Maslow's work was instrumental to McCarroll's agenda at the AHA, and the two of them became friends.[40]

San Francisco and Buffalo

The version of humanism that was emerging from McCarroll's work was ill-suited to the tiny Midwestern town of Yellow Springs, Ohio, where the national headquarters was located. When McCarroll was hired and moved there, Humanist House was new—a comfortable, modernistic office building that had only recently been completed with moneys from a major multiyear fund drive. After three years, however, it was becoming clear that the kinds of activities that he favored—namely, developing an experimental humanistic community—could not be managed in sleepy Yellow Springs. Moreover, McCarroll was not especially interested in editing the journal. He was an organizer by nature, and his goals required sustained attention to local activities. The magazine got in the way. Dissatisfaction with this state of affairs within the organization led to a major restructuring. In 1967, a new editor was picked, and for the first time the headquarters and the editorial office were split up. They sold the building in Ohio and moved to San Francisco, while the editor, a young philosopher named Paul Kurtz, worked from his office on the Amherst campus of the State University of New York at Buffalo.

A few board members lobbied hard against a move to San Francisco, favoring a large East Coast city like Washington, DC. Corliss Lamont, in particular, was unhappy with a move west. But McCarroll prevailed in the end by making a persuasive case that the cultural transformations effervescing in California would make it an ideal place to build the movement.[41] Humanist House opened its doors in San Francisco in a huge Tudor-style mansion overlooking the Presidio, with a magnificent view of the Golden Gate Bridge.[42]

The transition from small-town Ohio to one of America's largest cities—at the heart of the most radical countercultural community in the country—brought further changes to the movement.

Once in San Francisco, McCarroll turned his attention to local community building and experimented with a host of new techniques and ideas, many of them allied with humanistic psychology. One group that McCarroll hooked up with was Synanon. Part encounter group, part commune, part rehabilitation center, this Southern California creation became well known for its unconventional but highly effective drug and alcohol rehabilitation program. McCarroll viewed it as an example of the kind of humanizing effort that the new counterculture was producing.[43]

Meetings and encounter groups regularly took place in Humanist House. As many as fifty people might show up for a single event. During his first two years in San Francisco, McCarroll organized two research programs that used encounter groups to explore issues like personal growth, self-awareness, and "interactional sensitivity." McCarroll experimented with family-based sessions that allowed parents and children to work out their difficulties in a controlled setting. It was at this time that he also hosted the nation's first interracial black/white encounter group, an event that made the national news.[44]

Life at Humanist House was chaotic at times. Only a few miles from Haight-Ashbury, people of all sorts were constantly in and out of the house. The Diggers—a radical activist group known for their street theater performances—began to hang out there during the day. That said, McCarroll did not want the place to become simply a hippy crash pad. He wanted to test and create programs that would foster personal growth. Thus, McCarroll's Humanist House assumed a distinctive form. Aloof from the freewheeling counterculture, it was, instead, a place driven by a contemplative and directed concern for the psychological well-being of the person. Those looking for love-ins and stoner parties had to go elsewhere.[45]

Many humanists disliked what they saw happening in San Francisco. McCarroll was experimenting with ceremony in ways anathema to many of the old guard. For instance, he incorporated robes and candles in his encounter groups. This sent up a red flag for those who feared he was becoming a cult leader whose Catholic upbringing was beginning to resurface.[46] When word got out that nude encounter groups were being held at Humanist House, some people hit the roof.[47]

All this was happening at a difficult time in McCarroll's personal life. He was watching his wife Claire succumb to diabetes. She had been his longtime companion and a partner in a lot of his social activism going back to their days together in Portland, Oregon, when they set up a Planned Parenthood office together. The tragedy of his wife's illness, along with his focus on local events, took a toll on his leadership of the AHA. He neglected administrative duties essential to a national organization: membership lists fell into disarray, and communication between the headquarters and members dropped off. These changes worried humanist board members who had little understanding of McCarroll's new agenda.[48]

This precipitated a revolt, and a group of old-guard humanists joined forces to remove McCarroll as executive director. They managed to elect a new president to the AHA's governing board in 1969, Lloyd Morain, a California businessman. Like McCarroll, Morain had been involved in running a local Planned Parenthood with his wife. Both husband and wife were long-time humanists and quite active in the organization. This was actually Morain's second stint as president, having served for several years in the early 1950s. His first act as the new president was to fire McCarroll and take over the executive offices, installing a new executive director to help clean up the organizational mess.[49]

The intense period of collaboration between the AHA and the third force movement left its mark on humanism, but the heady and experimental days of the humanist counterculture were over. In its place, a more intellectual and political humanism asserted itself. The focus turned to organization building once more. Although Maslow remained on the magazine's editorial board and many humanists remained committed to humanistic psychology, the focus on human potential was relegated to the sidelines of the organization. Many people had come to associate humanism and third force over the years, and this turned out to be a liability for those who believed that humanism should be open to different schools of thought. This was especially true for the behaviorists and positivists who had joined the movement and brought in opposing points of view.

It was in the editorial office in Buffalo, New York, where the tide began to turn. There Paul Kurtz pushed the *Humanist* in a radically new direction. Kurtz devoted himself to humanism with a boundless energy that allowed him to rise quickly in the movement, and as editor, he stood as a counterweight to McCarroll, with an editorial policy that gave a strong voice to those scientific humanists who had been alienated by McCarroll's efforts.

The threat to the rationalists and scientific humanists in the group had seemed quite real at the time. An article by Dreikurs published under Mc-

Carroll's editorship is a case in point. Entitled "The Scientific Revolution," Dreikurs's article began with a discussion of recent ideas in theoretical physics that, he claimed, were leading to "revolutionary changes in epistemology." We needed to rethink the nature of science and its role in the world. Quantum mechanics was calling into question the classical notion of causality, as well as the age-old distinction between the observer and the thing observed. Between quantum physics and humanistic psychology, scientific objectivity would have to be rethought.[50] As radical as this idea was, Dreikurs's final step went too far for the rationalists: "What is objectionable," he said, "is the abuse of the scientific method, the deification of scientific findings, the disregard of scientific limitations, and the dependency of science as the only source of knowledge."[51] For a movement that was continually having to fight for the acceptance of science in society, Dreikurs's caveats were raising people's hackles.

The article hit a nerve with Hermann Muller, who responded in a scathing letter to the editor. The ex-AHA president rejected Dreikurs's fundamental assumptions because they represented "a view contrary to that of scientific humanists and of the great majority of humanistically minded scientists." Dreikurs had claimed at one point that biological mutations were an example of the breakdown of cause-and-effect mechanisms in biology, but Muller knew better. As a world-renowned geneticist, he explained that mutations showed no such thing, and Dreikurs was reintroducing ideas and theories that had been abandoned by science long ago.[52]

Kurtz was on Muller's side in this debate. A protégé of Sidney Hook as an undergraduate, Kurtz completed his PhD in philosophy at Columbia and considered himself to be a Deweyan in outlook. It was as a student that he discovered religious humanism and joined the Unitarian Church, though his philosophical disposition and his family's irreligious background placed him firmly in the rationalist wing of the movement. Later in life, he called himself a "third generation freethinker."[53]

Kurtz would become a dominant force in late twentieth-century humanism. His enthusiasm for the project, his desire to connect with people outside of academia, his aggressive personality, and his savvy political sense made him an extremely effective leader. Some considered him overly controlling, and others charged that he was unethical. Nonetheless, it is hard to find any humanist today who doesn't recognize his effectiveness in building up the movement, in terms of both its membership and its wider cultural influence.

Kurtz had a knack for popularization and knew what topics would interest his readers; he liked to use controversy to highlight issues and make points. And he did this to great success in humanist dialogues with both Marxists and Catholics. Early on, he worked to bring more prominence to the movement by hosting a syndicated TV series called *The Humanist Alternative* that was broadcast on public television stations around the country. In the series, which ran for three seasons, he interviewed noted writers and activists with a humanist bent, and he often put them in dialog with nonhumanists in order to explore the contrasts. Several of his initiatives, such as Humanist Manifesto II with its high-profile list of signatories, made page one of the *New York Times*.

Humanism beyond Freedom and Dignity

Kurtz was also a behaviorist and believed that the humanist movement had recently come to be too closely associated with humanistic psychology: the work of McCarroll and Dreikurs at Humanist House, the Humanist of the Year awardees, even the magazine's editorial board, which included Maslow. So when Kurtz took charge as editor, he sought to redress the imbalance. He wrote to B. F. Skinner and encouraged him to submit an article—"our pages are open to you any time," he said—and also added Skinner to the publications committee as a counterweight to Maslow.[54]

Skinner had been approached by McCarroll a couple of years earlier to write a response piece but was decidedly cool to the request: "I really believe that if humanism is going to depend on the third force, it is doomed. Sometime I will explain why."[55] He responded similarly when Kurtz first approached him to join, complaining that the articles in the magazine often seemed "to rule out the usefulness of a scientific approach to human behavior."[56] Kurtz, sympathetic yet insistent, explained his own views: "It is important that we dissuade anyone who simply identifies humanism in *The Humanist* with humanistic psychology. As a matter of fact, humanism in this country drew much of its inspiration from the behavioral revolution. I would consider myself as [a] behaviorist <u>first</u> in methodology, and I find obscene, some of the nonsense that humanistic psychologists advocate (though I do accept some of the values of growth and creativity)."[57] If there was a point at which the humanist movement began to tack away from the third force influence and toward behaviorism and positivism, it would be here.

In the same way that Kurtz would use the IQ debates to highlight controversy and tease out differences of opinion, so too he wanted to publish arguments over behaviorism in order to show that it was not antithetical to humanism. And just as we have seen in the IQ debates, Kurtz was not impartial. If it hadn't been for his advocacy, the recognition of Skinner as Humanist of the Year could hardly have happened.

The fight over whether behaviorism was a form of humanism began in earnest in 1970, when Kurtz received an article by Floyd Matson, president of the Association for Humanistic Psychology. Matson urged humanists to once and for all sever their allegiance to behaviorism. Matson wrote that humanists should "become the active conscience" of the field of psychology and that they should be "searching out and exposing—and condemning—each and every dehumanizing, depersonalizing and demoralizing force." The behaviorist's dreams, he asserted, were "the humanist's nightmares."[58] With its over-the-top rhetoric, Kurtz thought that Matson's article could spark a healthy debate, so he penned a quick letter to Skinner, asking for a rejoinder. In the end, Kurtz published Matson's piece along with responses from two of Skinner's colleagues, followed in the subsequent issue by a short comment from Skinner himself, all of which, he thought, "set the record straight" by affirming that "behaviorism is an authentic form of humanism."[59]

The basic argument that the behaviorists made was that scientific control of behavior was not bad. Like any other technology, it enabled control and thus gave people greater choices. Using techniques developed by behaviorists, it became easier to control one's own actions. It could also lead to many beneficial social transformations. Behaviorism, they thought, was just another field of science that ought to be employed, like the rest, to help mankind better understand and master nature—the only difference was that in this case the nature in question was *human* nature.[60] The fact that behaviorism was utilitarian and deterministic was an advantage, they thought, not a detriment. Skinner summarized this position by saying that "behaviorism *is* humanism. It has the distinction of being effective humanism." This was the standard argument that scientific humanists had been making for years: science provided rigorous and utilitarian control that gave humans the means to make their world better.[61]

Within a few months, Skinner's long-awaited book *Beyond Freedom and Dignity* came out, and suddenly Skinner was everywhere in the news. He appeared on the cover of *Time* magazine and was featured on talk shows across the country.[62] *Beyond Freedom and Dignity* was his most controversial book. In it, he attacked conventional social and political thought, so much so that both conservatives and liberals were outraged. The central argument claimed that the ideals of freedom and dignity, so central to Western political and social thought, were simply fictions of the mind. Freedom, according to Skinner, was simply the name people give to a *feeling* and had no real bearing on whether a person was being controlled or not. When control is aversive, as in the case of slavery, people *felt* that they were being controlled, but when control was nonaversive, as when someone was paid to perform work, they felt more comfortable about it, even though the controlling force—the employer— remained. Under slavery a person worked out of fear, whereas in a wage system a person worked for money, yet in both systems the worker was controlled and made to perform tasks. The only difference, according to Skinner, was that in the latter system people don't feel bad about it. Arguing that control was both ubiquitous and unavoidable, Skinner proposed that mankind adopt a technology of environmental conditioning guided by scientific principles. This technology would create an environment in which people perform social functions according to a plan without feeling coerced in any way. By eliminating the feeling of coercion, the citizens of a society could live happy and fulfilling lives.[63]

This was not quite what Kurtz had expected. In a strikingly honest and blunt review, he questioned the ethical foundations of Skinner's science: "I must confess an earlier sympathy with Skinner's behaviorism. I had thought that humanistic psychologists were overly critical of it, not recognizing its important contributions toward making possible a new science of man by freeing us from archaic conceptions. Humanism, in my view, should be rich enough to allow for a diversity of beliefs, and it should be able to encompass both behaviorism and humanistic psychology. After mulling over Skinner's new book, however, I must qualify my earlier generous estimation of his position."[64] He reflected on his own earlier "disquietude" after having read *Walden Two*, Skinner's novel about a utopian community controlled entirely by behavioristic technology. Both in that work and in this new one, Kurtz noticed "the lurking totalitarian implications" of behavioristic science. Two very different possibilities emerged from Skinner's thought: a world in which we all shared in the decisions about how science was to be used, or one ruled by "a group of scientific technocrats" with unlimited powers. He concluded his review with a strong request: "Is it not time for Professor Skinner to make his position unequivocal? . . . Is Skinner a democrat or not?"[65]

Unlike Kurtz, some humanists were untroubled by Skinner's recent assertions, and the awards committee listed him as their first choice in a list of three top candidates for Humanist of the Year. The AHA Board of Directors affirmed their choice. Considering the central role that academics had been playing in the movement, the lack of academic leadership in this regard is notable: the president of the AHA at the time was Lloyd Morain, a businessman, and the chairman of the awards committee was Robert L. Erdmann, an IBM human factors engineer trained in psychology. Erdmann's work involved the application of behavioral methods in education, man-machine interfaces, and management. In other words, the deliberations over the Skinner award were led in both committees by men involved in practical business applications rather than theoretical work.[66]

Erdmann's kind words for Skinner ("it is especially fitting that we publicly express our abiding respect and affection for him as humanist and active supporter of the American Humanist Association") emphasized his humane approach to the treatment of people.[67] The latter point has frequently been downplayed by Skinner's detractors who focus on the notions of control and human freedom. His supporters, however, often highlight the fact that he has, in Erdmann's words, "consistently opposed the use of punishment in the control of human behavior."[68] Indeed, Skinner's own reply to Kurtz printed in the *Humanist* pointed out how this respect for individual well-being could and should be applied to matters of governance. What Skinner called nonaversive behavioral technologies have been applied in mental institutions, schools, childcare, industrial settings, and prisons and have created better conditions in all of these places. They could be used in government "for the good of the individual, for the greatest good of the greatest number, and for the good of the culture or of mankind as a whole." He added, "These are certainly humanistic concerns, and no one who calls himself a humanist can afford to neglect them."[69] This was not a declaration of democracy, by any means, but neither was it inherently totalitarian.

The fight over behaviorism in the humanist movement was a philosophical dispute that had become institutionalized. On the one side was a mode of thought centered on the scientific spirit, characteristic of the philosophers in the Dewey circle. They were holists at heart and believed in a scientific method that could be applied pragmatically in many different areas of thought. The language of the scientific spirit had been enormously useful in capturing the kind of arguments that made early religious humanism possible. This way of thinking still resonated with a large number of humanists, and it was echoed by humanistic psychologists and the many sympathizers with the counterculture.

On the other side, the defense of behaviorism by high-profile humanists reinforced the ideas of the scientific humanists with their narrower view of scientific method that was associated with reductionism and positivism. The implications of this change were significant. Positivism restricted science to a rigid methodology involving mathematical, experimental, and empirical study. Broader statements of value and emotion were de-emphasized in the philosophy as scientifically unverifiable. Those who saw science as a means of control found this outlook especially useful. The debate over Skinner's radical behaviorism was perhaps the greatest test over the viability of scientific humanism. If Skinner could be considered a humanist even after all he had said in *Beyond Freedom and Dignity*, then the scientific humanists need not worry. That success was made possible by the continued effort by Kurtz and others to establish that behaviorism could be compatible with humanist values.

This battle over behaviorism and positivism was being fought at the same time that the IQ-race debates were taking place, as discussed in the previous chapter, and it revolved around the same basic issue: was this reductionistic account of human beings acceptable to humanists even when it contradicted long-standing humanist ideals about such things as equality and freedom? The successful incorporation of behaviorism into the movement suggested that it could be. Indeed, the next decade would make it clear that other elements of positivism would also be able to thrive. Science would be put to use in a more aggressive way to combat weakness in human thinking, and this new formulation of science would become a hallmark of a very different form of humanist activism.

WEST COAST HUMANISM. *Top*, the move of Humanist House to San Francisco in 1967 from Yellow Springs, Ohio, brought vast changes to the AHA, not least of which was a prominent mansion overlooking the Presidio where new kinds of humanistic social experiments took place. *Bottom*, executive director Tolbert McCarroll (on the right) helps raise the UN flag there, a sign that the movement still retained its allegiance to its postwar internationalist agenda. Humanist House, San Francisco, California. Photo courtesy of the American Humanist Association. Tolbert McCarroll raising the UN flag. Photo courtesy of the American Humanist Association.

COUNTERCULTURE HUMANISM. From the moment Tolbert McCarroll assumed editorship in January 1965, the content of the *Humanist* shifted sharply to an engagement with the existential and social concerns of the emerging cultural revolution. Rudolf Dreikurs, who authored the article on experimental schools in the issue shown here, highlights the movement's strong ties to humanistic psychology and its interest in the social agendas of the new left. Cover of the *Humanist*, January/February 1965. Photo courtesy of the American Humanist Association.

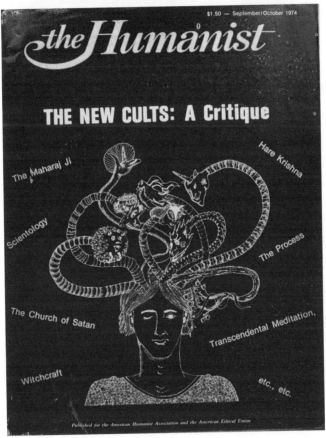

$1.50 — September/October 1974

the Humanist

THE NEW CULTS: A Critique

The Maharaj Ji

Hare Krishna

Scientology

The Process

The Church of Satan

Transcendental Meditation,

Witchcraft

etc., etc.

Published for the American Humanist Association and the American Ethical Union

AGAINST THE OCCULT. *Top*, under the editorial direction of philosopher Paul Kurtz, the *Humanist* assumed a much sharper tone. Both Kurtz and his mentor Sidney Hook (on the right) were driven by a concern that we might be witnessing in modern times a culture-wide failure of nerve. *Bottom*, the *Humanist* cover from September/October 1974 marks this critical turn in the movement, as the rationalist wing of humanism began to warn of the dangers of mystical thinking among the so-called New Age cults. Paul Kurtz and Sidney Hook. Courtesy of the Center for Inquiry Photo Archives, Center for Inquiry Libraries. Cover of the *Humanist*, September/October 1974. Photo courtesy of the American Humanist Association.

the Skeptical Inquirer
THE ZETETIC

The Moon and the Maternity Ward
Good-bye to Biorhythms / Foiling
a "Cold Reading"/ Students and the
Paranormal / High Sorcery in Colombia

Published by the Committee for the Scientific Investigation of Claims of the Paranormal
VOL. III NO. 4   SUMMER 1979

THE SKEPTICS MOVEMENT. Encouraged by widespread publicity—both positive and negative—after a brash public attack on the practice of astrology, Paul Kurtz and a team of scientists, magicians, philosophers, and laymen launched the Committee for the Scientific Investigation of Claims of the Paranormal in 1976. The frequently sardonic *Skeptical Inquirer* would be its face to the world as it attacked pseudoscience and irrational thinking in ever-new ways. Cover of *The Skeptical Inquirer*, Summer 1979. Courtesy of the Center for Inquiry Photo Archives, Center for Inquiry Libraries.

THE STARBABY AFFAIR. The tactics of the skeptics made them many enemies, and even some fellow rationalists were disillusioned by what they contended were unfair attacks. This snarky article by Dennis Rawlins, an ex-CSICOP board member, in the occult magazine *Fate* remains a widely cited exposé of the group by an ex-insider. Cover of the offprint of the article "sTarbaby" by Dennis Rawlins, © 1981. Used by permission of *Fate* magazine, with thanks to Jim Lippard for the image.

BIOLOGICAL DETERMINISM AND RACE. *Top*, the scientific debate over IQ and race in the January/February 1972 *Humanist* highlighted the challenges of biological determinism for an organization whose core principles stemmed from left-wing progressive thought. *Bottom*, Norm Allen, founder of African Americans for Humanism, expressed several decades later a more nuanced approach to the authority of science as a foundation of humanism, in large part because of its potential for abuse. Cover of the *Humanist*, January/February 1972. Photo courtesy of the American Humanist Association. Norm Allen. Courtesy of the Center for Inquiry Photo Archives, Center for Inquiry Libraries.

DEFENDING EVOLUTION. *Top*, Bette Chambers, president of the AHA during the 1970s, embraced the humanists' scientific agenda with a passion. With a degree in marine biology and three young daughters, she was an early herald of the dangers posed by creationists to the teaching of evolution in the public schools. *Bottom*, Harvard paleontologist Stephen Jay Gould (on the right) was one of the most prominent public figures to join the humanists in fighting creationism. Here, Gould and fellow scientists Christian B. Anfinson and Francisco Ayala announce the submission of an amicus curiae brief in a 1987 Supreme Court case. The brief, which arose out of a California skeptics group, argued that creationism was not a scientific idea but a religious one. Bette Chambers. Courtesy of Jan Sharar. Christian B. Anfinson, Francisco Ayala, and Gould. Courtesy of the Center for Inquiry Photo Archives, Center for Inquiry Libraries.

Power Books

Tim LaHaye

You are
engaged in

The Battle for the Mind

a subtle warfare

THE FUNDAMENTALIST ATTACK ON SECULAR HUMANISM. Tim LaHaye's 1981 best-selling book *The Battle for the Mind* repackaged a decade's worth of conspiracy theories, claiming that humanism was an exceptionally powerful movement that posed great dangers to America because the country's core traditions were founded on Christian, not humanist, values. Cover from *The Battle for the Mind* by Tim LaHaye, © 1980. Used by permission of Revell, a division of Baker Publishing Group.

THE DEBUNKING SPIRIT. *Top*, "the Amazing" James Randi became one of the staples of evening entertainment at CSICOP and other humanist conferences, where he would perform shows that demonstrated the gullible nature of the human mind. In this photo, Randi (center) manages an escape in front of an audience at one of these events. *Bottom*, Randi's most important humanist stunt, however, was probably his exposé of Peter Popoff's faith healing ministry, as shown on the cover of *Free Inquiry* in the summer of 1986. "The Amazing" James Randi doing an escape act. Courtesy of the Center for Inquiry Photo Archives, Center for Inquiry Libraries. Cover of *Free Inquiry*, Summer 1986. Courtesy of the Center for Inquiry Photo Archives, Center for Inquiry Libraries.

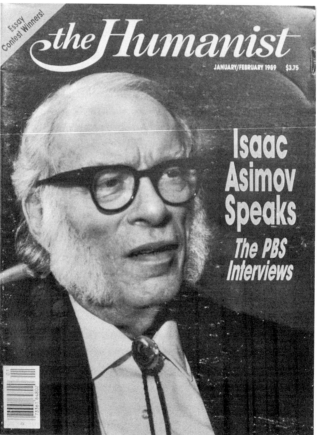

POPULARIZING SCIENCE IN A HUMANIST MODE. *Top,* by the 1980s, the humanists found allies among a number of celebrity scientists. Carl Sagan, by far the most prominent of them, presented a worldview that integrated humanistic ideals into his message about the importance of science. In this photo, Sagan is being honored as Humanist of the Year for 1980. *Bottom,* Isaac Asimov, science fiction author and popularizer and Sagan's close friend, identified proudly as a humanist and served as president of the AHA for nearly a decade. Carl Sagan receiving the Humanist of the Year award. Photo courtesy of the American Humanist Association. Cover of the *Humanist,* January/February 1989. Photo courtesy of the American Humanist Association.

CHAPTER NINE

Skeptics in the Age of Aquarius

I'm trying, in my way, to bring society to a rationalist point of view, to get people to deal with things as they actually are and not as we'd like them to be and not as some misinformed people would like us to believe they are. It's my firm conviction that scientists investigating claims of paranormal powers are making a gross error in assuming that they are qualified to observe cases in which chicanery could be occurring. . . . Scientists are known for thinking logically and in a linear fashion, therefore they are easily deceived.

James Randi (1976)

In an interview for the *Humanist* magazine in 1976, stage magician James Randi offered his critical assessment of modern parapsychology. Randi was one of a new league of debunkers who had recently banded together under the auspices of the American Humanist Association. He was already well known for exposing the trickery of many self-proclaimed occultists and was especially famous for debunking Israeli psychic Uri Geller. Geller's shows—on television and in person—seemed to demonstrate an amazing ability to read minds and to mentally manipulate physical objects. Randi believed that all these phenomena involved trickery. He was critical of the media for playing along with these entertainers and was particularly concerned about the popularity of parapsychology and its effect on the scientific enterprise. Even some scientists were being bamboozled by people like Geller. Randi believed that more scientists needed to stand up and expose these hucksters: "There are few more serious things for scientists to do than to deal with this kind of thing and put it out of business permanently."[1]

Randi's aggressive promotion of science and reason was very attractive to a group of humanists led by Paul Kurtz. Kurtz was still editor of the *Humanist* at this point and had also started his own rationalist publishing house, Prometheus Press. Kurtz's attack on pseudoscience and the occult emerged naturally out of his humanistic philosophy, but this was distinct from earlier forms of humanism. Kurtz, Randi, and the other debunkers promoted a type of rationalism that was grounded in a philosophical perspective very different from earlier religious humanism. It was an outgrowth of scientific humanism, which had been growing since the 1950s, and humanists who embraced the debunking spirit adopted a distinctly different set of assumptions about science than those who did not. This new trajectory was aligned closely with positivism and the Anglo-American analytic tradition in philosophy, in marked contrast to the Deweyan pragmatism of humanism's early years. Kurtz understood these philosophical differences clearly. As an editor and publisher, he played an instrumental role in shifting the philosophical orientation of American humanism, creating a new movement that was more confrontational and that focused on knowledge and truth as opposed to human experience.

Though Kurtz himself traced his academic genealogy back to John Dewey and American pragmatism, he was more comfortable in the company of logical positivists and other analytic philosophers.[2] We have already seen how different these two philosophical orientations were: pragmatists and positivists had sharply opposing views on scientific reductionism, especially regarding its use in the human sciences. The pragmatists approached the world holistically, whereas analytic philosophers believed in the utility of reductionism. The pragmatists had been especially good at integrating science into the lived experience. By contrast, the analytic philosophers, by focusing narrowly on logic, mathematics, and empirical fact gathering, believed that the scientific method was a means of correcting our error-prone senses and thought processes. In a nutshell, Deweyans trusted human experience, while skeptics, with their more positivistic mind-set, did not; they emphasized human fallibility and the need to apply rigorous filters on any claims about perceived phenomena. Randi's deep skepticism of human perception fit well into this mode of understanding. These two diametrically opposed views of human experience widened an already deep division within humanism.

Analytic philosophy presented a more hardened view of science, making it possible to use it as a weapon against illogic and fantasy. In this respect, it seemed to fit the needs of the current moment better—at a time when humanists now felt more embattled in the surrounding culture—than did the uplifting rhetoric that had prevailed since the Humanist Manifesto days. The popularity of Kurtz's debunking project proved the key to success. More than any scientific humanist before him, Kurtz was able to solidify this new strain of humanism, and this would be part of his lasting legacy. He did this not only by showing how science could be put to use in the culture wars but also by attaching this new outlook to the idealism that had sustained the humanistic ethos and its Enlightenment foundations. The result was a humanistic splinter group of self-declared *skeptics* with their own social and cultural agenda. This chapter charts the rise of this new group and treats the difficulties it encountered both inside the humanist movement and within society at large.

The New Cults Symposium: Debating the New Age

The immediate impetus for the skeptics movement was a newly perceived threat to American culture—and to humanism in particular. It was the challenge of the New Age, an eclectic amalgamation of ideas and practices that flourished in the counterculture of the sixties and early seventies. New Age proponents opposed the hyper-rationalism associated with modern science and the bureaucratic state. The movement consisted of new religious and esoteric beliefs, ranging from the spiritual traditions of India and the Far East to the magical practices of old Europe. It also included idiosyncratic approaches to scientific theories, as well as unusual and controversial discoveries. The very term *New Age* came from an astrological belief that humanity was on the cusp of a major civilizational change related to a 26,000-year cycle of the earth's orbit around the Sun.

Like the counterculture, the New Age was far from antiscientific. Indeed, it would not be wrong to characterize many of its thinkers as intoxicated by the revolutionary potential of science. But what was science for some people was hogwash for others. The New Age practitioners held views of science that contradicted much of what was being taught in the academy. Indeed, many of their ideas were embraced precisely *because* they were revolutionary.[3] Thus, the skeptics found themselves in the unusual position of defending the status quo of the academy against the unorthodox ideas of cultural radicals.

In 1974, Kurtz put together a special issue of the *Humanist* devoted to what he called "the new cults."[4] The magazine's cover set the tone: It was pitch black with a white line drawing of a medusa-like figure staring at the reader, its right eye closed. Serpents with grotesque faces twisted in the air above this head, labeled in red with the names of popular cults they represented: "Hare Krishna," "Scientology," "The Maharaj Ji," "The Church of Satan," "Transcendental Meditation." The list concluded with the words "Etc., etc."

In his opening editorial, Kurtz characterized all of these activities as "cults of unreason." His tone was academic and critical. He believed that people's attraction to cults was a reaction to the alienation they felt toward modern technological society. The cults that were responding to this alienation encouraged people to take "a romantic flight to other worlds of dependence and hope."[5] The cults offered hope to people in despair, but it was a false hope—one based on myths. Humanists needed to find a way of cutting through the false myths and offer a more realistic response to the uneasiness of modern life. Humanism was well suited to this task, since it was both rational and passionate: "Humanism, in emphasizing critical reason, [does not] leave out the aesthetic and religious dimensions of experience. . . . Passion and poetry are as much a part of life as logic and observation."[6]

Kurtz also put the phenomenon into historical perspective, comparing the challenges of the present day with what happened in the ancient world. The cults today "are not unlike the mystery cults that emerged in post-Athenian Greece or the post-Roman period." And like those older cults, the ones today are arising at a time when "there is uncertainty and despair in society" and a sense of "impending doom." If this argument sounds familiar, it should. Kurtz was echoing his mentor, fellow humanist Sidney Hook, who had attacked orthodox religionists during the Second World War for their own failure of nerve.[7] Both Hook and Kurtz borrowed this historical trope from classicist and humanist Gilbert Murray, who had written earlier in the century. Rational thought—according to Murray, Hook, and Kurtz—was a courageous step forward into the unknown, the path taken by a confident and self-reliant people who cast away the chains of old dogmas and superstitions. Classical civilization had died because people became afraid and retreated into mysticism and comfortable otherworldly faiths rather than soberly facing the problems of the here and now. The moral was clear for modern Western society: if the populace loses confidence in human institutions and human effort, civilization might also be lost.[8]

Using a tried-and-true formula, Kurtz put authors with opposing points of view side by side to highlight the controversy in the academy.[9] Several authors writing in a humanist mode reiterated the "failure of nerve" motif. In a piece entitled "The Yinning of America," writer Ethel Romm talked about the psychology of cultic belief, arguing that the widespread turn toward the occult was the result of the comfort those beliefs gave to people searching for "some sense of control over chaotic forces."[10] In another article, Ethical Culture leader Joseph Blau lamented that modern institutions "are no longer sensed as parts of an order, a cosmos." He thought that Americans were losing faith in their scientific civilization.[11] The authorities of the past could no longer give people a sense of certainty or security. Again and again, the writers highlighted the theme of courage against the seductive attractions of the occult.

In an opposing view, University of Michigan's A. Theodore Kachel saw these new social movements as an *encouraging* sign that pointed toward an age of liberation and freedom. The "spiritual awakening" that came from participating in some of these religious forms provided people with beneficial psychological resources by expanding their horizons and allowing them to experience the world anew. Movements of all sorts could provide this kind of liberation, everything ranging from Catholic neo-Pentecostalism to Transcendental Meditation. Participating in political activism could also result in similar kinds of liberation: the Black Power and women's liberation movements were energizing in this way. The overall thrust of his essay was to explore the idea of what he called mythic freedom, an "escape from the overly rationalized Western consciousness." For Kachel, this mythic freedom had an enormously beneficial effect. It "keeps one constantly in touch with the holy, [with] what it really means to be human, 'the really real.'"[12]

Philosophically, the most significant part of the New Cults issue was an exchange between two philosophers, Marjory Clay and Richard T. Hull. Clay had just finished a master's degree from Northwestern, and Hull was a younger colleague of Kurtz at the State University of New York at Buffalo. The two authors knew each other and had worked together. Although they had quite different responses to New Age thinking, they believed that these disagreements, so apparent in the symposium, could be bridged.[13]

Clay argued that modernity was faced with a dilemma in having to choose between scientism on the one hand and occultism on the other. Scientism was an outlook that overemphasized objectivity, while occultism overemphasized subjectivity. On the whole, she sympathized strongly with the New Age criticism of scientific rationality and its characterization of the limits of science. She thought that the mechanistic worldview associated with Newton and Descartes "too often ignores the human subject," and she singled out B. F. Skinner's behavioristic psychology as the most egregious and dehumanizing variation of it.[14] Quoting the romantic poet Percy Bysshe Shelley, she stated that "man, having enslaved the elements, remains himself a slave."[15] In this respect the occult outlook could help restore a balance by reasserting ways of thinking that had been lost. Occultism was more organic and, in this respect, more human centered than scientism. It could remove the chains of "functional rationality" and help re-enchant our lives with an appreciation of wonder and mystery. She acknowledged, though, that on its own it was deeply flawed by exaggerated and impractical subjectivity.

Hull defended mainstream science in his piece and criticized occultism as epistemologically unsound: the cults asked you to "believe in order to understand," which was precisely the reverse of careful scientific study. Hull acknowledged, though, that the scientific method was limited and not capable of handling all aspects of human experience. In fact, he proposed that scientists pay attention to reports of occult phenomena since they might reveal new phenomena to be studied. Occult claims might be a precursor to interesting areas of scientific investigation.[16] Indeed, both Clay and Hull advocated a collaborative interaction between the two groups—scientists and occultists. They even identified several possible areas of study: extrasensory perception, biofeedback, and acupuncture. Clay had already noted that nearly 70 percent of the scientists in a questionnaire "professed varying degrees of belief" in ESP, and even Hull took notice of the work of ESP researcher J. B. Rhine, stating that he has done "some of the finest examples of scientific investigation" on the topic. In Hull's words, "these excursions of science into occult phenomena represent . . . a rejection of the presumptions of scientism."[17]

Despite that apparent rapprochement between these two philosophers, it was clear which side Kurtz and the *Humanist* were on. The Clay-Hull proposal of a comfortable collaboration was wishful thinking. The very layout of the magazine pointed to trouble: a quarter-page advertisement for two anti-occult books was placed right in the middle of Clay's article. The ad by Kurtz's Prometheus Press also sent a not-so-subtle message. The blurb describing a book entitled *Science, Reason and Religion* called it "a critique of the new cults," and the ad for *The Art of Deception* claimed that it would teach you how to "recognize a fallacy" and to "see through deception."[18]

The New Cults issue drew a flurry of letters to the editor. Most of them were by humanists who lauded it and approved of its critical tone. One reader congratulated Kurtz and urged him to consider republishing the articles in paperback form. Another reader, humanist Harvey LeBrun of Berkeley, California, was delighted that the *Humanist* was standing up for science. He explained that he and many other rationalists "watched in amazement and incomprehension" when they saw friends and associates turn away from rational thought to "LSD, Transcendental Meditation, occultism, mysticism, one or another of the pseudo-science fads, or whatnot."[19]

Stars Aligned against Astrology

Kurtz's next project would utterly reject the idea of any kind of collaboration between scientists and occultists. Teaming up with an astronomer, he set out to publicly attack astrology as an ineffective and possibly dangerous pseudoscience. It was an easy target, and an appropriate one for his broader agenda, since it held such a prominent place in the rhetoric of the counterculture. The New Age, after all, was named after an astrological event: the shifting of the vernal equinox from the constellation Pisces to that of Aquarius because of the wobble of the earth's axis, a change that many people found portentous.[20] A popular song—the Fifth Dimension's "Aquarius"—immortalized the notion in 1969, representing this shift in the most positive terms: "peace will guide the planets / And love will steer the stars." Some of their lyrics were especially occult and psychedelic ("Mystic crystal revelation / And the mind's true liberation"), making it an enthusiastic paean to the trippier aspects of the counterculture.

Even on a mundane level, rationalists were confronted by what they considered astrological nonsense every time they opened the newspaper. Most of the papers in the country ran a horoscope column (and had done so for years), and it was at this seemingly innocuous practice that Kurtz and his team took aim. Bart Bok, the Dutch-American astronomer who was working with Kurtz, had been attacking astrology as far back as 1941. A well-known figure in his field, this former Harvard professor and director emeritus of the University of Arizona's Steward Observatory denounced astrological belief not only because it was so overtly unscientific but also—and probably more importantly—because it was a morally bankrupt belief system and psychologically debilitating.[21]

The psychological analysis was what convinced Bok that his crusade was important. Astrology, Bok believed, fostered an attitude of dependence and a flight from responsibility, a failure of nerve. In his earlier campaign, Bok had enlisted the support of the humanistic psychologist Gorden Allport to describe the logic behind this. Allport's report, endorsed by the Society of the Psychological Study of Social Issues, had argued that all forms of occultism were psychological crutches, which flourished in times of crisis when people sought out coping mechanisms. Astrology, Allport concluded, encouraged "an unwholesome flight" from real life's problems by allowing a person to "evade taking responsibility."[22]

Lawrence Jerome, a California science writer who was then working on a book critical of astrology, joined the other two men, and together they produced a short statement that summarized their main criticisms. In a nutshell, astrology was unscientific, having arisen at a time when people thought that the earth was stationary and at the center of the universe; it relied on magical forces, not naturalistic ones; and people accepted it on faith, like a religion rather than a science. Like Bok, they also spoke of the false reassurance offered by astrology: "In these uncertain times, many long for the comfort of having guidance in making decisions. They would like to believe in a destiny predetermined by astral forces beyond their control." Finally, it enjoined people to aspire to higher standards and take responsibility for their lives: "We must all face the world, and we must realize that our futures lie in ourselves, and not in the stars."[23]

Bok's previous efforts had been frustrated. He had been unable to enlist support from the astronomical community because the scientists believed astrology to be a frivolous pastime that was not a serious threat to science. Kurtz was more effective. Astronomers weighed in this time, and the final draft received nearly two hundred endorsements from members of the American Astronomical Society and the National Academy of Sciences.[24] The statement, alongside articles by Bok and Jerome, appeared in the September 1975 *Humanist*, whose cover announced that "186 scientists, including 18 Nobel Prizewinners, disavow astrology."

In an audacious move, Kurtz sent a copy of the statement to hundreds of newspapers around the country along with a letter urging them to discontinue their daily horoscope column or at least to publish a consumer protection warning against the "pretentious claims of astrological charlatans."[25] So at the beginning of September, just as the *Humanist* hit the newsstands, newspapers all over the country published reports of the humanists' attack on horoscopes. News stories, editorials, and syndicated columns commented on it. Eric Sevareid even gave it a sympathetic nod on *CBS Evening News*. The conservative pundit George Will agreed that "astrology is preposterous" but feared that believers "are not apt to pay attention to what real scientists say." Harriet Van Horne's syndicated column urged people to "give thanks for the 186 scientists." Some newspapers, such as one in Dover, indicated that they had been considering for some time canceling their horoscope column, since it was just "hokum."[26]

Of course, many people took the opposing side. One writer in Dayton defended the astrologers, saying that the humanists' statement merely indicated the elitism of the scientists, who did not see that "scientific rationality" was the problem. An editorial in the *Chicago Tribune* called the scientists spoilsports, and another newspaper article claimed that the statement itself was "the most futile verbal broadside of recent memory." Astrologers responded as well, sending letters to the *Humanist* to explain their position. Kurtz printed several long replies. One called the recent attack scurrilous and said that science was "the New Superstition" while astrology was an ancient study that had the weight of such eminent figures as Newton, Kepler, and Galileo behind it. Kathleen Russo and Carolyn Bermingham, the heads of the Federation of Scientific Astrologers, wrote a more nuanced letter. They said that "the serious student of astrology does not condone the use of horoscope columns" but that astrology, nevertheless, could still be studied rationally "as an instrument of developing a more substantial understanding of each individual . . . and his own potentials." One astrologer from the American Federation of Astrologers in New York challenged the 186 scientists to a test, claiming that under controlled circumstances he could astrologically deduce an appendectomy patient out of a group of nine other healthy men and women. Overall, the astrologers believed that the scientists were mostly ignorant of the actual practice of astrology and believed that their own astrological work did, in fact, have a scientific foundation.[27] All in all, the response was impressive. Kurtz's strategy worked so well that even he was surprised. "Very frankly," he said, "we never expected such an enormous reaction: we must have hit a funny bone."[28]

Organizing Skeptics

The difference between the "Objections to Astrology" statement and the earlier dialogue between Hull and Clay over the New Age was vast. Clay and Hull had proposed a collaboration between scientists and occultists, a mutual interaction that would benefit both sides. The anti-astrology campaign in the *Humanist*, on the other hand, offered no pretext of anything other than outright debunking. From the beginning, its aim was to refute astrology's claims and demonstrate how absurd and foolish it was.

Debunking the occult was certainly not new. Nineteenth-century scientists like Michael Faraday had taken it on, and in the 1920s stage magician Harry Houdini had teamed up with *Scientific American* to expose psychics and mediums. The popular science writer Martin Gardner published a book in the 1950s entitled *In the Name of Science*, where he described and discredited many pseudoscientific beliefs, ranging from the incredible (modern arguments to prove that the earth was flat) to the popular (dowsing, UFOs, chiropractic, and the island of Atlantis).

Similar efforts were taking place elsewhere as well. Just a year before the "Objections to Astrology" statement, Cornell astronomer Carl Sagan had organized a scientific panel at the American Association for the Advancement of Science to study the sensational astronomical ideas of Immanuel Velikovsky, who explained biblical miracles in terms of planetary collisions. Around the same time, James Randi began attacking Uri Geller. A Belgian committee of scientists had also become interested in studying the occult, and they had already published one response to an unusual form of statistical astrology. The genius of Paul Kurtz was that he was able to turn these disparate debunkers into a coherent organization with a clear mission.[29]

In the spring of 1976, at the annual conference of the AHA, Kurtz held a symposium on "antiscience and pseudoscience." Speeches at this conference exposed pseudoscientific claims of all kinds: UFOs, doomsday predictions, ESP, and psychic powers. Some of the talks were quite academic, explaining the philosophy of skepticism and its relation to the scientific method. The eminent philosopher Ernest Nagel was even at the conference—a leading light in the logical positivist movement. Many of the participants were not academics, though, and only a few were rank-and-file humanists. Philip Klass was an engineer and science writer who had investigated UFOs and written two books arguing that the sightings could be explained without resorting to extraterrestrial spaceships. Daniel Cohen was a science writer interested in doomsday prophecies. The conference even had themed entertainment: Randi performed one of his magic shows in which he exposed the trickery of the spoon benders and mind readers.[30]

The most significant outcome of the conference was the formation of the Committee for the Scientific Investigation of Claims of the Paranormal, which came to be known as CSICOP. The off-putting acronym, pronounced *sigh-cop*, was a fitting name for an organization that would adopt the role of a science police, who would expose the fraudulent claims of pseudoscientists, cranks, and scam artists.[31] The committee was chaired by Kurtz and had a governing board and list of fellows that included many well-known scientists, philosophers, writers, and entertainers. The group's main function would be to publish a journal, originally called the *Zetetic*, devoted to a new debunking enterprise; this awkward title was soon changed to the more accessible name *Skeptical Inquirer*.

The pages of the *Skeptical Inquirer* covered all sorts of unusual phenomena, like perpetual motion, dowsing, biorhythms, and pyramid power. There were articles and brief notes describing claims of unusual creatures and mystical beings, ranging from angels and fairies to "cryptozoological" beasts like the Loch Ness Monster and Bigfoot. Some articles explored quite exotic claims, such as the idea that "ancient astronauts" from other planets had helped build the pyramids in Egypt and make rock drawings in Peru. One popular book that was exposed by the journal argued that the Old Testament book of Ezekiel was actually a report of a close encounter with extraterrestrials. As time went on, the journal expanded to include attacks on claims made by religious believers involving psychic healing, spirits and ghosts, near-death experiences, the Shroud of Turin, and creationism. There were also attacks on subjects that one would not ordinarily associate with the paranormal: encounter groups, yoga, and organic gardening.[32]

The tone of the journal started out quite academic, but within a few issues it quickly shifted to cater to a more popular readership. The editors clearly enjoyed poking fun at some of the more outlandish beliefs, especially in short reports coming from various media outlets. The "Psychic Vibrations" column, for example, over the years noted that the rumors of Bigfoot were being joined by a number of similar but less known sightings: Littlefoot, Bighead, and even Skunkfoot, a foul-smelling creature prowling around Chesapeake, Virginia.[33] Both the sections for book reviews and letters to the editor encouraged engagement between writers and readers, with responses by authors printed alongside reviews and critical letters. The meat of the journal was, of course, its articles. Some of them were quite long and technical. They discussed statistical or other evidential and analytical matters but punctuated this narrative with firsthand accounts of debunking expeditions. Many articles concerned the psychology of perception and credulity, focusing on how imprecise and error-prone the human mind is. There was also a great deal of attention to the media's uncritical reception of pseudoscience and an especially serious concern about the spread of unscientific ideas in higher education.

Behind the group's utilitarian mission to debunk unscientific claims, one finds a deeper naturalistic worldview grounded in the Anglo-American analytic philosophical tradition. That tradition centered on epistemological questions, especially related to the goal of understanding the nature of truth and finding ways to verify knowledge claims. The 1976 *Humanist* issue that introduced CSICOP contained three articles defending this tradition: one by Kurtz; another by a colleague of his at Buffalo, Marvin Zimmerman; and the third by world-renowned logical positivist Ernest Nagel. The presence of two other internationally influential analytic philosophers on the list of CSICOP Fellows—Antony Flew and Willard Van Orman Quine—reinforced this philosophical bias. The articles denounced anti-intellectual and antiscientific attitudes that were common in the culture. The philosophers attacked the idea that science was inherently limited because it could not deal with transcendent values, they rejected the claim that logic and emotion "denuded" human experience, and they argued that the embrace of subjectivity was philosophically unsound. Nagel's article focused its attack explicitly on "the ranks of philosophers and historians of science" who criticized the idea that "scientific inquiry is the most effective way to obtain warranted knowledge."[34]

While the philosophers were quick to defend science and differentiate it from poetry and art, they did attempt to connect this austere scientific outlook with Enlightenment values typical of mainstream humanism. Zimmerman, for example, claimed that the scientist was by nature open-minded; the "commitment to science also suggests commitment to tolerance toward any and all ideas, no matter how wild they appear to be." He concluded his article with the injunction that it was necessary to both expose irrational beliefs and defend people's right to believe in them.[35] Kurtz went further, claiming that "man does not live by reason alone" and acknowledging that the whole realm of human experience must be part of any viable humanistic philosophy. "The arts are the deepest expression of our 'spiritual' interests," he said. "We need poetry and symbol, creativity and drama, passion and emotion, love and devotion. We need to appeal to the whole person, not simply his cerebellum. We need to celebrate life and its potential goods, to find joy and happiness as part of it, and to satisfy the quest for meaning."[36] In this way, Kurtz sought to marry the old-style humanist embrace of human experience with the narrower epistemology of scientific practice.

The project took off quickly, but there were signs of internal strife from the outset. CSICOP members were not all of the same mind about the organization's purpose, and a major conflict emerged early on between Kurtz, the committee's chairman, and Marcello Truzzi, the first editor of the journal. Truzzi had been instrumental in the formation of CSICOP because he was already well known by both scholars and debunkers of the paranormal. He published a newsletter composed of bibliographies and synopses of scholarly books and articles dealing with the paranormal from a wide range of academic disciplines. In Truzzi's mind, this was an academic pursuit, and the more he worked on it, the more it seemed to him that a new interdisciplinary

scholarly field was on the verge of being born. Scholars who studied the occult and paranormal came from disciplines as varied as psychology, folklore,
and philosophy.[37]

As a sociologist, Truzzi's thinking about popular belief was fundamentally
different from Kurtz's philosophical point of view. For Kurtz, the issues involved methods of finding truth and establishing intellectual rigor. Social
cohesion in modernity was based on right thinking, so it was crucially important to teach scientific method and denounce nonsense. Truzzi's perspective
could not have been more different. Though he, too, was a naturalist and did
not accept the occult worldview, neither was he frightened by it. He was
simply fascinated, from a sociological point of view, by the beliefs of social
groups. These cultural developments were interesting in their own right. For
example, Satanism, in his estimation, was a playful diversion akin to the reading of science fiction, and the people who practiced it were not inherently
dangerous, either to themselves or to the foundation of society as a whole.
Truzzi's perspective was in part derived from his own childhood experience
growing up in one of the twentieth century's most famous circus families. His
father, Massimiliano Truzzi, was a Russian-Italian juggler who made his way
to America and performed with the Ringling Brothers circus. Marcello was
always fascinated by the odd and anomalous. By no means an enthusiast of
the New Age, he was intrigued by it as a sociological phenomenon and found
its acolytes to be thoughtful and interesting individuals. As a sociologist, he
adopted a disinterested approach to the occult revival but remained sympathetic to the people he studied. Philosophically, he was a naturalist through
and through, but he did not expect others to share his perspective.[38]

Had Truzzi remained focused on his original academic mission, it is unlikely that he would have teamed up with Kurtz in the first place. However, he
was already being pushed toward skepticism by several colleagues who were
or had been at one time involved in stage magic: the psychologist Ray Hyman,
the popular science writer Martin Gardner, and the magician James Randi.
These men were more deeply skeptical of occult claims because they knew
how easy it was to trick people. Psychics claimed that the astonishing feats that
they performed came from hidden forces or supernatural powers, but stage
magicians saw only illusion and misdirection, not much different from a con
game. Hyman, Randi, and Gardner saw nothing in the psychics' performances
that couldn't be explained using strictly ordinary deceits. If there were anything to the paranormal and the occult, it was going to be necessary to first
get rid of the pretenders.[39]

It quickly became apparent that Truzzi's tolerance of irrational beliefs would not be acceptable to either Kurtz and the humanists, on the one side, or his magician friends, on the other. As editor of CSICOP's journal, Truzzi had set a tone in the first few issues that irritated all of them. He was too neutral. He was also too critical of the debunking enterprise. Treating the debunkers just as he would any other social group, he noted that they too had their faults. As time went on, he came to be especially upset by what he saw to be simply closed-minded dogmatism. However, it was Truzzi's stance toward science that most bothered the skeptics: at one point he stated that he had "no special vested interest in protecting contemporary science's paradigms."[40] It was precisely the protection and promotion of those paradigms that Kurtz and most of his fellow skeptics believed to be the mission of CSICOP.

After the second issue of the journal, Truzzi resigned as editor. Ten years later, he continued to be a critic of the group: "By substituting horselaughs for syllogisms, we act to suppress dissent. It is for this reason that I am often more sharply critical of so-called 'skeptics,' with whose 'orthodox' conclusions I may in fact largely agree, than I am towards the 'maverick' scientists towards whom I have been accused of showing too much tolerance."[41] Truzzi had wanted to publish an interdisciplinary academic journal, while the rest of the committee envisioned a popular magazine. They wanted to fight pseudoscience, not study it. Upon Truzzi's departure, Kendrick Frazier, a science editor, took over the *Skeptical Inquirer*, and the magazine at once assumed a popular, critical, and frequently sarcastic tone.

The Mars Effect Controversy

Defending science by debunking the paranormal turned out to be more challenging than the skeptics first imagined. The focus on discrete facts and analytical methods central to the debunking enterprise proved unworkable in areas where there were basic disagreements over epistemology and metaphysics. One event in particular highlighted the pitfalls of this strategy: a controversy satirically termed by a former CSICOP member the "sTarbaby affair." The conflict arose out of a carelessly written passage in Lawrence Jerome's article in the "Objections to Astrology" issue of the *Humanist*. In the article, he characterized the different types of astrology as either unscientific magical thinking or shoddy attempts at science. As an example of the latter, he cited the work of the contemporary French researcher Michel Gauquelin, who claimed to have uncovered some statistically relevant astrological correspondences.[42]

Gauquelin and his wife Françoise had spent years collecting birth data on thousands of individuals with notable careers in different fields around Europe. The goal was to find out whether there were any correlations between a person's horoscope and his or her mental or physical characteristics. The Gauquelins' analysis revealed that, surprisingly, there seemed to be. The complicated statistical study they did showed correlations between the configuration of planets in the sky at the time of a person's birth and his or her athletic abilities. Mars turned out to be the key variable. Great athletes, the Gauquelins discovered, had a statistically greater likelihood than others of being born when Mars was in a particular region of the sky. This discovery did not echo any existing astrological schools of thought. However, the statistics implied that planetary configurations seemed to have a measurable influence on people's lives.

The claim had seemed patently ridiculous to Jerome, so he had written it off as a statistical error. It turned out that he had been too quick in his judgment, and the Gauquelins pointed out that Jerome simply had not correctly understood their analysis. Letters to the editor exchanged in the *Humanist* over the subsequent months turned into a complex and technical discussion over statistical significance and scientific methodology, a long, tedious, and probably irrelevant conversation for most readers of the magazine. Though the skeptics came to admit that Jerome's analysis was mistaken, they remained entirely unconvinced by the Gauquelins' arguments and believed that the statistics were somehow in error. The dispute concluded with a proposal that a group of skeptics test the Gauquelins' claims in a new way, as suggested by biomedical statistician Marvin Zelen.[43]

In what initially looked like a spirit of cooperative scientific investigation, in keeping with the ideals of Clay and Hull, the Zelen committee and the Gauquelins worked together. The Gauquelins collected the data, and members of the committee did the analysis. However, there were difficulties from the outset. In fact, it was questionable whether this was really a scientific investigation at all. The fact that there were no community-set protocols for such a study, no refereed disciplinary journal within which to publish, and indeed no disciplinary community for this study at all meant that the analysis was entirely detached from the usual scientific institutional setting. Instead, the entire public dispute took place in the pages of a popular magazine of opinion. Moreover, the skeptics even characterized the test not as a scientific study but as "a challenge." That public context influenced the entire nature of the test, putting enormous pressure on each side to win the dispute.[44]

Not surprisingly, the two sides arrived at entirely different conclusions and had no means to adequately adjudicate them. The new data and the new analysis showed a modest, albeit statistically significant, astrological signature, and the Gauquelins believed that their ideas were confirmed. The skeptics, though, were not convinced, so they reexamined the entire experimental setup and subdivided the sample by sex and location. When they made these changes, the entire statistical relationship was affected. Now, the subsets separately showed only negligible and inconclusive results.[45] The Gauquelins and their supporters were outraged. By redoing the analysis on the subsamples, the skeptics were changing the experimental setup that all had agreed to. The skeptics were being not just unfair but dishonest and unscientific.

That same conclusion was also reached by one particularly cantankerous member of the CSICOP executive board, an astronomer by the name of Dennis Rawlins. Even without a PhD or an established academic position, Rawlins had published in several prominent scientific journals. Moreover, he was an expert statistician. He was outspoken in his disdain for occultism—he thought that most believers in the paranormal were fools or liars and often said so—and he was a stickler for accuracy. It was Rawlins who had done much of the painstaking calculation on the Zelen test *before* Kurtz and Zelen subdivided the data. Rawlins, like the Gauquelins, was furious with his colleagues for manipulating the data in order to minimize results that clearly favored the astrological claim.

Rawlins's increasingly angry attempts to get his fellow skeptics to acknowledge their bias got him voted off the governing council and admonished to keep silent in public meetings. Indignant, Rawlins penned a witty, savage account of the affair that consumed 250 manuscript pages, which he sent to the occult magazine *Fate*. There, it was pared down and published as a thirty-two-page article.[46] The piece, entitled "sTarbaby," accused Kurtz and "the Committee" of dishonesty and blamed them for turning CSICOP into an advocacy group that cared little for the truth. He likened Kurtz to Brer Rabbit in the famous folktale about the tar-baby, in which all efforts of the rabbit's escape from Brer Fox's ingenious trap merely intensified his problems.

Rawlins believed that CSICOP was behaving in a manner antithetical to the scientific spirit—by fostering partisanship, promoting secrecy, and even telling lies. By contrast, the members of the committee justified their response as wholly rational. They saw the subsample analysis as a refinement, not a breach of protocol. Nonetheless, the episode had serious ramifications. A few reflective committee members were saddened by how this attempt at impartial analysis had deteriorated into a series of personal accusations, scheming, and secretive deliberations. This was not the scientific study they had anticipated. The many details of this acrimonious affair have been presented, explored, and rehashed in both published and unpublished documents and could fill up an entire filing cabinet.[47] The widespread publicity was fodder for critics who attacked the skeptics' claim to scientific neutrality, and it became clear that the skeptical project as set up through CSICOP could never do rigorous and sustained investigative work with paranormal proponents as collaborators. Their studies rarely convinced true believers, nor could the proponents of the paranormal effect any change in the minds of the skeptics.

CSICOP and Religion

The skeptics movement flourished beyond the expectations of Kurtz and its founders. Faced with what they saw as a wholesale assault on rationality in the culture, the self-proclaimed defenders of science found a large, like-minded cohort of readers and followers. Total paid circulation increased enormously in the first decade. By 1979, there were about five thousand subscribers; by 1982, it had doubled; by 1986, it had more than doubled again, eventually leveling off at about thirty-five thousand at the end of the decade.[48] Outside of the journal, local groups of skeptics began popping up all over the country. By the spring of 1986, on the group's tenth anniversary, the magazine listed twelve states as having one or more skeptical societies (Texas and California had seven between them). These local groups were never officially affiliated with CSICOP, in large part because of legal liabilities. Nonetheless, they were recognized as having "aims similar to CSICOP's." In that same anniversary issue, the inside back cover of the journal also gave a "partial list" of ten international committees in Australia, Belgium, Canada, Ecuador, France, Great Britain, Mexico, New Zealand, Norway, and Sweden, many of which had sprung up in the first couple of years of CSICOP's existence. The committee had by this time forty-four scientific and technical consultants, mostly academics, in fields as diverse as mathematics, chemistry, astronomy, geology, pharmacology, nuclear medicine, materials science, anthropology, sociology, psychology, philosophy, and even biblical history.

Quite early on, CSICOP became institutionally separate from the AHA.[49] To begin with, Kurtz was fired as editor of the *Humanist*, due to allegations of financial mismanagement and disagreements about editorial control. Nonetheless, the split between CSICOP and the AHA did not appreciably change the humanistic nature of the skeptics movement. The two groups continued to have interlocking membership, and only a few years after his sacking by the *Humanist* Kurtz started a competing humanist journal called *Free Inquiry*. Though skepticism had a narrower focus and never explicitly embraced any religious, theological, or broadly philosophical worldviews, those bigger concerns motivated the work of the skeptics.

The relationship between skepticism and religion was complex, but most skeptics took a dim view of traditional and orthodox religion. Starting with the second issue of the *Zetetic*, creationism emerged as a frequent topic, where it was regularly portrayed as yet another form of pseudoscience. Geologist Norman R. King penned a scathing review of a recent book *The Creation-Evolution Controversy*, calling it deceptive—an argument for creationism under the guise of neutrality. A letter writer in the following issue approved and called the authors of all creationist literature "religious fanatics."[50] A cover story a few years later by anthropologist Laurie Godfrey took a more aggressive swipe at creationism, especially its appearance on college campuses. She denounced fundamentalists, evangelistic grassroots organizations, Bible research groups, and born-again Christianity. She did not distinguish this kind of religiously based pseudoscience from other sorts of paranormal belief: "Simultaneously, courses on scientific creationism appeared in college curricula, along with courses on astrology, Atlantis, the teachings of von Däniken, and so on. Indeed, many academicians have jumped on the paranormal bandwagon."[51]

The cover story of the spring 1979 issue of the *Skeptical Inquirer* was on the psychology of near-death experiences, something relevant to many religious traditions. Its author, professor of psychology James E. Alcock, explained religion in purely naturalistic terms, noting that "during the social evolution of human behavior, religions gradually developed that helped to assuage [the] age-old existential and ontological anxieties about living and dying."[52] The belief in an eternal soul, he continued, arose out of a functional need to alleviate anxiety. His overall assessment—after explaining that mystical and near-death experiences can be replicated with drugs, meditation, and other sorts of naturalistic techniques—was that religious experiences should be studied but that the metaphysical trappings that accompanied them must be abandoned.

Shortly after this, science fiction writer Isaac Asimov attacked the astronomer Robert Jastrow for putting science in the service of religious ideas. Jastrow's *God and the Astronomers* claimed that recent astronomical discoveries could be compared to some aspects of the Christian worldview—the Creation story, in particular, could be justified with big bang cosmology. Asimov attacked Jastrow for "abandoning his faith in the power of reason" and asserted that science and religion have simply nothing to say to each other. Similarities between the claims of believers and scientists were simply coincidental.

Asimov leveled the same attack on the physicist Fritjof Capra, for writing a popular physics book, *The Tao of Physics*, that made comparisons between ancient Eastern ideas and quantum mechanics. For Asimov, the ideas of early Christians or ancient Taoists should never be justified through science. And certainly science should never be intertwined with these religious points of view. Asimov concluded on a familiar note. Fearing that we were witnessing a period in which "rational minds . . . lost their nerve," he explained that "there has been at least one other such occasion in history, when Greek secular and rational thought bowed to the mystical aspects of Christianity, and what followed was a Dark Age."[53]

In all of these cases, skepticism was used as a means of attacking outdated and unscientific ideas and explicitly or implicitly attacking religion in the process. As an offshoot of rationalism, the skeptical mentality was an abridged humanism that focused primarily on scientific claims without any discussion of theology or metaphysics. Nor did it present any comprehensive life philosophy in the mode of traditional humanism. The skeptics' indebtedness to positivism and analytic philosophy caused it to see the social and psychological dimensions of human experience as impediments to truth rather than positive features of human life. For the skeptics, science was utilitarian rather than inspirational, a tool for separating truth from poppycock.

Curiously, by adopting such a position, the skeptical project remained accessible to some people in a way that conventional humanism was not—namely, individuals who valued rationalism yet who were not themselves atheists. Since Kurtz recognized this, he never allowed the atheists in the group to define skepticism as simply out-and-out rationalism. This posture of neutrality was in large part due to the presence of Martin Gardner on the CSICOP executive committee. Though Gardner was a debunker with a regular column in the journal entitled "Notes of a Psi-Watcher," he was not a humanist. Gardner called himself a philosophical theist because, although he followed no institutionalized religion, he did believe in a personal God, an afterlife, and prayer.[54] Nor was he alone among people of faith in finding common cause with the skeptical program. There was cautious support for this kind of debunking even among some orthodox Christians because it helped them rebut the threat of the New Age, which they considered religious superstition. The mainstream evangelical publication *Christianity Today*, for example, ran an article that made this point explicitly: "Although *The Skeptical Inquirer* is highly critical of Christian orthodoxy, it nevertheless contains valuable exposés of alleged occult encounters, New Age healings, and other stories that promote paganism. Christians ought to make common cause with its authors to share what is true, presenting a Christian alternative when needed."[55]

Skepticism proved to be a viable offshoot of humanism that grew and developed rapidly in its first few decades. Its narrower focus gave it a clearer mission and made it especially effective in its activism. This also made it attractive to those who were unwilling to call themselves humanists because of all of the metaphysical baggage that came with that term. Nonetheless, a strong thread of humanistic values remained embedded in the skeptical temper, and its aggressive, confrontational, science-based activism would play a continuing role in some of the humanists' future battles.

CHAPTER TEN

The Fundamentalist Challenge

One conclusion that the [Rafferty] report draws is particularly ominous; for it maintains [that] *all* forms of secular humanism which are taught in our public schools—in science, history and elsewhere—are to be taken as a form of religious indoctrination.

Paul Kurtz (1969)

In the late 1960s, California's conservative superintendent of education Max Rafferty appointed a committee to respond to a perceived "moral decay" in the culture. In Rafferty's opinion, teenage behavior was out of control: there was increased drug use, widespread sexual promiscuity, a rise in illegitimate births, and a surge in violent crimes. With the blessing of conservative governor Ronald Reagan, a new Moral Guidelines Committee convened with a mandate to fight a statewide epidemic of immorality. Grounded in a fearful right-wing conspiracy narrative, the report talked about "incursions" into the schools and ideas "alien to our heritage." It identified secular humanism as the culprit. The report was a long (seventy-four pages), poorly edited typescript that asserted that the American Humanist Association was the hub of a nationwide conspiracy. It denounced John Dewey as "the high priest of progressive education" in America and noted Dewey's endorsement of the Humanist Manifesto (which the report reprinted in full).[1]

One of the main concerns of the report involved the state's sex education curriculum, which was then being attacked by a group of conservative Christian parents who objected to course materials that they considered pornographic. Indeed, they charged that the materials encouraged kids to have sex and used techniques that undermined parental authority. The report took that objection and manufactured an outlandish conspiracy to go with it. Among its more sensational claims, the report stated that the humanists were preparing to use both behavioristic psychology and sensitivity training to condition and indoctrinate school children in a way not unlike what the Nazis had done.

According to the report, the Germans had sponsored sex orgies for teenage boys in the Hitler Youth in order to establish psychological control over them by fostering base instincts and thereby short-circuiting their rational and moral sensibilities. "This conditioning through emotional, animalistic responses has been developed by the Communoid forces," it stated, "who apply these techniques to control group behavior."[2] The link to humanism was through the sexologist Lester Kirkendall, an AHA board member, who was one of the founders of a new national group called the Sexuality Information and Education Council of the United States (SIECUS).[3]

A central claim of the report was that teaching Christian morality was a legal right and that the recent takeover of education by secularists was unconstitutional. "The American Republic was, and is, established upon a firm belief in divine providence," the report asserted. It justified this claim on the idea that human freedom was possible only in a world with a divine moral sanction, that in a godless cosmos, where mankind was on his own, true liberty was impossible: "man's blessings—all his freedoms—stem from a source that is higher than man." The atheistic direction of modern education, and especially the exclusion of Christian teaching in the schools, which was now established in a number of recent Supreme Court rulings, threatened to undermine the basic foundations of the country by creating a new generation that rejected religion altogether. The strategy moving forward was to demonstrate that so-called secular humanism was really a type of religion and that much of the new secular public education was unconstitutional. "The state is forbidden to promote a Godless religion," it stated, "just as it is forbidden to promote any one sect."[4] Humanist-sponsored curricula were "tantamount to indoctrination in a religion and contrary to public policy."[5]

This report, with its over-the-top claims and its Christian legal agenda, emerged from a number of forces brewing in California at the time. In recent years, an American right-wing movement had formed out of a merger between anti-Communists, especially the conspiracy-prone John Birch Society, and ultraconservative Christians, both Catholic and Protestant. This new politicized fundamentalism achieved a fair amount of clout in California and exerted influence on state politics. Rafferty himself had been elected state superintendent as a lifelong conservative, whose political career was fostered in large part by the ultrareactionary John Birchers.[6] The connection to extreme groups within the evangelical camp was clear as well. The Rafferty report echoed claims made by the Oklahoma-based Christian Crusade ministries, one of whose publications was entitled *Is the School House the Proper*

Place to Teach Raw Sex? All in all, these political and religious conservative activists pushed an agenda that included Christian orthodoxy, American patriotism, family values, and conventional gender roles. Their anti-Communist connections encouraged their suspicion of fifth column traitors and promoted wild conspiracy theories. It was in this context that the idea of a humanist conspiracy involving subterfuge and mind control came about.[7]

Church and State in Humanist History

The fundamentalists' tendency to blame the humanist movement was understandable in certain respects. Secularism and church-state issues had topped the humanists' agenda in the second half of the century as the country rode a wave of secularization that affected the culture, the law, and especially education. During that time, the country witnessed radical changes in how religious faith was expressed in the public sphere. Religious classes, school-day Bible reading, prayers in the classrooms, and the swearing of oaths for public office were all contested and declared unconstitutional. Although the 1950s is often considered to be a period of revived religiosity in the country—it was then that the word *God* was added to the pledge of allegiance and to the currency—it is equally right to see the time as one of increasing secularism, in which traditional religious beliefs and practices lost influence in many sectors of the public sphere.[8]

The secularizing of public institutions did not happen on its own; it was abetted by activists who challenged the legality of public support for religious practices.[9] Beginning in the 1940s, individuals and groups like the American Civil Liberties Union (ACLU) initiated court cases that argued that the First Amendment made it unconstitutional for public institutions to provide any support for religion—separation of church and state had to be absolute in order to protect the religious freedom of all citizens. The AHA watched the cases closely and provided reports and updates in the journal and in their member newsletter. The plaintiffs in these cases were sometimes already affiliated with humanist groups, but not always. It was sometimes only later that individuals in these cases became members. A common pattern emerged in which litigants who had won cases were actively recruited by the humanists, and their national prominence often launched them into leadership positions.

One of those people was Vashti McCollum, whose involvement in a landmark 1948 Supreme Court case brought her widespread notoriety and eventually led to her election as AHA president in 1962. The case revolved around McCollum's objection to an Illinois state law that allowed released time religious instruction for all public school students. Released time was a relatively popular program that set aside an hour each week during the public school day for a faith-based class on the religion of the family's choice: Protestant, Catholic, or Jewish. This left unbelievers and liberals without an option other than simply declining instruction altogether.[10]

Having grown up in a liberal Protestant family, McCollum had initially enrolled her son in the class for Protestants, but the teaching had been so conservative that she eventually removed him. The boy was taunted by his classmates, and McCollum sued the school district.[11] The religious community was sharply split on the issue, with liberals, Jews, and even some conservative Baptists strongly supportive of a strict separationist position, while many traditionalists argued that the nation had a responsibility to provide moral education in the form of religious teaching. Many supporters of released time interpreted McCollum's position as an attack on religion outright. She was promoting atheism, they thought, and one person even called her "a wicked godless woman, an emissary of Satan, a Communist, and a fiend in human form."[12] The relatively young AHA (less than seven years old at the time) decided not to publicly support the case, feeling that it might be counterproductive and could prejudice the case against McCollum because the group was "too far left religiously."[13]

The court ruled in favor of McCollum against released time instruction, affirming a principle established in a decision from the previous year in which Supreme Court justice Hugo Black had stated that "the First Amendment has erected a wall between church and state [that] must be kept high and impregnable." School districts that lent support to a partisan religious curriculum breached that wall and violated the First Amendment.[14]

After the decision, McCollum joined the AHA, where she served on the board for more than ten years, during which time she was elected president. McCollum's rise to president in 1962 was prophetic. That year and the following, the Supreme Court heard two more church-state cases, *Engel v. Vitale* and *School District of Abington Township v. Schempp*. The first concerned school prayer, and the second grappled with the question of Bible reading in the schools.[15] The *Engel* case arose when several New York families challenged the practices of a state-mandated school prayer because the specific prayer was antithetical to their faith. As Unitarians, agnostics, Ethical Culturists, and Jews, they objected to the idea that humanity was servile to an all-powerful personal God, as the prayer clearly affirmed. The second case on Bible reading involved Unitarians and atheists, and one of the litigants, Madalyn Murray (later Madalyn Murray O'Hair), would eventually gain international notoriety as America's foremost atheist. She, like McCollum, also sat on the board of the AHA some years later, albeit for a much shorter tenure. The court in both cases struck down laws as violations of the Establishment Clause. Public entities such as schools could not be allowed to establish any one religious position, they said.[16]

Paul Blanshard, the AHA's legal watchdog, had followed both cases closely in his regular *Humanist* column, where he defended the AHA's uncompromising stance on the separation of church and state. Of course, even as he praised the court rulings, he urged humanists to be vigilant: "We [humanists] must be the American alarmists," he declared, echoing the words of James Madison. It had been fifteen years since the McCollum case, and things had changed a lot. The AHA was no longer worried that its presence would prejudice the court against it and had submitted an amicus brief in the *Schempp* case, where, indeed, all the plaintiffs were humanists.[17]

Two other Supreme Court decisions were instrumental in shaping the legal environment of the 1960s with regard to the status of humanism, and they did so in ways that were less obviously beneficial to humanist concerns. In the case of *Torcaso v. Watkins* (1961), the atheist Roy Torcaso had objected that the state of Maryland required all state employees to take an oath affirming belief in God, which excluded him from holding public office. The other case, *United States v. Seeger* (1965), involved the religious exemption for conscientious objector status, a concern that was especially relevant during the increasingly deadly and ever more controversial Vietnam War. The court sided with the plaintiff in both cases.[18]

The AHA's executive director Tolbert McCarroll, a lawyer by training, had written the AHA's amicus brief for the *Seeger* case, and in it he explained that the term *religion* should refer to "any system of ideas which gives perspective to life, the emotional involvement with a philosophical point of view, or simply the search for meaningful ethical standards." McCarroll went so far as to say that even those nonbelievers who objected to the term *religious* could be considered religious in the legal, philosophical, and anthropological sense.[19] This was precisely the way that Dietrich and Reese had used the term since the first years of their movement.

The *Torcaso* decision turned out to be especially significant because it affirmed that a religion need not assert belief in the existence of any supernatural entity, stating that the government must not "aid those religions based on a belief in the existence of God as against those religions founded on different beliefs." In a widely cited footnote, it listed several religions of this sort, including both "Ethical Culture" and "Secular Humanism" by name.[20] That sentence gave the religious humanist cause a major boost, and it had the effect of pushing the AHA toward a new legal status. The group had been founded in 1941 as an educational association, not a religious one, so it held an educational tax exemption. After the *Seeger* decision, McCarroll spearheaded a successful effort to have the organization redefined as religious. One reason for this was that, during the Vietnam War, this change made it possible for the AHA to support the petitions of pacifist unbelievers who sought to get conscientious objector status.[21] The redesignation also put the organization fully in line with McCarroll's other efforts to promote humanist counseling and various ceremonial practices like weddings and funerals.

It was a good strategy at the time that fulfilled the goals of some religious humanists, but the characterization of the AHA as a religious organization was increasingly out of step with the majority of its members' beliefs as they became ever more secular.[22] Nonetheless, for the next two-plus decades, the group's tax designation as a religious entity remained on the books, lasting until the early 1990s.[23] The other fallout from these two cases resulted from the way that they were used to help frame church-state law in ways that the fundamentalists believed they could exploit if they could convince others that secular ideas and practices were ultimately just as religious as Christianity.

The Evangelical Antihumanist Crusade

Seven years after the Rafferty report, Arizona Republican John Conlan warned his colleagues in the US House of Representatives of the secular humanist threat to American education. He echoed Rafferty's claim that secular humanism was an insidious religion that had infiltrated the American public school curriculum and was undermining the country's Judeo-Christian foundations. To counteract this threat, he proposed an amendment to a funding bill that would explicitly prohibit support of educational programs "involving any aspect of the religion of secular humanism."[24] Although the amendment passed in the House, it died in a House-Senate conference.

The notion of a humanist conspiracy was developed further in a 1977 book-length exposé. Written by evangelical Claire Chambers, *The SIECUS Circle* returned to one of the big concerns of the Rafferty report, namely, sex education, which Chambers believed was at the center of the conspiracy. It was through the SIECUS network that humanism infiltrated American institutions with its socialist, internationalist agenda. Her book documented the interlocking boards of America's most powerful liberal and cultural institutions, beginning with the AHA and SIECUS and then tracing connections to such places as the ACLU and the National Education Association. She even implicated the ecumenical National Council of Churches.[25]

One of the most scholarly versions of this argument was an article in the *Texas Tech Law Review* in 1978 written by Representative Conlan and John W. Whitehead. Whitehead was a young lawyer from Tennessee who had just finished a book attacking federalism from a Southern Christian states' rights perspective. The point of the *Texas Tech* article was that the search for a neutral secular public sphere was impossible. All efforts by the courts to insist on neutrality were doomed to failure because when Christian influence was removed, anti-Christian influences inevitably took their place. The Supreme Court, which claimed to be on the side of neutrality and tolerance, was in fact siding with the humanists and liberal religionists. In so doing, it gave preference to liberal and humanistic values that were, by definition, opposed to Christian ones.[26]

A 1980 publication, titled *Battle for the Mind*, continued the argument that humanism was an insidious anti-American movement. Author Tim LaHaye, who would later become famous for his *Left Behind* novels, reiterated the conspiracy theories of Rafferty and Chambers. However, he simplified the arguments against humanism—sometimes to the point of caricature—so that they would appeal to a wider audience. His basic contention was that the humanist worldview was promoted by an elite ruling class. The humanists' power arose from their control of education at all levels, including the media, the federal government, and a panoply of liberal activist groups. Its worldview produced an unhealthy obsession with sex, self-indulgence, and rights without responsibilities that in the end was a "rebellion at God, parents, [and] authority." The well-being of the young was especially compromised when they grew up lacking "skills, self-worth, purpose, [and] happiness."[27] LaHaye illustrated the power relationship that drove this conspiracy by drawing a schematic image showing ordinary Americans ("religious pro-moralists," born-again Christians, and so forth) chained to a small group of humanists who were dragging the rest of the country downward away from God.[28]

All of the above attacks were public and political ones, and many were written explicitly for and by laypeople, but there was a deeper theological framework that lay behind these popular political arguments. The religious attack on humanism and secular culture derived many of its ideas from three influential Calvinist Presbyterians: Francis Schaeffer, Rousas J. Rushdoony, and Cornelius Van Til. These three men helped lay the intellectual foundations of late twentieth-century evangelical Christianity. Their cultural criticism, and especially their attacks on humanism, strongly influenced later conspiracy theories.[29]

Francis Schaeffer identified humanism as the central problem facing modern civilization but understood it as an intellectual crisis, rather than a conspiracy. In his view, secular culture was by definition humanistic and revolved around the concepts of materialism, autonomy, and freedom, which he believed were fundamentally misunderstood by secular thinkers. The materialist view of mankind, in its rejection of the spiritual soul, reduced men to animals, and ultimately to lumps of matter. Without anything special to differentiate human beings from the rest of the world, there was no way to determine right from wrong. Human autonomy led to isolation, which in turn led to an ethical vacuum, in which people lost all connection to a divine moral code. Further, although secularists believed that rejecting the myth of God gave men freedom, Schaeffer argued that it simply released the powerful from all ethical constraints. Ultimately, humanistic thinking resulted in a world where humanity itself was dispensable. It led to the Holocaust and Stalin's purges, to abortion, child abuse, and slavery. Humane behavior could not survive in a relativistic world without a religiously based moral framework.[30]

Rousas Rushdoony held similar ideas about humanism but analyzed the political implications of secularism more deeply, ultimately adopting a much more radical position. Rushdoony's parents had escaped from the Armenian genocide of 1915, when over one million Christian Armenians died at the hands of the Ottoman Turks. His reflections on this event made him always suspicious of the nature of state power. His 1963 book *The Messianic Character of American Education* argued for the absolute separation of school and state, a work that made him a major figure in the homeschool movement.[31]

For his part, the theologian Cornelius Van Til questioned the very possibility of a religiously neutral position. He described the conflict of worldviews between Christians and non-Christians as inevitable, since each side started from completely different presuppositions about the world. Because there was no neutral ground, there could be no rational means of discussion that bridged the gap between the two sides. They were simply incommensurable beliefs. The goal of all evangelical activity, according to Van Til, should be to expose the contradictions in the non-Christian worldview.[32]

One of those contradictions, according to both Schaeffer and Rushdoony, was to be found in the scientific control of human nature, epitomized in the behaviorist psychology of B. F. Skinner. Behaviorism was the epitome of the philosophy of humanism, they contended, and it clearly demonstrated the inevitable outcome of materialistic philosophy. Although humanists claimed that their philosophy increased human freedom and compassion, the internal logic of materialistic humanism contradicted that claim. Instead, humanism led to a deterministic and totalitarian ideology. Rushdoony found a similar problem with John Dewey and his ideas on education: "It is questionable whether liberty can long survive under a continued onslaught of Deweyism," he said.[33] Rushdoony feared that the state would use social-behavioral technologies to control and thus dehumanize people. He concluded that only an unwavering transcendental faith could preserve human dignity and freedom. Christianity was the only way.

The Humanist Response

As fundamentalist activism crescendoed after the defeat of the Rafferty report in California, it achieved a remarkable success in the establishment of the Moral Majority as a powerful, national right-wing political organization that helped elect Ronald Reagan to the US presidency. Throughout this period, the evangelicals honed their attacks against the stealth religion of secular humanism, using the phrase as a shibboleth and a call to arms. It signaled a rapid politicization of religion that blindsided unsuspecting liberals.

The humanists were more aware of this new politicized religious threat than most liberals and mobilized, both in print and on the ground. In the case of the Rafferty report, Kurtz published counterattacks and updates in three issues of the *Humanist*.[34] Confronted with the real possibility that California might enact this reactionary educational agenda if the report's guidelines were adopted, humanists also joined forces with other liberal groups to lobby the State Board of Education. This coalition included Unitarians, Ethical Culture members, and the Council for Liberal Religious Thought. During the State School Board hearing, AHA president Lloyd Morain defended the humanist position, speaking out in a room packed with right-wing activists in patriotic garb. The humanists were on the barricades in this battle and sighed with relief when moderation prevailed and a completely different set of guidelines was adopted.[35]

It was at about this time that the humanists also established their own lobbying organization in Washington, DC, joining forces with Unitarians and Ethical Culturists to form a group called the Joint Washington Office for Social Concern.[36] The focus of this group was liberal social causes: church-state issues, women's rights, abortion rights, and similar topics. In addition, individual humanists often joined other organizations that targeted one or more areas of fundamentalist right-wing political engagement: People for the American Way, Americans United for Separation of Church and State, the ACLU, and similar groups advocating for liberal political and legal causes.

The humanists also used successful media strategies pioneered by Kurtz to publicize the humanist point of view. The mid-1970s witnessed a flurry of public statements of this sort. After the "Objections to Astrology" statement in 1975, there was a "New Bill of Sexual Rights and Responsibilities" in 1976, a "Statement Affirming Evolution as a Principle of Science" in 1977, and a "Statement on the Family" in 1980.[37] Of greatest importance during this period were two major manifestos of humanist principles, the Humanist Manifesto II of 1973 and the Secular Humanist Declaration of 1980, both of which were spearheaded by Kurtz and both of which garnered front-page stories in the *New York Times*.[38] The Humanist Manifesto II was not directly aimed at fundamentalism, but the Secular Humanist Declaration was—adopting, identifying with, and defending the very phrase *secular humanism*, which had been so vilified by the fundamentalists. This declaration was announced at the same time as a new humanist journal, *Free Inquiry*, and a new humanist organization, the Council for Democratic and Secular Humanism. It was the culmination of Kurtz's decade-long effort to forcefully rebut the assault on humanism and humanistic values.

In the seventies many of the AHA's Humanist of the Year awards had an explicitly political purpose. In 1970, they honored octogenarian A. Philip Randolph, an African American union organizer and political activist. Then, in 1974 and 1975, the committee made the unusual choice to honor two people each year, bringing the total number of awards to four. Mary Steichen Calderone was recognized for her work in advancing sex education; Joseph Fletcher, author of *Situation Ethics*, was a pioneer in nonreligious ethics; Betty Friedan helped launch the feminist movement with her confrontational book *The Feminine Mystique*; and Henry Morgentaler, a Canadian doctor, was working to liberalize abortion law in his country. Morgentaler's award was especially noteworthy, given that he was in jail at the time of its presentation. All of these honorees were promoted for the agendas that were being attacked by evangelicals and conservative Catholics: sex education, secular ethics, feminism, and abortion.

The articles in the *Humanist*, the newsletter *Free Mind*, and the new magazine *Free Inquiry* provide a glimpse into the humanists' understanding of the evangelical mind-set. They also highlighted the humanists' defense of modernity. The *Humanist* and *Free Inquiry* printed multiarticle symposia at times of peak fundamentalist activity when there were high-profile assaults on humanism. And beginning in 1980, there was a steady outpouring of articles in response to the formation of the Moral Majority by the Baptist minister Jerry Falwell.[39]

A 1977 *Humanist* symposium on fundamentalism illuminates the variety of ways, which ranged from pragmatic to academic, in which humanists came to understand this surprising rise of fundamentalism in American culture. The symposium was a response to a Gallup poll that showed that a vast number of Americans now identified with beliefs that were strikingly orthodox; indeed, over a third of the population considered themselves born again and believed in the literal interpretation of the Bible. In a two-issue symposium on "The Resurgence of Fundamentalism," Kurtz responded to this poll by asking a number of humanists and ethical culturists to comment.[40]

The most searching essay in attempting to understand why America was so conservative came from the psychotherapist Albert Ellis, the 1971 Humanist of the Year. Ellis asked, "Why do so many intelligent and well-educated people today fall back on devout dependency on supernatural forces, which their own parents and teachers had largely abandoned in the 1920s and 1930s?"[41] He found a variety of causes: "the human biological tendency to irrational and crooked thinking,"[42] people's innate need to find certainty in their lives, and propaganda from both the religious community and some sectors of the scientific community that promoted sloppy thinking. In another article, humanistic philosopher Walter Kauffman laid the responsibility on educators who had become lax in their intellectual expectations.

Humanists differed widely in their level of concern about fundamentalism and what they believed should be done about it. The rationalists in the 1977 symposium expressed shock at the continued superstitiousness of one of the world's most advanced scientific societies and found this very troubling.[43] The two Ethical Culture writers were more sanguine, however, and considered such beliefs unsurprising: it was natural, they believed, that people would find solace in religion.[44] Even the usually aggressive Sidney Hook asserted that the beliefs themselves must be considered entirely acceptable in a pluralistic society as long as the norms of democracy were not being undermined.[45]

When it came to the evangelical claim that all forms of secular culture were simply manifestations of the religion of secular humanism, the symposium participants were in close agreement with Kurtz's view that the entire ethical foundation of the West was rooted in a deeply humanistic value system. "The fact is," wrote Kurtz, "that humanism—in the broad sense—is the deepest current in Western civilization, predating the Judeo-Christian era and having its roots in Greek-Roman civilization." Among the values that Kurtz and others defended as humanistic were compassion, freedom, equality, dignity, justice, creativity, and tolerance, none of which were in any way specific to a particular religious outlook. Moreover, humanists emphatically defended the secular public sphere and its centrality to American law. As Hook put it, "Secularism per se is not a religion at all but a view about how church and state should be related. . . . The argument for the separation of church and state in the United States is cogent and uncontested."[46] Leo Pfeffer, a humanist and constitutional lawyer who had argued several of the separationist cases in the Supreme Court, followed this up by claiming that it was the Conlan amendment, not any supposed humanistic educational material, that was unconstitutional.[47]

All in all, the humanist dispute with the fundamentalists was never an intellectual dialogue. The evangelicals' wild conspiracy theories and John Bircher rhetoric were simply not amenable to intellectual engagement, with their hyperbolic language and imprecise and erroneous claims. The humanists saw their aggressors as simply ignorant and anti-intellectual. Of the Rafferty report, for example, the humanists used phrases like "poor scholarship," "illogic," "mishmash of misinformation," "smear tactics," "confused," "childish," "dishonest," "shocking," "prurient," and "rigid."[48] The attack on humanism a few years later by Representative Conlan, a Harvard law graduate, was considered intolerant and bigoted, and they characterized his attack as a "new inquisition."[49] They did not consider these attacks as part of an intellectual agenda to be debated; rather, they were political stratagems to be opposed. In this way, a wall formed between the two sides, and each viewed the other as wholly alien.

The concerns of the humanists only grew stronger as the years passed. By 1980, the humanists saw the evangelical resurgence as something of great concern. "If we try to ignore the attacks of the Moral Majority and the New Right," said Maxine Negri, who was chairperson of the Commission to Defend Humanism, "we may be ploughed under. We may not be able to match their dollars, but we can try to match their persistence and determination. Toward this end every humanist and every friend of humanism needs to be willing to participate. Now is the time to stand up and be counted."[50]

A Manifesto and a Declaration

The humanist response to fundamentalism can be gleaned from a new humanist manifesto published in 1973. The first humanist manifesto of 1933 was never intended to be an eternal or definitive statement of humanism. The writers expected that the movement would evolve as time passed, and even before the decade was out, Bernard Fantus had already started drafting revisions. But time passed, and no new manifesto appeared. By 1973, one was long overdue. Edwin Wilson and Paul Kurtz teamed up to write the new document for the AHA. By this time, Kurtz was hosting a TV show, *The Humanist Alternative*, which aired on commercial and educational stations in the United States and Canada. Kurtz was learning how to use the various media to discuss and promote ideas effectively. In late August of 1973, he held a news conference to announce the publication of Humanist Manifesto II.[51]

It was a significant moment for the movement. The *New York Times* gave the event front-page coverage in its widely read Sunday edition. Flanking the article were two photographs, one of Kurtz and the other of the late John Dewey, who had signed the first manifesto. Liberally quoting from the manifesto and listing all of its signers, the *Times* highlighted the stature of the people involved. It was an important event for humanism by dint of its being featured so prominently in the country's most prestigious newspaper.[52]

Nonetheless, the media response was not all positive. Conservative columnist Garry Wills wrote a sardonic attack on humanism, decrying its ugly utilitarian value system: "They live in a funny world, these humanists—one where tribalism is a human experience, but prayer isn't. Abortion seems to be one of their favorite human experiences, and their rage for civil liberties makes them demand the right to suicide, divorce, euthanasia, and the other happy sacraments of science."[53] The final twist in that statement regarding the "sacraments of science" provides a strong indication of how closely humanism was being typified by its opponents, even those who were not fundamentalists, as a religion of science.

In the year leading up to the publication of the Humanist Manifesto II, there were extensive discussions in the pages of the *Humanist* about how to best present the movement to a new generation. A draft was circulated to prominent humanists worldwide. The resulting document was both more cosmopolitan and more secular than the first manifesto. The new manifesto was also more than three times as long. It was divided into six main sections, which dealt with religion, ethics, the individual, democratic society, world community, and humanity as a whole.[54]

Science was discussed in the new manifesto, but perhaps not as much as one might expect. This manifesto characterized science in very utilitarian terms; it was simply a way to understand nature and a path toward modern technology. The document began with a long discussion of the fruits of science as a useful tool: "We have virtually conquered the planet, explored the moon, overcome the natural limits of travel and communication; we stand at the dawn of a new age, ready to move farther into space and perhaps inhabit other planets. Using technology wisely, we can control our environment, conquer poverty, markedly reduce disease, extend our life-span, significantly modify our behavior, alter the course of human evolution and cultural development, unlock vast new powers, and provide humankind with unparalleled opportunity for achieving an abundant and meaningful life."[55] Noticeably absent in this discussion was any talk about the scientific spirit or the ethos of science. Science was not an inspirational force, nor was it tied to any ethical framework. This was a more positivist conception of science, divorced from values and emotional engagement.

The sarcastic analysis of Garry Wills notwithstanding, the 1973 manifesto was also somewhat less sanguine about science than its predecessor: "Science has sometimes brought evil as well as good," it stated. It also acknowledged that "in learning to apply the scientific method to nature and human life, we have opened the door to ecological damage, over-population, dehumanizing institutions, totalitarian repression, and nuclear and bio-chemical disaster."[56] This is quite a change from the first humanist manifesto, where the scientific method was represented as a pathway to human ethics. In the new manifesto, humanist ethics were seen as something outside of science, to be imported into it—something that would be able to tame science and guide it.

Religion was also treated more extensively in the 1973 manifesto. The statement on religion began by saying that sometimes religion could be entirely consistent with humanistic values: "In the best sense, religion may inspire dedication to the highest ethical ideals. The cultivation of moral devotion and creative imagination is an expression of genuine 'spiritual' experience and aspiration."[57] Most of the discussion of religion, however, was negative. The manifesto focused on what it called traditional religions, which tended to be dogmatic, authoritarian, and obscurantist. Traditional religions, it stated, were also unscientific, positing a supernatural being for which there was no evidence. The old arguments against supernaturalism were presented: supernatural religions were escapist and encouraged dependence and obedience; they also diverted people's attention from social injustices, by asking people to concentrate on the next world, rather than this one. "No deity will save us; we must save ourselves," it explained. A new argument emerged as well, one taken directly from humanistic psychology, namely, that supernatural faiths hampered people's self-actualization.

Unlike the original manifesto, this one did not attempt to redefine religion, and it did not characterize humanism as a religion. Instead, the document explained that not all humanists considered themselves religious; many more described their humanism with adjectives like *scientific, ethical,* and *democratic.* In line with the liberal tradition, the manifesto explicitly stated that this version was "not a final credo or dogma, but an expression of a living and growing faith." Acknowledging that new statements will supersede it as times changed, the document stressed that current signatories need not agree with every point.

The new manifesto also sharply separated ethics from religion. The goal of human life was to be creative, enrich one's life, and strive for happiness. "Ethics," it said, "is autonomous and situational needing no theological or ideological sanction." Humankind should be guided by reason and critical intelligence balanced with compassion and empathy. On the topic of sex, the manifesto promoted toleration of all practices that did not harm others and were nonexploitative: "individuals should be permitted to express their sexual proclivities and pursue their lifestyles as they desire."

The other sections of the manifesto described humanist ideals related to political, economic, and social relationships. In those areas it focused on the importance of human autonomy and free will. It strongly supported the freedom and dignity of all people, embraced the ideals of an open and democratic society, encouraged equality and the end to discrimination, and supported universal education. Finally, it turned to the world community and urged nations to work together to build a peaceful international order that managed natural resources, tempered poverty, and encouraged communication and cooperation.

The list of signatories at the end of the document provides a further contrast with the previous manifesto and illustrates how the movement had become more secular over the years. The ratio of ministers to academics was flipped: scientists, philosophers, and scholars overwhelmed ministers by more than three to one. Among the list of notables was science fiction author Isaac Asimov, codiscoverer of the DNA double helix Francis Crick, British philosopher Antony Flew, dissident Soviet physicist Andrei Sakharov, and Harvard psychologist B. F. Skinner. There were 120 signatures in all—a far cry from the thirty-three individuals who had signed the first one.

Seven years after Humanist Manifesto II, another statement appeared, "A Secular Humanist Declaration." Since Humanist Manifesto II, the fundamentalist assault on secular humanism had crescendoed to a fevered pitch, and this document reflected that shift. It also reflected a schism within the humanist movement itself that was related in part to the intellectual issues discussed in earlier chapters but related most directly to institutional conflict that emerged within the AHA. Kurtz had been fired as editor of the *Humanist* over disagreements about editorial policy and had even been attacked for financial mismanagement. The nasty affair left both parties feeling badly burned.

As a result, Kurtz left the AHA and created a new group called the Council for Democratic and Secular Humanism (CODESH). Because this was his own organization, Kurtz was able to impose a more combative tone than was possible while working within the AHA. It was clear to him that the times required a more aggressive response to religion in general, and especially toward fundamentalism. In that vein, Kurtz started a new bimonthly magazine, *Free Inquiry*, where on page one of the first issue the Secular Humanist Declaration was announced.

Like Humanist Manifesto II, the declaration received page-one coverage in the *New York Times* in an article entitled "Secularists Attack 'Absolutist' Morals: 61 Scholars and Writers Denounce the Rise of Fundamentalism."[58] Published just three weeks before the US presidential election, the article highlighted the way in which the evangelical attack on secular humanism had been incorporated into campaign speeches: "Preachers on the stump and on television repeatedly rail against what they see as an atheistic plot to stamp out religion." The *Times* article focused on the differences in ethics and morality between humanists and fundamentalists, contrasting the evangelical outlook founded on biblical authority with the scientifically based morality of the humanists. It defined secular humanism as "a philosophy that favors exclusion of religion in making moral and political decisions." It explained the sharp political and moral divisions between the two sides, the role of science in humanism, and the strongly atheistic and skeptical outlook that drove it.

The Secular Humanist Declaration of 1980 solidified the humanist movement's response to fundamentalism. LaHaye's *The Battle for the Mind* had been published earlier that same year, and religious politicization had now become entrenched in American culture in ways that were hardly imaginable ten years earlier. Meanwhile, fundamentalism was transforming international politics. The Iran hostage crisis, led by Islamic fundamentalists, had at least as much influence in swinging the election to Ronald Reagan as homegrown Christian fundamentalism. The growth of fundamentalism both at home and abroad deeply concerned Kurtz. Modernity and the foundations of a free and democratic world were on the line. Rationalism and Enlightenment values were being opposed all over the world: "We are today faced with a variety of antisecularist trends: the reappearance of dogmatic authoritarian religions; fundamentalist, literalist, and doctrinaire Christianity; a rapidly growing and uncompromising Moslem clericalism in the Middle East and Asia; the reassertion of orthodox authority by the Roman Catholic papal hierarchy; nationalistic religious Judaism; and the reversion to obscurantist religions in Asia."[59] Fundamentalism was dangerous in whatever guise, for it always contained an antimodern outlook that opposed progress, science, and democratic values and ultimately pointed the way to barbarism. When it came to religion, then, the Secular Humanist Declaration stood apart from the previous two manifestos. The strong opposition to authoritarian power formed the core of the declaration. Although the document acknowledged that religion could sometimes be a force for good, most of the time the words *religion* and *church* were used in a negative sense. Churches, it claimed, were often run by "religious bigots" who instituted tyrannies and restricted freedom of thought.[60] The term *religion* and its cognates appear far more often in the Secular Humanist Declaration than in the Humanist Manifesto II, and almost always in the negative sense.[61]

Science was also discussed in more detail than in the Humanist Manifesto II, primarily in terms of its utility. Science was "the most reliable way of understanding the world," the declaration announced; it was opening up "exciting new dimensions of the universe" and extending our understanding of human behavior. The declaration used some of the same cautionary language that had been present in the Humanist Manifesto II—"we are aware of, and oppose, the abuses of misapplied technology and its possible harmful consequences"—but that language was overshadowed to a great extent by the document's strong defense of scientific research: "we urge resistance to unthinking efforts to limit technological or scientific advances."

A whole section of the declaration was devoted to evolution, defending the theory and explicitly opposing the teaching of creationism: "it is a sham to mask an article of religious faith as a scientific truth and to inflict that doctrine on the scientific curriculum." Evolution, the document warned, could be a critical issue for all of science: "if successful, creationists may seriously undermine the credibility of science itself."[62]

The declaration opposed the fundamentalist worldview in nearly every way. It declared that ethics was entirely autonomous from religion and based solely on critical intelligence. It asserted the goal of an absolute separation of church and state and claimed that this separation was essential to a democratic civilization.

Politically, the declaration also departed from the humanist norm, a result of Kurtz's own political trajectory. As a friend and confidante of Sidney Hook, Kurtz tacked rightward during the decades leading to the 1980 election. Indeed, in this watershed election in which religious politics was so evident, Kurtz himself astonishingly voted for Reagan, rather than the born-again Democrat, Jimmy Carter. As we have previously seen, Kurtz was also opposed to New Left cultural politics, which he feared could be just as dangerous and authoritarian as Christian fundamentalism. In the end, he adopted a libertarian point of view, and the declaration reflected this. The rhetoric of the declaration praised the ideals of freedom, individualism, and self-government in ways that would have been familiar to many on the right. It talked about "economic freedom," "competing in the marketplace," and avoiding "undue influence by centralized political control." It even stated that "the right to private property is a human right without which other rights are nugatory." This was, by far, the most economically conservative of the manifestos.

As for the new humanist organizations, Kurtz's journal *Free Inquiry* was designed to be "unabashedly intellectual" but not academic: its tone was popular rather than scholarly. It drew on the rationalist and freethought strand of secularism that had flourished in the nineteenth century rather than on the genteel ministerial tone of Unitarian religious humanism. It was meant to engage the fight against the enemies of the Enlightenment, not to sound like a Sunday morning sermon. The chief aim of CODESH, the organization that published the journal, was to "influence opinion leaders and the media."[63]

Kurtz's editorial commentary in the first issue of *Free Inquiry* explained that the entire project was grounded on the need to defend Enlightenment values against "a massive resurgence of fanatical dogmas and doctrines." Fanaticism came in many forms. Christian and Islamic fundamentalism joined antiscience ideas and paranormal beliefs of the New Age as threats to Enlightenment. Even more dangerous were the many authoritarian political ideologies that plagued the modern world: from Nazism and Stalinism in midcentury to the Communist regimes and terrorist groups of his day. The traditional left-right divisions did not apply; authoritarianism could be found on both sides of the political spectrum and in many other areas of human engagement. Humanists needed to be willing to fight for democratic individualism with a "commitment to scientific evidence and reason" and a "value of individual freedom and dignity."[64] This new rhetoric was more pragmatic and activist than in earlier humanism, grounded more in political and combative language than in religious and aspirational ideals.

One area of that activism, which predated Kurtz's secular humanist agenda by decades, was the fight to defend the theory of evolution. It is here that we find humanists involved in a variety of struggles against Christian fundamentalists, struggles that assumed quite different forms over the century. Since this agenda played a singularly prominent role in humanist history, it deserves a chapter of its own.

CHAPTER ELEVEN

Battling Creationism and Christian Pseudoscience

Periodically in public school administration, the ghost of William Jennings Bryan rises from the mephistic swamps of the antievolutionist camp to threaten the factual teaching of modern biological science. . . . It hardly needs to be said that this amounts to arguing that the theory of the stork be given "equal treatment" in sex education texts!

Bette Chambers (1972)

Bette Chambers was one of the American Humanist Association's most committed members, a supporter of humanism from her days as head of the Humanist Chapter of Minnesota in 1962 until her death in 2015. She was the second female president of the AHA and a strong influence in the association during the 1970s. She edited the organization's newsletter *Free Mind* and raised three children during those same years, declaring herself to be "a happy, not politically contented, feminist [in] an all Humanist family."[1] She was trained in marine zoology, and her husband was a biologist. She took special interest in the teaching of evolution in the public schools—or, rather, the lack thereof. As she put it, "This intricate and beautiful principle of modern biology is taught almost nowhere without extensive apologetics or having first been filtered through a sieve of nervous religious disclaimers."[2] Chambers, as president of the AHA, coordinated a major statement on the teaching of evolution that was sent to school districts around the country. In this and several other ways, she was a major player in the humanist effort to counteract the creationists in their efforts to get their curriculum into the public schools.

Chambers's work in this area is especially notable because she was a humanist mother concerned about public school science education. We have talked about the scientific spirit as an aspirational outlook adopted by philosophers, scientists, and religious thinkers, but it touched the general public as well. It motivated parents and activists, just as it did scientists. The theory of evolution became especially important in America as a focal point in the public perception of science more generally. For many people, evolution represented science itself; the fate of evolution was inseparably linked to the fate of science. Discussion of the theory therefore often invoked an intense, visceral response and became a touchstone for political ideals like democracy, as well as for personal character, reflecting on a person's honesty or courage. The public debates over evolution were a flashpoint in America not least because they pitted humanists against fundamentalists.

The struggle over the teaching of evolution was a persistent aspect of twentieth-century American culture.[3] Indeed, it has become an iconic fight, seen by many as an example of the inevitable war between science and religion. In reality it tells us a lot more about how different factions of the American public fought over social power. In the case of humanism, their struggle with creationists changed how they thought about the use and defense of good science. The aspirational aspects of the scientific spirit were, in these battles, replaced by confrontational rhetoric. Science came to be a weapon in the culture wars.

This chapter follows the previous one in exploring how fundamentalist Christianity influenced late twentieth-century humanism. In it, we look back at several earlier episodes in which humanists confronted antievolution activism and then return to the period of heightened conflict beginning in the early 1970s when humanists challenged Christian efforts to teach creationism in the public schools. Humanists responded to this very specific attack on modern scientific thought in several ways. Moreover, though creationism posed the most sustained challenge to science, humanists began exposing other pseudoscientific claims promoted by Christians, including one instance of outright fraud. Despite the many differences among humanists in philosophy and ideology, they all agreed that fundamentalism held an outdated epistemology and, as a result, posed a major threat to science. In the end, the humanist-fundamentalist battles shaped the way that humanists used scientific knowledge, reinforcing its value as a weapon against conservative and reactionary religiosity.

Early Battles over Evolution

Humanists were engaged in the battles over evolution from day one of the movement. The University of Wisconsin philosopher Max Otto was one of the first humanists to find himself defending evolution against attacks by conservative Christians, but this was not his choice; he was dragged into it. In 1910, even before he completed his PhD, Otto was given a course to teach at the university entitled "Man and Nature." He decided to center the course around the topic of human evolution and its implications. Having been an evangelical himself not that many years before, he knew how controversial this topic would be, but he also knew that controversies engaged students and would make them think.[4] His course turned out to be extremely popular—and indeed, the class size eventually reached about six hundred. But that popularity brought challenges.

Not long after Otto had begun teaching the class, a local Catholic clergyman denounced the university for allowing a professor to teach dogmatic materialism. Otto had gone to great lengths to ensure that there was an open, nondogmatic environment in the class by, for example, hiring teaching assistants who were Christians. However, the religious attacks continued. One Baptist pastor called on the president of the school to either discontinue the course or offer another one that presented a theistic perspective. Sometimes the objections became explicitly political, with the claim that Otto was teaching Communist propaganda. In fact, during one election season, a Republican candidate for the Senate demanded that the university halt atheistic courses such as Otto's and that the faculty swear an oath of patriotism to the United States. The case was newsworthy enough that the antievolutionist firebrand William Jennings Bryan alluded to it in one of his popular cross-country lectures on the "menace of Darwinism."[5] In the end, Otto's course was never touched. Until Otto's retirement, the university's presidents always defended him and thwarted his critics.[6]

At the same time that Otto was engaged in these local battles, the flamboyant humanist minister Charles Francis Potter brought the debate to the national stage. A maverick minister and excellent promoter, he wrote and lectured prolifically. He is considered by some scholars to be one of the three founders of religious humanism and sits alongside Curtis Reese and John Dietrich as a major force in the movement. A signatory of the first Humanist Manifesto, he founded his own independent church in Manhattan—the First Humanist Society of New York—from which he officiated humanist weddings, funerals, and coming-of-age ceremonies.[7]

Potter's fame arose in large part from a series of nationally advertised debates in 1923 and 1924 with Baptist preacher John Roach Straton, head of New York City's largest fundamentalist church. The first Potter-Straton event seized national attention, and the two men agreed to meet several more times over the next few months, butting heads over the central dogmas of Christianity. The debates drew huge crowds—over twenty-five hundred people attended the first debate at Straton's Calvary Baptist Church, and thereafter the two men were booked in Carnegie Hall. Radio broadcasts of the events and newspaper coverage spread the news around the country. The second of those debates explored the proposition "That the Earth and Man Came from Evolution."[8] In keeping with this theme, Potter held an "Evolution Day" celebration before the final debate. This event included the unveiling of a statue by the artist Carl Akeley depicting a young man emerging from a gorilla. At the event, Potter gave a sermon in which he explained evolution in terms of its spiritual significance to mankind. There was even music along with the reading of a verse by poet-scholar William H. Carruth.[9]

Potter continued his attack on antievolutionism during the infamous Scopes "Monkey" Trial in 1925, when he was asked to participate as "librarian and Bible expert for the defense." In that capacity he helped Clarence Darrow draw up a list of factual inaccuracies in biblical scripture. Potter's list was key to Darrow's strategy of embarrassing his famous opponent William Jennings Bryan by showing that Bryan had ridiculous views and accepted Old Testament stories literally, like the one about Jonah and the whale. During the trial, Potter also encouraged an alliance between scientists and liberal ministers to oppose the fundamentalists. The ideological divisions in the country were stark, but the divisions, he argued, did not pit science against religion as much as they pitted a liberal, scientific outlook against the old-time, superstitious forms of Christianity.[10]

Humanistically minded scientists also found themselves engaged in the struggle to defend evolution. At around the same time that Otto and Potter were fighting their battles, the two young scientists Julian Huxley and Hermann Muller, both of whom would become major figures in the humanist movement, discovered how little the American public knew about the science of evolution. It all started when the young British grandson of T. H. Huxley took a teaching post at Rice University in Houston, Texas. While there, he encountered American antievolutionism firsthand. As a newcomer to the country, he was flabbergasted to find so much resistance to the theory so long after it had been established as a principle of science. The American reaction was so unlike anything in Britain. During the heady years leading up to the Scopes Trial, Huxley wrote a long editorial to a London periodical where he lambasted a Kentucky antievolution bill that was then under consideration. He acknowledged the great cultural divide between Britain and the United States and then tried to explain the amazingly backward views of so many rural Americans. "To the average educated Englishman, the fact will doubtless seem unbelievable," he noted.[11]

When Huxley vacated his position at Rice, he got his friend Muller to replace him. Even though Muller was an American and thus more familiar with this country's religiosity, he, too, was shocked at the strength of antievolution sentiment in the southern states. In a 1918 letter to Huxley, Muller explained, "A mountain threatened to fall on my head this fall when I gave three public lectures on evolution—I carried the audiences perfectly (if I do say so), but these little old fogies in out-of-town hamlets started to send in letters of 'criticism,' and then too some ministers association took affront at 'Darwinianism.' Just think of a place where you can't speak of evolution! That almost disproves evolution in itself, doesn't it?"[12]

The memory of this event remained with him, and he would continue to promote efforts to educate the public and defend the teaching of evolution until the end of his life.[13] In the early 1960s, having recently served a term as president of the AHA, Muller became active in a national effort to strengthen biology education as part of a steering committee for a new series of high school biology textbooks. The new series made evolution the central framework for understanding biology, so schools that adopted these books could not avoid teaching it. "A hundred years without Darwin are enough," Muller declared.[14]

At about this same time, Muller was asked to participate in a local debate sponsored by a group of fundamentalists in Arkansas, who claimed that even the scientists themselves didn't believe in evolution but only promoted it as a smokescreen to salvage their atheism. Not one to ignore an opportunity to defend science even when it was unlikely to change many minds, he accepted the challenge and sent two bright and articulate graduate students to Little Rock in his stead. One of those men was Carl Sagan. These two thirtyish New York Jews arrived in a region quite foreign to them. According to a later account, they both felt quite uncomfortable in a city notorious for its segregationist policies, and Sagan (for a reason that he never disclosed) found himself mortally threatened and took refuge in a Roman Catholic monastery.[15] The debate, however, went on as planned against a fundamentalist who had a PhD in biology. It was here, in Little Rock, that we can see the young Sagan gaining a gradual understanding of the larger role of the scientist as a public intellectual, and we see him engaged in popularization in ways that many of his scientific colleagues were not and would never be.

All in all, Huxley, Muller, and Sagan were shaped as spokesmen for science at least in part by the evolution debates that all three participated in as young early-career scientists. As they developed in their fields, their larger public agendas coincided closely with that of the humanist movement—indeed, all would eventually be named Humanist of the Year, and both Huxley and Muller, as we have seen, were active advocates within the movement. But one finds a larger theme here as well. The humanist engagement with Christian fundamentalists over the theory of evolution goes back to the early years of the humanist movement, which points to the surprising resilience of this debate in American culture. The antievolution disputes were forums in which both humanists and Christian fundamentalists were engaged in types of public activism. The two groups were fighting over influence in American culture, especially in the educational system, which had become, as we have seen, a flash point in the culture wars. The humanists were responding to an onslaught of activism by Christians who were learning how to deploy political power at all levels of the culture. When these activist Christians targeted evolution, they were often directly targeting humanism as well in order to protect themselves from what they considered to be mortal dangers to their religious heritage.

The Fundamentalists' Antievolution Agenda

To understand the humanist efforts after the 1960s, we must first understand how the fundamentalists' agenda was molded by some very specific circumstances. First, the antievolutionists met a shattering defeat in the Supreme Court when it declared state laws that banned the teaching of evolution on religious grounds to be unconstitutional. Second, a new form of creationism arose that adopted the form and language of science so that it was no longer dependent on scriptural evidence to make its case against evolution. Third, fundamentalists increasingly believed that the teaching of evolution was part of a larger assault on Christianity in America.

In the first case, the legal status of teaching evolution flipped in 1968 with the Supreme Court decision in *Epperson v. Arkansas*. In that case, the court held that states could not restrict the teaching of evolution on religious grounds. The Scopes Trial forty years earlier had been inconclusive. The constitutionality of the Tennessee law barring the teaching of evolution was never taken to a higher court. This meant that the flurry of state laws that had been enacted in the wake of Bryan's antievolution campaign remained on the books in many places. Arkansas, for instance, had a statute that forbade teaching "that mankind ascended or descended from a lower order of animals." The court struck down that law on grounds similar to those in the school prayer and Bible-reading cases earlier in the decade, stating that it infringed on the Establishment Clause of the First Amendment. States could only make laws for a secular purpose, and when Arkansas forbade evolution simply because it conflicted with certain religious beliefs, the state was overstepping its constitutional authority.[16] This hit fundamentalists hard. It was yet another dagger in the back, another attack on Christians in the public schools.

The second major change in this period arose in creationist apologetics. Opponents of evolution began promoting what they believed was an alternative scientific view that supported their biblical ideas, claiming that there was empirical evidence for a young earth and the creation chronicle of Genesis. Although a lot of Christians, including quite a few evangelicals, dismissed "creation science" as a crank idea, some fundamentalists got behind it and founded an institute, the Creation Research Society. This was composed of scientists with academic credentials who argued for the Genesis account of the origin of the world. Their goal was to call into question existing science by criticizing such things as radiocarbon dating. They also presented alternative theories that could explain in scientific, materialistic terms how a biblical-style creation could have occurred. Reinterpretation of geological forces helped them argue that rock formations considered to have been created over millions of years actually could have been created quite quickly and catastrophically. Standard scientific ideas about the fundamental laws of physics were even reevaluated. Those laws, the creationists argued, must have changed over time, allowing for rapid transformations in the early moments of the universe, 10,000 years ago. These creationists used scientific terminology and cited peer-reviewed papers. They avoided scriptural quotations, so their work more closely followed current scientific publication conventions. The group even published a school textbook written from this perspective.[17]

Finally, evolution came to be linked directly to the growing set of cultural changes that threatened fundamentalists. This was the time in which Max Rafferty commissioned his report on the moral state of California public schools, which attempted to connect all of the secularizing tendencies together into a single massive humanist conspiracy. Evolution joined sex education, secular morality, and behaviorism as planks of a humanistic agenda to undermine religion.

Given all of these changes, fundamentalists adopted a new legal strategy. No longer able to lobby for laws that would restrict evolution, they now began to talk about "equal time" so that creation science could be taught alongside evolution.[18] The equal-time agenda appeared almost immediately after the *Epperson* decision was handed down, when it was mentioned in the Rafferty report discussed in the previous chapter. The report linked evolution to immorality and called it "an a priori assumption of the Humanist religion."[19] There was no definitive evidence for the theory, so it should be treated as tentative and theoretical, rather than factual. Moreover, the court should demand complete neutrality, meaning that it should not favor the secular theory of evolution over the Christian theory of creation. Both, they argued, were religious tenets. "If the origins of man were taught from the point of view of both evolutionists and creationists, the purpose of education would be satisfied," stated the report. "By concentrating on only one theory and ignoring others, it is tantamount to indoctrination in one special religious viewpoint."[20]

Evangelicals began making ever more aggressive attacks on the teaching of evolution. Battles arose in state educational bodies, and for a while in the 1970s, California became ground zero for the conflict. Attacks on the teaching of evolution crescendoed through that decade and spread around the country. By 1980, right-wing evangelicals strengthened their political influence and ever more aggressively pursued the equal-time agenda. By the election year of 1980, the fundamentalists had been able to organize creationists at the state level around the country.[21] Grassroots coalitions of activists lobbied their legislators in at least fifteen states to propose equal-time bills, and Reagan himself announced support for them. By the end of 1981, two states, Louisiana and Arkansas, had enacted these laws that followed a formula proposed by a young Yale Law School graduate, Wendell Bird, an affiliate of the Institute for Creation Research. Bird sought to establish complete parity between "creation science" and "evolution science," as he called them, making the by-now-familiar claim that the former was just as scientific as the latter, and that the two were equally tied to religious doctrines.[22]

Defending Evolution: Two Strategies

The creationist equal-time agenda was getting started in the early 1970s at a time of change in the leadership in the AHA. With that change came new strategies for defending evolution. These new leaders, laywomen and men without academic or religious credentials, utilized the organizational capacity of humanism to build a more robust response to the creationist threat.

Their often more personal perspective encouraged their activism in ways different from the scientists and philosophers.

The first of these lay leaders was Bette Chambers, who was the mother of school-age children. Her husband's job had taken the family back and forth across the country, so she had developed a good sense of the public school science curricula all over the United States. She was dismayed at the impoverishment of biology teaching everywhere she went. In Spokane, Washington, for instance, one of her daughters came home after school one day asking whether God had programmed the spider's brain to build a web. Chambers discovered that the Spokane public schools were showing science films created by the Moody Bible Institute, expensive productions that promoted a creationist message. From that point on, she began sounding the alarm, discovering, like Muller, that too many scientifically minded people simply didn't worry enough about the issue to make a fuss. For many of these well-educated men and women the Genesis account of creation was so ridiculously absurd that they assumed it would remain an insignificant movement and eventually die out on its own. Yet, as Chambers knew well, creationism was no small movement, nor was it confined to the South; it was a large and growing national problem that required serious attention.[23]

After she became president of the AHA, Chambers was well placed to provide wider exposure to the creationist threat. The California State Curriculum Committee took up an equal-time recommendation in 1972 and asked the Board of Education to adopt only textbooks that taught both evolution and creationism. The AHA jumped into action, appointing a team to lobby the state board and vowing to carry their fight to court if necessary. Chambers wrote a regular column in the AHA newsletter *Free Mind* to keep the membership apprised of the fight as it unfolded.[24]

She also pressed for a new proactive response. Noting that the strategy Paul Kurtz had used for publicizing the second humanist manifesto and the anti-astrology statement was extremely effective—both of which had received national news coverage—Chambers started working on a similar project. The idea was to publish a statement on the matter endorsed by a long list of prominent supporters and experts. Hermann Muller had tried this earlier. In the months before his death, he had drawn up a 750-word statement on evolution and circulated it among his colleagues. The statement, with 177 names attached, was published in the *Bulletin of the Atomic Scientists*.[25] Chambers believed that the *Humanist* could do better, so she put together a board of advisors that included Isaac Asimov, the chemist Linus Pauling, and the biologists Hudson Hoagland, Chauncey Leake, and George Gaylord Simpson, prominent scientists and popularizers all. They rewrote Muller's earlier statement and once again distributed it for signatures.[26]

The AHA "Statement Affirming Evolution as a Principle of Science" was shorter than Muller's original and more pragmatic. Unlike its predecessor, this statement directly addressed school boards, textbook publishers, biology teachers, and educational agencies. It emphasized evolution's extremely long time frame ("thousands of millions of years"), it called out the unscientific elements of creationism ("creationism is not scientific; it is a purely religious view"), and it encouraged activism (asking people to resist equal-time laws and support classroom teachers who taught evolution). Finally, it explained that evolution was neither a religious doctrine nor a tenet of "secular humanism."[27]

Given the strongly activist intent of the statement, Chambers was worried that scientists might be reluctant to sign for fear of jeopardizing their reputations as disinterested scholars. But "not to issue this statement," she said, "strikes me as intellectual cowardice."[28] In the end, her fears were unwarranted. All but three people who were asked to sign the statement did so, and many more names were added by way of scientists who learned about the project through the grapevine. All told, a total of 177 experts endorsed the statement, precisely the same number that Muller had gotten. In January of 1977, the *Humanist* printed this as its cover story, and the statement was simultaneously mailed to major school districts across the United States.[29]

The cover of the magazine showed a staged photograph of a mother with a Bible in hand pulling her child out of a classroom where the teacher was lecturing on human evolution. It was a twist on a familiar scene out of the film *Inherit the Wind*. The text read "Evolution vs. Creationism: The Schools as Battleground." The statement was published alongside three other articles. One of them, by Preston Cloud, a research scientist from California, explained that "fundamentalist creationism is not a science but a form of anti-science, whose more vocal practitioners . . . play fast and loose with the facts of geology and biology."[30] Scientifically literate people ignored the creationist movement at their peril, he proclaimed, because it is run by savvy political activists. A second article was by William Mayer, the current director of the Biological Sciences Curriculum Study. He likened creationists to King Canute, who commanded the tide to retreat: they peddled unbelievable ideas that had the same scientific credibility as the writings of the Flat Earth Society.[31] And Chambers herself wrote a short piece that explained why she had pushed for this statement in the first place. That same issue contained a symposium exploring the rising religion of Christian fundamentalism. The country's new religious politics was by this time unescapable, and the humanists were among the loudest voices raised against it.[32]

The *Los Angeles Times* picked up the story, quoting Chambers as saying, "It becomes imperative to state that [creationism] is rubbish lest science education in America become the laughingstock of the civilized world." Yet to Chambers's dismay, the evolution statement received scant notice elsewhere. Not many other newspapers covered it.[33] It was the second time in a decade that scientists had called out creationism in this way, and there was still not much response. Chambers blamed Kurtz, who, she claimed, never got behind the issue and refused to activate his publicity machine.

In 1980, Fred Edwords, a local San Diego humanist leader, spearheaded a new journal called *Creation/Evolution*. Edwords was only thirty-two at the time. Though he had not completed a college degree, his intellectual curiosity had led him into humanism, where he and two other members became engaged in the growing antievolution movement. Edwords's status as a layman was central to his response to the evolution-creation struggles. Attending some public debates, he and his fellow activists paid close attention to the social and cultural dynamics during the event and began to see problems with the standard responses. In particular, they criticized the tendency of the people on the evolution side to present only academic responses to antievolution.

The magazine *Creation/Evolution* was not meant to be a humanist journal; it was designed for anyone who was concerned about the attacks on evolution, and it sought to be a resource for the "network of contributors, advisors, and debaters from around the country." People came to the table with great passion, but few of them really understood much about the creationists, about what they thought, or about the arguments they used. This was especially true of the scientists, who approached the public debate just as they would if they were trying to convince their colleagues. As a result, well-meaning evolutionists would enter into debates with creationists simply assuming that their greater understanding of science would give them the winning edge. Unfortunately, since they did not have any familiarity with the creationists' arguments or objectives, the evolutionists were frequently quite ineffective at persuading an audience. As one editor of *Creation/Evolution* put it, "Sound science, poorly articulated loses out to well-packaged pseudoscience."[34]

Creation/Evolution was important in helping activists network. Up to that time, most people who sought to debate and rebut creationism were working alone. "Each one had to start at the beginning with his research and preparation. Each had to learn the hard way the debate and instructional tactics of contemporary creationists." The idea for the new magazine was to establish a more knowledgeable and better-organized movement.[35]

This grassroots strategy differed markedly from the public statements and manifestos. Signed, authoritative statements of the sort Muller and Chambers had published were seen as elitist and counterproductive. According to one activist, writing in *Creation/Evolution*, these projects were "largely a waste; and the more prestigious are the names attached to a statement, the less effective it is likely to be." To fight local battles, one needed local participation, persistence, dedication, and a good network of both experts and regular citizens. The creation-evolution campaigns had to be fought on the ground, community by community, using tactics that political organizers had been using for decades to win elections. You had to circulate educational literature, write letters to local newspapers, talk to neighbors, and speak up at school board meetings.[36]

Edwords's organizational prowess was apparent to the leaders of the AHA. He had also helped organize a local conference for the AHA and was a known quantity. At about the same time as he was starting *Creation/Evolution*, he became the main administrative officer for the AHA, moving from California to the headquarters in Amherst, New York. Edwards worked in that capacity for the next two decades, soon assuming the role of executive director. For years, he continued editing *Creation/Evolution*, which was being published under the organizational framework of the AHA. (This lasted until a new organization, the National Center for Science Education, took over the journal in the early 1990s.) For most AHA members who were not *Creation/Evolution* subscribers, Edwords published a regular column in each issue of the *Humanist* to update its readers on the state of play of the antievolution movement, similar to what Chambers had done the previous decade.

The creationist fight was more or less inseparable from the larger battle against a rising tide of fundamentalism and its attacks on science and rationality. The AHA renewed its commitment to science and scientific knowledge in several ways. Most visible was the election of Isaac Asimov as AHA president in 1984. With a PhD in chemistry, Asimov was one of the country's foremost popularizers of science and one of its most renowned science fiction authors. Even before his election as president, Asimov had worked closely with the group. He had helped Chambers, for instance, redraft Muller's "Statement Affirming Evolution." As a chemist, he was repulsed by the misuse of science across the culture and was one of the founding fellows of the Committee for the Scientific Investigation of Claims of the Paranormal. As an outspoken atheist and an immigrant Russian Jew, he was no friend of Protestant fundamentalism, which he saw as an enemy of religious liberty. He found fundamentalists' naivete on scientific issues to be outrageous.[37] As a popular writer, Asimov had a knack for communicating arguments and ideas in accessible ways to an audience of nonintellectuals.

Humanist of the Year and subsequently AHA president, Asimov gave the group popular name recognition on AHA stationary and in the masthead of its publications, though he was only marginally active in the organization itself. Chambers acted as his liaison and took over all of the main administrative duties, contacting him only for specific endorsements. She, Edwords, and humanist board members periodically asked Asimov for statements of support on key issues, to help in publicity and fundraising. In this fashion, Asimov endorsed both *Creation/Evolution* and its new affiliate, the National Center for Science Education.[38] He also supported other groups allied with the AHA, including Americans for Religious Liberty, which participated in legal fights over church-state issues.[39]

Beyond Antievolution: Attacking Fundamentalist Pseudoscience

CSICOP and the Council for Democratic and Secular Humanism, both founded by Paul Kurtz and based just across town from AHA headquarters in Amherst, New York, also embarked on crusades against creationism. The overarching agendas of these two groups, as we have seen, were somewhat different from that of the AHA, and so too were the rationale and methods of their activism. In 1980, Paul Kurtz was in the process of building a more aggressive, in-your-face response to fundamentalism, one that appealed to the feistier group of self-proclaimed secular humanists. Combatting creationism was an integral part of his early agenda, but so was attacking the very premises of fundamentalist epistemology. In the next several years, the projects that grew out of Kurtz's initiatives blossomed in different ways as they utilized science-based arguments to attack and undermine fundamentalist thinking. The positivist, skeptical temperament turned out to be especially useful in this crusade, since it was by its very nature intellectually combative. These two groups went after their targets in a variety of ways, through publications, conferences, committees, legal briefs, and sting operations.

The groups attacked creationism on two fronts: on moral grounds for its authoritarian and antidemocratic activities, and on intellectual grounds for its unscientific claims and epistemological deficiencies. The intellectual arguments were key to many humanist attacks. Kurtz's "Secular Humanist Declaration" of 1980 had denounced scientific creationism, calling it "a sham" and stating, "We deplore the efforts by fundamentalists (especially in the United States) to invade the science classrooms, requiring that creationist theory be taught to students and requiring that it be included in biology textbooks." Not only did fundamentalist efforts threaten academic freedom, but they also endangered "the integrity of the educational process." The paragraph concluded by claiming that "if successful, creationists may seriously undermine the credibility of science itself."[40] Alongside the declaration, in the first issue of *Free Inquiry* was a three-page article specifically devoted to the defense of the principle of evolution and attacking creationism in the biology classroom.

The following summer *Free Inquiry* published another discussion of creationism, which highlighted the national threat: there were by this time over fifty creationist associations around the United States. It analyzed Wendell Bird's legal arguments for equal time and published a set of axioms of scientific creationism.[41] Shortly after this, there were more articles in this vein, including an account of "the continuing Monkey Wars" in the American courtrooms and a snarky article by a Unitarian minister entitled "Three Cheers for the Creationists!"[42]

Though the concern about creationism mostly revolved around primary and secondary education, Kurtz recognized that there were threats to the academy as well and published an article by British philosopher Antony Flew entitled "The Erosion of Evolution: A Treason of the Intellectuals," a scathing attack on scholars in the academy who played into the hands of the creationists. Flew began the article by recounting a case in which staff members at the Natural History Museum in London, one of Britain's premier scientific institutions, published a brochure that publicly doubted the strength of the scientific evidence for evolution. One reason for this, he contended, was that people looking for academic support for evolution ran into scholars who seemed to be actively attacking evolution. For example, fellow philosopher Karl Popper had claimed that there were no decisive experiments or observations that could disprove evolution, and this disqualified it from being a legitimate scientific theory. Not surprisingly, creationists trumpeted Popper's criticisms as confirmation of their own views.[43] And Popper was just one of many academics who were dangerously promoting ideas that were "demoralizing" to those who were on "the front lines, defending science and enlightenment."[44]

These two aggressively secular and skeptical organizations were clearly concerned about defending evolution, but gradually a larger agenda became clear: the protection of science from the threats posed by religion (which usually meant fundamentalism). One of the first places we see an extended discussion of this is in a 1982 CODESH-sponsored conference. This was a mostly academic discussion at the campus of the State University of New York at Buffalo. The occasion was the celebration of Charles Darwin on the centennial of his death; according to Kurtz, it was the major commemoration of this event in North America. The conference, entitled "Science, the Bible, and Darwin," brought eighteen speakers to campus from various fields, including scientists, philosophers, and religious studies scholars. A Unitarian minister and James Randi also participated. The two-day event included talks on Darwin and his influence, on biblical criticism, on ethics and religion, and, of course, on the creationist challenge to evolution. The papers on evolution and creationism explored the topic from several angles—philosophical, scientific, and historical—with an emphasis on the conflict between rational thinkers and fundamentalists.

More significant than the Darwin session were the papers dealing with the study and critique of religion. The speakers in one session recounted the conflict—more than a century old by now—between modern biblical scholars and fundamentalists, whom the former decried as naive and literalistic. They pointed out that most fundamentalists were still objecting to the scientific studies in linguistics, history, and archeology that had transformed our understanding of the development of Christianity. Another session on ethics and religion argued an even more radical thesis, namely, that the study of ethics should be wholly autonomous from religious influence. The philosopher Joseph Fletcher, an AHA Humanist of the Year, made the point most clearly in his paper "Why Ethics Should Avoid Religion."[45] Finally, as was becoming usual in meetings organized by Kurtz, the entertainment for the evening included a show by Randi in which he "pointed out the magical basis of religious miracles" and showed its similarities to paranormal claims.[46] The conference resulted in a special seventy-page issue of *Free Inquiry*. The takeaway from these papers was that creationism and antievolution were structurally part of a broad epistemological attack on science and scientific method. Evolution was just the tip of the iceberg, the entire scientific worldview needed support, and religion in all of its aspects had to be scrutinized closely.

This agenda was nowhere more evident than in the Religion and Biblical Criticism Research Project, chaired by Gerald Larue, professor emeritus of archaeology and biblical history at the University of Southern California. Larue was a signatory of the Secular Humanist Declaration and would go on some years later to win a Humanist of the Year award. The impetus for this new project was to focus hard-nosed scholarly analysis on the institution of religion itself, the claims and actions of which were too often unexamined. The power of scientific investigation should be put to use to directly contest the claims of religionists when they were out of line with current scholarship. This was "not the work of agnostics, atheists, and nonbelievers," according to Larue, but rather "products of the continuing study of the Bible by Jewish and Christian scholars."[47]

Although this new committee called itself a research project, Larue was not interested in producing new research; instead, he sought "to disseminate the results of the many investigations carried on over the past century that have not been shared with the public."[48] In other words, the program would provide a public face for the widely shared scientific understanding of the Bible. Its political concerns were evident the following year when the group sponsored a conference in Washington, DC, entitled "Religion in American Politics." This was in response to a congressional resolution signed by President Reagan that declared 1983 the "Year of the Bible."[49] The seminar included academics as well as liberal politicians.

CSICOP, for its part, was especially well set up to test certain kinds of claims by religionists. Though the group did not attack religious belief as such, the magazine did go after what it considered to be religious pseudoscience and efforts to justify supernatural faith with supposed scientific evidence. The Shroud of Turin was one subject that periodically turned up in the pages of the *Skeptical Inquirer*. The shroud is a long rectangular cloth housed in the Turin Cathedral that has markings of a human form on it in the likeness of a bearded man. Legend has it that it was the burial cloth of Jesus of Nazareth and that the image had a supernatural origin. In 1982 the *Skeptical Inquirer* did a special issue critiquing the myth, and a few years later it reported that recent radiocarbon tests had shown the cloth to be a medieval production, not nearly old enough to have been wrapped around Christ's body.[50]

Creationism was a common topic in the *Skeptical Inquirer*, a perfect example of religious pseudoscience, and thus fair game for the skeptics. In one issue, for example, a paleontological excavation in Texas was written up. The creationists who were doing the digging claimed to have found evidence that humans and dinosaurs lived at the same time. On private land right next to Glenn Rose State Park, famous for its well-preserved dinosaur tracks, the group had unearthed a limestone stratum that seemed to show human and dinosaur footprints side by side. The skeptics—who included high school teachers and students, as well as out-of-state experts in paleontology, geology, and archaeology—denounced the excavation as spurious. When a documentary film was made about this investigation, *The Case of the Texas Footprints*, it was funded by, among other groups, the AHA and CSICOP.[51]

In one especially notable instance, this debunking mode of attack was directed at a popular evangelical preacher and led to one of the most sensational exposés of the decade. The target was Peter Popoff, whose faith healing ministry was burgeoning as he traveled around the country. His revivals brought in hundreds of thousands of dollars each month in contributions because he seemed to be able to summon the power of God to heal the faithful.

Enter the Faith Healing Investigation Project, a new CODESH subcommittee, led with remarkable inventiveness and subterfuge by James Randi.[52] Randi knew that Popoff's rallies were all a con job and believed he could uncover the trickery that the evangelical was using to fool people. Aided by volunteers from local skeptics and rationalist groups, Randi was able to infiltrate one of these events. He planted skeptics in the audience posing as sick people and managed to get an electronics team fitted with radio gear unnoticed into an empty room in the building. With that setup, they were able to pick up and record wireless transmissions sent by Popoff's wife to her husband on stage wearing an invisible earphone receiver. As she read off information from cards that the revivalists had filled out when they entered, her husband called out the names of people and had them come up to be healed. From the perspective of the audience, it looked like Popoff had divinely inspired knowledge. God was actually talking to him. After the event, Randi revealed what was actually happening to a national TV audience in a sensational episode of Johnny Carson's *Tonight Show*, where he replayed the tapes he had recorded. The sting operation landed Randi a prestigious MacArthur Fellowship.[53]

One last project helps to bring this discussion to a close. Returning full circle, we find once again a court battle over evolution and an old strategy to fight it—a document in defense of science affirmed by world-class authorities. The case at issue was the 1987 Supreme Court case *Edwards v. Aguillard*, in which a constitutional challenge was brought against an equal-time law passed by the state of Louisiana that required teachers to teach creation science alongside evolution. In the lead-up to the case, an amicus curiae brief was filed by over twenty state and other scientific organizations and seventy-two Nobel laureates. The project had gotten its start when a member of the Southern California Skeptics group contacted Nobel physicist Murray Gell-Mann, who was himself a member of the board. Over the course of the next five months, a small team that included Gell-Mann and a couple of young skeptics managed to get a brief drafted and distributed for signatures to the Nobelists, the most elite group of scientists in the world. It was a twenty-three-page brief that provided a definition of scientific method and explained why creation science was not scientific. Then, with great fanfare, on August 18, 1986, the well-known paleontologist Stephen Jay Gould, joined by two other noted scientists, gave a press conference announcing the remarkable brief at the National Press Club in Washington, DC.

The *Edwards* case over equal time was an especially important one for defenders of evolution, and the Supreme Court sided with them in a crucial defeat for the creationists. As it had done earlier, the court ruled that the obvious religious purpose of the Louisiana law made it unconstitutional.[54] Humanists celebrated, and fundamentalists fumed. One more phase in the history of this long-standing controversy had ended (though it would not be the last, by any means).

Throughout the century, humanist groups responded to attacks on evolution and other scientific claims by fundamentalist Christians with a wide variety of strategies, publishing articles, collecting signatures, and investigating fraud. The strategies changed over the years as different people became involved, but the fight against the fundamentalists' antievolution campaign united humanists of every stripe. The differences among humanists paled in comparison to their united concern about the fundamentalist misuse of science. Humanists were, of course, not the only ones to defend evolution. Throughout the twentieth century, many groups stood up against both antievolution and creationism, including many people of faith. But humanist activism stood out. The forcefulness of the humanist response, as well as its persistence across the century, points to the singular importance of this aspect of the humanist agenda. The aspirational language of scientific spirit does not appear often. Instead, rationalism and skepticism flourished in these circumstances. The fight against fundamentalism reinforced a hard-line secularist stance, so that when humanists were forced on the defensive, they reached for a positivistic epistemology that demanded a sharp distinction between science and religion.

CHAPTER TWELVE

The Humanist Ethos of Science
in Modern America

The Cosmos is all that is or ever was or ever will be. Our feeblest contempla-
tions of the Cosmos stir us—there is a tingling in the spine, a catch in the voice,
a faint sensation, as if a distant memory, of falling from a height. We know we
are approaching the greatest of mysteries.

Carl Sagan (1980)

When the astronomer Carl Sagan received the Humanist of the Year award
in April of 1981, his television series, *Cosmos*, had just ended, and millions
of Americans had experienced a tour through the universe and through the
history of civilization. The inspiring and poetic language of the series captured a sense of wonder about the natural world and a fascination regarding
the human quest for knowledge. This was the same message that humanists
had been advancing in sermons, lectures, books, and manifestos for over a
century. In a country where an overwhelming majority of men and women
still maintained a comfortable belief in God, Sagan's opening words in *Cosmos*, quoted above, challenged that faith. Nonetheless, the show was an unmitigated success, a testimony to the scientist's skill at weaving a fascinating
story of mankind's journey at the intersection of religion and science.

Aspects of humanism resonate strongly with Americans of many faiths—
its optimism, its democratic values, its strong emphasis on science as a means
to know and understand. These coins of America's cultural currency were
core elements of humanism, and Sagan's beautifully produced series conveyed
the excitement of scientific exploration in ways that touched on all of these
ideals. Sagan's show was appealing in part because he told a captivating story
of the power of human imagination, a poetic tribute to science throughout
the ages. As a result, nonhumanists could overlook the religious differences
separating Sagan's worldview from theirs. It was an amazing feat to accomplish in a period of resurgent fundamentalism and the aggressive liberal reaction against it.

Two Figureheads of Scientific Humanism

Sagan was one of a number of science popularizers in the latter half of the twentieth century whose work embodied a strong humanistic ethos. Isaac Asimov was another. Asimov understood and popularized science in much the same way as did Sagan, and the two had become close friends over the years. They had quite similar worldviews; both were New York Jews who went on to become extremely visible public promoters of science. The American Humanist Association named Asimov Humanist of the Year three years after Sagan and appointed him president that same year, a position Asimov held until his death eight years later. Having completed a PhD in chemistry, Asimov identified himself frequently as a scientist, although he worked exclusively as a writer of both fiction and nonfiction books. His prolific production—reaching over five hundred books by the time of his death and including some of his generation's most well-known science fiction stories—made him a pop culture icon. It was through both science fiction and science popularization that he promoted his humanistic values.

Each man had connections to multiple humanist groups. Both, for example, in addition to being AHA awardees, were on the board of the Committee for the Scientific Investigation of Claims of the Paranormal. They were not activists in the organizational sense. Instead, they contributed by allowing themselves to be used as figureheads. (Indeed, Asimov's presidency was more titular than real; Bette Chambers did all of the actual administrative work, contacting Asimov only infrequently, mostly to have him sign off on publicity campaigns.) When the AHA asked Asimov to serve as president, he at first demurred, suggesting they call on Sagan instead: "Carl Sagan is as strong a humanist as I am; is younger; travels freely; is well-known to the public through his television work; is better-looking and has greater charisma."[1] Sagan, however, was not interested; he never allowed himself to be labeled easily. "Well, I'm certainly a human. And I'm certainly in favor of humans," he stated evasively when asked about this. "If that makes me a humanist, then I plead guilty." He was simply not a joiner: "I try to think for myself," he explained. Indeed, Sagan, more than Asimov, valued his status as an academic who needed to maintain an objective stance, and he did not want to jeopardize his credentials by putting himself too much in one ideological camp, even though his ideas made him an exemplar of humanism.[2] By contrast, Asimov advertised his support for many liberal causes, including such explicitly political ones as the American Civil Liberties Union, Americans United for Separation of Church and State, and the Freedom From Religion Foundation.

Asimov was not a bad choice for AHA president. He, like Sagan, had been within the humanist circle for years before his election. As early as 1959, Asimov's card was in the Rolodex of Edwin Wilson and was being tapped for speaking slots at local meetings. He was a founding member of CSICOP in the 1970s and a strong supporter of the ongoing campaigns against creationism. He was even married by Ethical Culture leader Edward Ericson.[3]

Sagan and Asimov really captured the scientific spirit of American humanism in the last quarter of the century. Their writings, both fiction and nonfiction, provided a romantic engagement with nature as revealed by modern science and technology. It was through science fiction (especially in Asimov's case) that a lot of this work was done. This genre was an ideal forum for the humanist message, and we can see its influence going back to the 1950s with AHA president Hermann Muller and his devotion to it. Asimov's stories always took readers into a world that worked according to a coherent logic and whose plots revolved around the collision between scientific rationality and mankind's irrational nature. His stories tended to be space operas that dealt with the grand sweep of future history and mankind's reach for the stars. His epic Foundation series, for example, turned a lens on the coming centuries of humankind as it colonized star systems throughout the galaxy. His stories forced readers to rethink their place in the cosmos. They were optimistic accounts, portraying a techno-utopian future where mankind was in control of the forces of nature. Toward the end of his life, Sagan, too, penned a science fiction novel, the widely heralded *Contact*, which soon became a motion picture. *Contact* speculated about our first meeting with alien life and the ramifications of understanding a universe that was far vaster than most earth-bound inhabitants could ever imagine.[4]

In their popular nonfiction as well, the two men humanized science in an effort to overturn its stereotype as a cold and impersonal enterprise. All successful popularization does this to some extent, making science more appealing and more personal, and touching on ethical ideals, all of which was true for these men.[5] Their popularization reflected their social idealism. A scientifically and technologically literate population, Asimov argued, was crucial to our future survival. Popularization encouraged support for science, helped to justify national funding, and inspired the career goals of young people. Science was even necessary in order to sustain a robust and more equal society: "We are, in the future, running the risk of creating a world of science-educated 'haves' . . . and science-uneducated 'have-nots,'" he said. "If one believes, as I do, that a strongly stratified society has the seeds of instability and destruction in it, then we should labor toward making science-education as broadly-based as possible."[6] Sagan echoed this sentiment. By educating people about science, we would "assist the machinery of Jeffersonian democracy."[7] In an essay celebrating the two hundredth anniversary of the Bill of Rights, he went even further by linking American democracy to science directly: "The Constitution was, in a way," he wrote, "a product of the scientific revolution."[8]

Their humanist prejudices arose most clearly when they discussed religion. On the one hand, Sagan endorsed the idea of a spiritual connection to nature. Sagan wrote *Cosmos* as a "personal odyssey" that touched on these deep, personal, and existential questions. Though he did not consider his work as having explicitly religious motivations, he used religious-like rhetoric and imagery. He emphasized the awe and reverence he felt when he contemplated the universe, something he believed to be a normal consequence of "any deep encounter with nature." One could not escape these strong emotions in the face of "the intricacy, beauty, subtlety, and elegance of the way the universe is constructed."[9] Many viewers of *Cosmos* were also struck by Sagan's use of religious imagery, which was sometimes strikingly conventional. The most evocative and memorable example of this was the set of his fictional "spaceship of the imagination" with its cathedral-like appearance. In numerous scenes, we see Sagan standing at the helm of this vessel, awestruck in front of an inconceivable expanse of stars and nebulae. The religious implications are hard to escape.[10]

On the other hand, the more frequent and explicit references to religion were negative. They depicted science triumphing over religious dogmatism, a common trope that historians term "the warfare thesis." This black-and-white view of history was all the more persuasive in the years after the rise of the politicized evangelical Christian right. The fundamentalists' antimodern reading of the Bible and their supernaturalist epistemology were a threat to both mainstream science and humanist values. Hence, the writings of these two popularizers abound in passages that echo the radical Enlightenment narrative of science freeing humanity from the bondage of the church.

Asimov, who considered himself a second-generation atheist, wrote many stories in this vein. The most famous example of this can be seen in his 1941 short story "Nightfall," a tale about a civilization on a planet with six suns and an orbit that keeps it in constant daylight, except for once every two thousand years when the suns and a moon align and the planet experiences true night. During the hours of darkness, the inhabitants go mad and destroy civilization, burning it to the ground in search of illumination. The central conceit of the story pits scientists against religious leaders. The clash is made explicit by the head astronomer Aton: "Our results contain none of the Cult's mysticism. Facts are facts, and the Cult's so-called mythology has certain facts behind it. We've exposed them and ripped away their mystery. I assure you that the Cult hates us now worse than you do."[11] The end of the story—as the scientists themselves succumb to madness and the conflagration begins—highlights another of Asimov's ideas, namely, that science and civilization are intertwined, rising and falling together.

Sagan emphasized that point as well. An especially important episode of *Cosmos* showed Sagan narrating the story of the destruction of the ancient Library of Alexandria and the murder of its librarian Hypatia, a mathematician and astronomer, at the hands of a Christian mob. He explained with deep regret how religious fanaticism at that time had destroyed centuries of human culture and learning.[12] Similarly, the rationalism of science was again pitted against the irrationalism of religion in the novel *Contact*, a story that was as much a personal meditation on religion as a science fiction adventure. Throughout the book we see two very different ideas of religion arise: one is organizational and political, while the other is personal and emotional. As one character explains, "I think the bureaucratic religions try to institutionalize your perception of the numinous instead of providing the means so you can perceive the numinous directly—like looking through a six-inch telescope. If sensing the numinous is at the heart of religion, who's more religious would you say—the people who follow the bureaucratic religions or the people who teach themselves science?"[13] It is the institutionalized religions, of course, that pose one of the greatest dangers in *Contact*, and the hero, the radio astronomer Ellie Arroway, a modern-day Hypatia, again and again comes up against the irrationalism of the religious mob.

The humanistic ethics of Sagan and Asimov also tapped into America's democratic spirit, which valued the importance of self-reliance and human ingenuity. When asked about the definition of humanism, Asimov answered a reader's fan letter: "If there are problems to be solved, human beings will have to solve them, or they will remain unsolved." In other words, humanism was simply an ethic that promoted problem-solving through science and reason. But it also promoted self-reliance—our salvation rested squarely with us: "We recognize our own limitations, but there is nothing and no one to make up for any shortcoming we may have."[14] This was language that could have been taken directly from one of the sermons of Unitarian humanist John Dietrich at his Minneapolis pulpit in the 1920s.[15]

All in all, Sagan and Asimov offer us a glimpse into an extremely popular late twentieth-century ideology of science that reflected the humanist values of the past century. By using science to encapsulate elements of humanistic thinking, they appealed to readers and viewers from a wide range of backgrounds and were considerably more influential than any of the humanist manifestos, with their numbered planks and philosophical assertions. Science became more than a utilitarian vocation; it represented the spirit of curiosity and inquiry and offered courage and an optimistic faith in the ability of mankind to succeed on earth. At the same time, as we have seen, it could be weaponized against fuzzy thinking and supernatural beliefs, whether they were based on traditional faiths or New Age mysticism.

I've focused on Sagan and Asimov here, but they were not alone. The prolific Harvard paleontologist Stephen Jay Gould, who also participated in a variety of humanist and skeptic events, was a third. And even before all three of these men had passed away, others like Richard Dawkins and Neil deGrasse Tyson added their voices to the chorus. All of them have been noted for their popular influence as well as their religious prejudices. One recent book takes issue with these "oracles of science," complaining that they used their towering status as celebrities to promote a narrow, science-based atheism, one that was explicitly competing with traditional religions and covertly inserting their secular prejudices where they did not belong. There is some truth to that, but the real story here, I suggest, is these men's talent at harnessing widely shared values that often crossed religious boundaries and integrating them into their popularizations of science. Contrary to many of their critics, there was much more to what these scientists did than simply tear down ancient faiths and comforting convictions. Rather, it was their willingness to personalize and embrace an ethical and visionary stance that made them so successful.[16]

Scientists and Humanism

This book has charted the path that led to the humanists' embrace of these scientist-popularizers in the late twentieth century. The movement that welcomed them began as a radical religion and gradually secularized. It started with a largely pragmatist worldview and became more positivist in outlook later. Those shifts did not happen overnight, nor were they consistently or universally embraced. The reason that the popularizers proved so successful within the humanist movement was that they synthesized different and often conflicting elements of humanism's complex history.

I have argued that the intellectual transformation within the humanist movement was linked both to an internal demographic shift and to cultural changes. It is to the demographic shift that I want to turn for the moment. To get a better feel for that shift, I have collected and categorized the signatories and awardees of the most visible humanist projects over the years: the two Humanist Manifestos of 1933 and 1973, the Secular Humanist Declaration of 1980, and the Humanist of the Year awards. In addition, I have also included the Academy of Humanism, a small group started in 1983 by the Council for Democratic and Secular Humanism. Each project had a different purpose, but all helped define humanism in the public eye.

The nature of the projects and the size of the samples preclude a true statistical analysis, so my analysis here should be taken as an impressionistic survey. My categories lump people together based on their occupations and sometimes their primary role in the humanist movement, which results in making occasional difficult choices. (Asimov, for example, though he was a popular writer, is listed as an academic because of his PhD in chemistry, which he frequently advertised.) We must be particularly wary of drawing firm conclusions related to the first manifesto and the Academy of Humanism since the absolute numbers there (thirty-four and thirty people, respectively) are so small that the bias of the creators of those documents is likely overwhelming any true statistical result. Nonetheless, I still think that the results are instructive—allowing us a bird's-eye view of the prominent supporters of humanism. Moreover, the data comport with the story I have told in these pages based on other evidence.

To begin with, the secularization of the movement is glaringly apparent in these data. Between the Humanist Manifesto of 1933 and the Humanist Manifesto II of 1973, the number of signatories who were ministers dropped from 50 percent to 21 percent, and it fell again to only 5 percent in 1980 with the publication of the Secular Humanist Declaration (figure 1). (Of course, the declaration was so aggressively irreligious that one might wonder why any clergymen signed that document—but these are humanists, after all.) Looking at the other two categories—academics (which includes practicing scientists outside the academy) and others (mostly businessmen, lawyers, entertainers, and writers)—another big change stands out. There is a sharp rise in the percentage of academic support for humanism over the years: it is

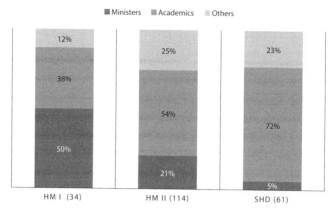

Figure 1. MANIFESTO SIGNATORIES BY PROFESSION. Using the signatory lists from each manifesto—Humanist Manifesto of 1933 (HM I), Humanist Manifesto II of 1973 (HM II), and the Secular Humanist Declaration of 1980 (SHD)—it is easy to see how the humanist movement secularized over time. The percentage of clergy declined between 1933 and 1980 in direct proportion to the increase in credentialed academic professionals. (Numbers in parentheses below each column refer to number of signatories to each document.)

above 50 percent in 1973 and above 70 percent in 1980. The message here seems clear: the main source of authority in the humanist movement shifted from ministers to university professors and scientists.

Given the importance of academic authority in humanist circles, we can see more clearly which disciplines were on top in figure 2. Most significantly, philosophers dominate in this area across all manifestos, accounting for between 43 and 52 percent of that category in each list. This is an important result because it highlights the one long-term source of stability in the movement. Humanism was a movement embraced and in large part built by philosophers across the century. That insight comes alongside an interesting surprise: natural scientists—people who worked in physics, biology, and medicine—constituted a relatively low percentage (between 15 and 18 percent) of academics in all the manifestos. By contrast, we find instead that human scientists (social scientists and psychologists) came into their own in the late twentieth century. Their numbers thrived in the later years, and they did so at the expense of the humanities scholars. After the 1933 manifesto, the percentage of signatories in the human science category rose in direct proportion to the decline in the percentage of scholars in the humanities, an important point that I will return to shortly.

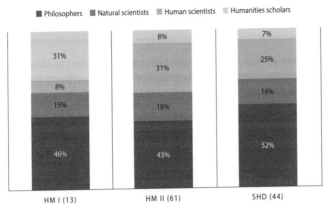

Figure 2. ACADEMIC AND SCIENTIFIC SIGNATORIES OF THE MANIFESTOS BY DISCIPLINE. Using the same signatory lists as in figure 1, but focusing only on the academic professionals, we can see how different sciences and academic disciplines have had shifting importance over time. The steady consistent dominance of philosophers is evident here, as is the relatively low but consistent participation by natural scientists. The major shift over time shows the rise of the human scientists (psychology, in particular) as scholars in humanities decline. (The numbers in parentheses at the bottom of each column refer to the number of academic signatories to each document.)

The Humanist of the Year awards given by the AHA provide another window into the public face of humanism, as shown in figure 3. Since this award highlights the professional expertise of the individual and that person's lifetime accomplishments in the service of humanistic causes, the disciplinary divisions are instructive, telling us much about the thinking of humanist leaders (at least those on the awards committees). This list of awardees (which I have limited to the forty people who received these awards in the period from its inception in 1953 to 1990) tells a different story than the lists of manifesto signatories. In contrast to those lists, the natural, biomedical, and human sciences dominate, consisting of over two-thirds of the awardees. I have separated the biomedical scientists from the other natural sciences to highlight the importance of the human-based disciplines: taken together, biomedical scientists and human scientists from various disciplines, especially psychology, account for over half of all awards. Perhaps most startling, though, is the dearth of philosophers—the only philosopher to be given the award in this period was Joseph Fletcher in 1974, an ethicist. This result is especially unexpected given how prominent philosophers were in the signatory lists and in the movement's development generally. It suggests that these men were more

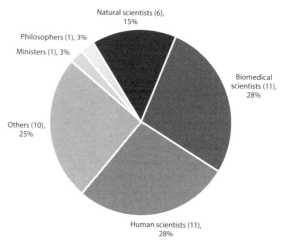

Figure 3. HUMANISTS OF THE YEAR 1953–1990 BY DISCIPLINE. This chart high-lights the composition of the awardees of the American Humanist Association's Humanist of the Year awards between 1953 and 1990. Unlike the signatory lists, it is the natural and human sciences that dominate the list of prize winners. Given the importance of philosophy throughout the decades, the almost total lack of philosophers in this list is notable. (Numbers in parentheses refer to the number of awardees in each category.)

important working behind the scenes than facing the public, where the visibility of prominent scientists has been much greater. Another notable point is that academic work counted higher than activism. Finally, less than one-quarter of the awards went to social and political activists.

A look at one last list is useful, I think. In 1983, as part of his new organization CODESH, Paul Kurtz established the Academy of Humanism "to recognize distinguished humanists and to disseminate humanistic ideals and beliefs to the world."[17] Kurtz celebrated the new academy in *Free Inquiry* as part of his announcement of the 350th anniversary of Galileo's trial, with an emphasis on "the vital importance of free inquiry to human civilization." Along the way, Kurtz listed a "humanist pantheon" of the world's most famous scientists, philosophers, and social thinkers of the past, including such figures as Socrates, Spinoza, Thomas Paine, Charles Darwin, Sigmund Freud, Marie Curie, and Margaret Sanger. The Academy of Humanism was described as an active forum of engagement with ideas and issues, but in the end it proved little different from the AHA's Humanist of the Year. It was not only an honor bestowed on well-known personages but also a way to bring publicity and attention to the council itself. The academy was highly international

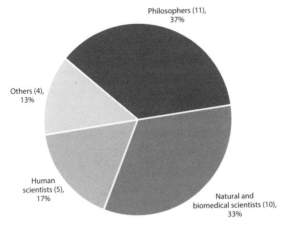

Figure 4. ACADEMY OF HUMANISM IN 1983 BY DISCIPLINE. This chart shows the disciplinary composition of the Academy of Humanism founded by Paul Kurtz. Here, philosophers are the dominant profession, followed by the natural sciences and then the human sciences. (Numbers in parentheses refer to the number of members in each category.)

and touted that fact by printing the list of its members organized by country. Among the thirty initial members, one finds a Who's Who of popular scientists and intellectuals: Francis Crick, Stephen Jay Gould, Karl Popper, Carl Sagan, Andrei Sakharov, and Edward O. Wilson.[18]

The chart in figure 4 gives us a means of comparing this group with the other lists of names, providing us with a perspective on the outlook of Paul Kurtz and his new companions in CODESH. Philosophy returns to play a dominant role, being the single largest disciplinary category (37 percent). Nonetheless, taken together, the natural and human sciences (at 33 and 17 percent, respectively) made up a much higher percentage than philosophy. There were no ministers at all in this group, though there was a theologian among the philosophers and a biblical studies scholar who was an archaeologist.

Science and Atheism

What can we make of those numbers? Are they unusual at all? I think that they do make sense of some of the trends in America in the twentieth century, based on what we know about scientists and faith. Although there is nothing about science that makes it inherently atheistic, secular, or humanistic, it is clear in the story of humanism that many contemporary scientists find humanism compelling and have been willing to add their names to humanist projects.

Historically this was not the case. If we go back to the sixteenth and seventeenth centuries, we see that nearly all the figures associated with the Scientific Revolution were believers, many of them obsessively so. Johannes Kepler, the astronomer who discovered the mathematics behind the planetary orbits, held a mystical view of God's plan for the cosmos. Galileo, for all his trouble with the church, thought of himself as a pious Catholic. Isaac Newton, we know today, turns out to have written much more on theology than he ever did on gravity, optics, and calculus. Examples in the eighteenth, nineteenth, and even twentieth centuries show many deeply religious scientists, for whom the study of nature was integral to their faith. These men and women believed that they were uncovering God's handiwork, and in that way, many of them considered scientific work to be performing a type of worship. Some considered it their religious duty to explore and understand God's book of nature.[19]

The past 150 years, however, have witnessed a sea change in the nature of science—in both content and practice. Considering those changes, are there now reasons to consider modern science fundamentally atheistic? Science has become more specialized now; scientists now see themselves as professionals, they now deal on a day-to-day basis with narrow technical problems, and it is entirely out of place to discuss God and religious matters as part of one's work. To be a believer and practice their profession, scientists do what other professionals have done: they privatize their beliefs, separating those beliefs from their work.[20]

Those changes seem to have taken a toll on belief. As long ago as 1914, it was already clear that scientists were very skeptical people as a group. In a study that year by the psychologist James Leuba, scientists were shown to be more skeptical than the average American: 42 percent did not believe in God, and another 17 percent were uncertain, which is remarkable in a country where the level of belief was at least 90 percent. But Leuba went on to find that among elite scientists, those at the top of their field, the level of belief was even lower; moreover, in a follow-up study nearly two decades later, he documented that the percentages had plummeted even further. By 1933, the data showed that only 15 percent of elite scientists were believers. At the end of the century, two scholars, Edward Larson and Larry Witham, repeated Leuba's survey and found that levels of belief in God among the rank-and-file scientists had not changed significantly, but when they examined elite scientists, they noted that the level had continued to fall below Leuba's numbers to a miniscule level of only 7 percent.[21] Over the course of the century, then, the top scientists in all fields had become an almost entirely secular community.

Given these results, it is not too surprising to observe that the community of elite scientists is becoming more and more humanist in orientation.

More recently, sociologist Elaine Ecklund has presented the results of a detailed survey that gives a nuanced picture of spirituality among scientists at elite American universities in the early twenty-first century.[22] She confirms the radically different level of religiosity between scientists and the overall population of the United States. Not only are scientists much more atheistic and agnostic, but they hold beliefs about the nature of religion that are far more negative. On the other hand, she also detected a strong strand of spirituality among them—even among those who identify as atheists or agnostics. But whereas the common use of the term *spiritual* connotes transcendence, these atheistic scientists use the term in a more Sagan-esque way that focuses on the natural world in its awe-inspiring magnificence.[23]

Secular Critics at the End of the Century

The humanist view explored in this book was not by any means universally embraced by secular thinkers at the end of the twentieth century, not even by those who called themselves humanists. In order to put the science-based humanism discussed here into a broader context, I've chosen to look at two critical perspectives that highlight some of the challenges it faced from other secularly minded people. In particular, I look at African American humanists, on the one hand, and postmodern theorists, on the other.

The humanist movement of the twentieth century was largely a white middle-class movement, despite its continued outreach to people of color. On an organizational level, the efforts of white humanists to engage black religious radicals were marginally effective at best. This was not because there was no one there to recruit. The African American population included nonbelievers and religious radicals who could easily have found a home in the humanist community.

Nor was the problem simply the group's resistance to change. We can see this by noting how women came to be integrated into the community over the course of the century. Though leadership in the movement had been largely male throughout the early part of the century, as time wore on, more and more women became involved and began to play key roles. The AHA had a female editor as early as 1953, a female president a decade later, and by the 1970s a host of women honored as awardees and sitting on the governing board. Beyond those especially visible indications, women also, by and large, were the ones who pursued the legal challenges in church-state cases where religious involvement in education was the primary concern. The women who were litigants in these court cases were soon joined by others who were activists, engaged in legalizing birth control, defending sex education, promoting equal rights, and fighting creationism. These issues all put women on the barricades in fighting for social causes and brought them into leadership positions in the AHA. (Incidentally, this general opening for women in humanism was emphatically not the case in the hardline skeptics movement, which remained, throughout the last quarter of the twentieth century, an almost uniformly male community, a gender difference that requires further exploration, more than what I am able to do here at this time.)

So what was it about the humanist movement that made it less attractive to people of color? African Americans have a rich tradition of atheism and religious radicalism. Contemporary black humanists point to freethinkers going back as far as Frederick Douglass in the nineteenth century. Such figures as W. E. B. DuBois and Zora Neale Hurston stand out as humanist figureheads in the twentieth century, as do numerous other intellectuals, artists, and activists. There were some Unitarian congregations that grew up in black communities, and a few of those had quite radical preachers. The Unitarian Lewis McGee was one of them, who preached a humanist message to his congregations. As we have seen, McGee worked side by side with Edwin Wilson in the 1940s in the executive office in Yellow Springs. That said, it was not until 1989 that writer and activist Norm R. Allen Jr. founded African Americans for Humanism (AAH) and edited its newsletter, the *AAH Examiner*, which discussed and promoted humanism in the black community, usually outside of the traditional church setting. Allen's movement was composed of both secular and religious humanists, including many Unitarian Universalists.[24]

Despite all this, the humanist movement had a much harder time engaging with the concerns of black radicals than with those of women. The challenges in bringing racial diversity to the humanist movement seem to have had a lot to do with the different roots of black and white humanism and the ideological trajectory of the white movement that this book has documented. It is not unrelated to the strong emphasis that humanists put on science. The experiences that gave rise to humanism among black thinkers did not engage the same intellectual concerns that drove the white humanist movement. The black scholar Anthony Pinn, who began writing humanist essays in the 1990s, explained the challenge for blacks. He was a follower of the black Unitarian humanist William R. Jones, and he describes how Jones understood the difference between the two groups: "The African American humanist project emerges not as a consequence of the Enlightenment but rather as a direct response to a unique set of circumstances facing African American communities in the United States."[25] The moral challenges that confronted the children and grandchildren of enslaved Americans in a manifestly racist society made many of the epistemological debates about science and religion almost superfluous to them. At the very least, epistemology was of secondary importance to a more pressing agenda: that of gaining equal status with other human beings in the society they lived in. Given the powerful force that Enlightenment rationalism had in framing the white humanist worldview, it is no wonder that the two groups diverged in their thinking.

I have already spent some time recounting the IQ-race forum of 1970 in chapter 7, but I want to return to this topic briefly. I do not mean to give too much emphasis to this event because it was only a single instance of surprisingly reactionary ideas; nevertheless, the event is telling in certain ways, and it helps us understand just how culturally specific the science-based culture of white humanism was and how that tended to limit its attraction to some people who might otherwise embrace an overall naturalistic worldview. With this in mind, we can see why the Jensen controversy played out as it did. While Paul Kurtz and Sidney Hook were focused on the debate between free inquiry and an open society, black humanists only saw further evidence of prejudice and intellectual domination. The supposedly scientific evidence of black inferiority could be considered nothing more than another effort to relegate the black "race" to subhuman status. The very concept of *race* in this controversy—which should have been one of the central points of debate—was simply glossed over and taken for granted by most of the participants, white men who wrote from the point of view of privileged members of the society and who could never feel threatened by the very existence of the discussion.[26]

Norm Allen took Pinn's critique further. Although Allen was a staunch supporter of evolution and worked with national groups to provide a black voice to defend it against creationism, he also warned against the uncritical embrace of science and the dangers of using scientific ideas to defend prejudice:

> Many white humanists embrace a dogmatic scientism. They believe that scientists are without biases, and they scoff at "conspiracy theories." Progressive Black humanists, however, are not so quick to scoff. Black humanists are much more likely to take note of the infamous Tuskegee experiment—*a genuine government conspiracy*—in which men were infected with syphilis and allowed to go untreated. . . . There are numerous instances of the use of science for evil

purposes. Black humanists are likely to be aware of this fact, and are less likely to rush blindly to the defense of science whenever controversial problems arise. Good science must be viewed as a means to an end—genuine progress—and not an end in itself.[27]

The arguments of Jones, Allen, and Pinn were by no means a rejection of science and rationality, but they express a more nuanced understanding of the concepts of scientific objectivity and scientific authority. Black humanists who fought for a rationalistic worldview did not conflate rationalism with science in the same way that white humanists did. This made them less confident about claiming that science was a cornerstone of humanist thought.

The one dramatic counterexample to the above point is the twenty-first-century popular astronomer Neil deGrasse Tyson, who was a protégé of the late Carl Sagan and ultimately took over his mantle as one of America's foremost science popularizers. Tyson's worldview is very similar to Sagan's, and the AHA gave him their Isaac Asimov Science Award in 2009.[28] Five years after that award, he hosted a major reboot of *Cosmos* that contained most of the same humanistic agenda. In the new *Cosmos*, the photogenic Tyson was able to convey that same emotional connection to science, the feelings of wonder and awe at the universe. He also reiterated the warfare thesis in new historical accounts that criticized institutional religion as an enemy of science and intellectual progress.[29]

One can conclude then that, despite an underlying appreciation for science and its ability to lift humanity out of antiquated belief systems, secular-thinking African Americans as a group were not especially attracted to the vision of the scientific spirit. Their concerns were so pressingly social that the idealism associated with Enlightenment rationalism did not provide the same kind of comfort that it did for white Americans. Indeed, African American literature is replete with examples of how white members of the society wittingly and unwittingly create rationalizations that obscure the kind of daily discrimination faced by people of color in this country. The ideology of science is no exception, as can be seen by the kind of abuses that went on in its name in the past, even within the humanist movement.

Turning to a somewhat different set of concerns, it is instructive to look at the postmodern critique of humanism. There we find a deeper, more tenacious, and incisive set of objections to science and our very ability to know the world objectively. In some respects, humanists were blindsided by the postmodern critique. After all, it arose out of the secular academy, an area of the culture that humanists had long had close ties to. But postmodern theorists were not scientists and were not enamored with rationalism like the humanists were.

Of course, challenges from the academy to a science-centered worldview were not without precedent. The midcentury Frankfurt School, a group of German émigrés, was especially worried about an excessive emphasis on scientific rationality. They argued that attempting to build a society on rationalism and a lifeless instrumentalist morality would lead to just the type of an authoritarian world that they had only recently escaped from. And we have already seen how Mortimer Adler hammered against the dangers of instrumental rationality in the 1930s and 1940s and how the humanistic psychologists of the 1960s vehemently opposed positivism on similar grounds.

By the mid-1980s, however, a major shift took place in the humanities. During that time, theorists in a wide range of fields made various arguments against humanism and Enlightenment thinking. Indeed, many postmodernists used the term *humanism* explicitly in their attacks to refer to the intellectual foundations of the whole of modernity. Though they seldom referred to the organized humanist movement we have been studying, nonetheless their attack was a direct assault on the philosophical framework that the organizations embraced.

For one they rejected any kind of universalism. Drawing heavily on French theorists like Michel Foucault and Jacques Derrida, they claimed there were no transcendent truths or other means of finding a fixed point of reference across all of humanity with its myriad of cultural and social differences. As a result, there were no fixed foundations upon which to ground any kind of knowledge claim or ethical precept. With regard to science, this meant that rational thought was not autonomous. Contrary to what many scientists and philosophers asserted, neither science nor rationality provided a privileged perspective that set it apart from any other assertion about the world. All claims to knowing were inextricably tied to power relationships, which meant that even science was justified only by a social network. There was nothing special about the way it worked and certainly no reason to elevate people of science to positions of cultural or political authority simply based on their occupation.[30]

According to postmodernism, science and rationality were not liberators, as humanists and others claimed, but rather sources of oppression. The rise of postmodernism soured many scholars on science both as an institution and as a source of knowledge, and they began to see the humanistic worldview of modernity and rationality as anathema. The fact that postmodernism captured some of the top scholars in the humanities helps to explain the continued striking imbalance in the academic support for the organized humanist movement, which was, as we saw above, almost entirely confined to the social and natural sciences. Within the humanities, the only group to give any kind of strong support for organized humanism was the Anglo-American analytic philosophers.

The postmodern critique of rationality, on the one hand, and the humanists' difficulties with race, on the other, are evidence of late twentieth-century secular challenges to humanist thought. The critiques took issue with humanism generally, but the strongest and most consistent objections were against what I have characterized as the positivist strain of humanism.[31]

The American Foundations of Humanism

One may ask what held humanism together over the decades. Those unfamiliar with the history of liberal religion in America will be tempted to see humanism as foreign and disconnected from mainstream American thinking—especially when one looks at America's high rate of belief in God. However, as I have shown, the humanist movement was an important part of the American scene; it fits easily within the broad tent of the American liberal religious tradition. Many of the humanists' ideas about modern science, individual freedom, and self-reliance were, in fact, distinctly American in character.

Like all social movements, humanism is an amalgam of ideas and motives. In this book, I have documented various influences on American humanism: the moral and intellectual spirit of the Enlightenment, the liberal religious theology of Protestant modernism, the secular outlook of liberal Jewish intellectuals, the Deweyan and later positivist traditions in American philosophy, and the visionary thinking of prominent American scientists. This diverse intellectual heritage has provided the humanist movement with many different resources to draw on, just as it has also given rise to internal conflict. A few words about each of these influences will be useful in better understanding their respective contributions.

First of all, humanism's Enlightenment heritage links this movement to the founding fathers of the republic. Indeed, one writer has called humanists "children of the Enlightenment." Since the eighteenth century, revolutionaries like Thomas Paine have urged people to look at the world without the distorting filter of traditional religion. At the same time that the country was experiencing a revolution in political ideas, rejecting monarchy and tyranny, many of its defenders were justifying those ideas on rationalist grounds. Proclaiming individual freedom and constitutional democracy, many eighteenth-century intellectuals urged a radical transformation in both social structures and moral commitments. The humanists have closely followed those thinkers in their belief that science and democracy are the two main pillars of modern thought.

The role of Protestant modernism places this movement within America's long tradition of liberal religion. The political influence of conservative Christianity today sometimes causes us to forget how dominant liberal theology was at the beginning of the twentieth century. It was taught in the seminaries and expounded by ministers throughout the country. Humanism traces its origins directly back to this theological tradition. Though it was radical insofar as it dismissed any talk of God, the early humanist outlook was merely an extension of religious liberalism, not a rejection of it. Modernism served as a kind of pivot point where these young ministers stood briefly before making their way into full-fledged humanism. In other words, these were religious men, who lived and breathed church air most of their lives.

Secular Jewish intellectuals who entered the movement served as a counterweight to the liberal Protestant tradition. Humanism would not have developed as it did had not many of its proponents emerged out of this non-Christian cultural milieu. In this, of course, humanism is no different from many other areas of American intellectual life where, over much of the twentieth century, Jewish intellectuals and artists reshaped American culture and imbued it with a distinctive understanding of what it meant to be an American. Primarily urban, and mostly New Yorkers, their outlook differed significantly from that of the humanists of Protestant heritage. Whereas the ministers held a generally positive view of the institution of religion, the Jewish intellectuals were deeply suspicious of it. They had little patience for the concept of religion largely because in America it remained inescapably associated with Christianity. For them, the humanist outlook needed to be separate from religion. John Dewey's protégé Sidney Hook and later Hook's protégé Paul Kurtz—both Jews—helped secularize the movement.

Of all professions, philosophers were among the most important influences on humanism. They shaped it in profound ways. If this book does anything, it helps to highlight the remarkable significance of this largely invisible group of academics. Few people have given thought to the complex social function that academic professors can play in the culture, but the story of humanism calls attention to these different roles. The extracurricular activities of these men, their ideas, their social networks, and their religious involvements were all crucial to the way that humanist history played out. Two schools of American philosophy fundamentally shaped its development. John Dewey has had an outsized influence on many areas of American culture, and it was true here as well. He held an almost grandfatherly place among humanists. His pragmatist vision supplied to humanism an optimistic faith in the potential of humanity to grow, thrive, and improve. Under the pragmatists, philosophy was not simply an intellectual enterprise; it was a tool for improving the world. In contrast to pragmatism, positivism and analytic philosophy made their way into humanism through philosophers and scientists alike. It was an approach to ideas that turned out to be well suited to an institution under siege. And by the second half of the century, humanism certainly was, with political enemies on the resurgent Christian right and attacks on rationalism among radical counterculture activists and New Age religionists on the left. Under these conditions, the sharp edge of analytic philosophy appealed to many humanists. With it, science could be repurposed as a weapon.

Visionary and politically active scientists built on this heritage. These scientists were attracted to humanism in large part because it gave them a popular forum where they could discuss political and moral issues that were not part of their professional activities. When they lent their authority and prestige to the humanist movement, they did so based on their desire to be ethical citizens. In the wake of the Second World War and science's transformative power in fighting and ending the war, many of the physicists, chemists, and engineers looked back and reconsidered their roles in the world. Their sudden rise as international spokesmen gave them a new, highly visible status. In that new role, they considered it to be their moral responsibility as cultural icons to speak out on behalf of the ideals of humanism, whose outlook closely mirrored their own. And humanists were eager to have them. From Muller to Sagan, from Maslow to Skinner, the scientists found that they were embraced by humanists. Not only did their visions align, but the scientists also brought with them an extraordinary authority and prestige. Their presence, like that of the philosophers, helped shape the movement and take it in new directions.

The history of the humanist movement in America is like a spotlight aimed at some of the most contentious areas of the nation's culture. It illuminates much about the nature of both science and religion in this country. Humanism, far from being an aberration in American history, epitomizes a set of ideas and outlooks that are quite common. It is one version of a distinctively

American experience, a product of the fertile soil of two centuries of American thought. And within the movement, the scientific spirit of American humanism was central. The relationship between science and religion in twentieth-century America cannot be fully understood without grasping it. Only by recognizing how that framework was so captivating for its adherents and so apparently dangerous for its detractors can we begin to comprehend the complexities of the American experience of religion and science.

Science and Millennial Humanism

The long history of American humanism and its entanglement with science told here covers most of the twentieth century. It ends about a decade before the turn of the new millennium, at a time when different strands within the movement had begun to mature into identifiable factions. As it stands, that narrative tells a coherent story that I believe both makes sense of what was happening within humanism and sheds light on larger changes in American culture and society up to that point.

Time moves on, however, and events since then have put humanism in a different place. As I write, we are now three decades beyond the conclusion of the story, and both humanism and the world around it have changed considerably—although, in some respects, conditions remain strikingly and stubbornly familiar. I want to end this book with just a few words about the directions that humanism has taken since then.

First of all, there seem to be indications that the previously very steady and strong religiosity in mainstream America is declining. Polls show that more and more people, especially young people, are rejecting traditional religious identification. Nearly 30 percent of respondents in recent polls indicate a very weak or nonexistent connection to traditional religiosity.[1] Of course, not all of these "nones" are atheists or agnostics, let alone humanists; nonetheless, a good fraction of them are nonbelievers, and one study recently characterized the rise in this population as an *awakening*. In a culture where it is increasingly acceptable to be a nonconformist in terms of race, ethnicity, gender, and sexuality, we should not be surprised to find religious nonconformity also more acceptable. And, indeed, atheists and humanists are certainly more comfortable these days in identifying themselves.[2]

There have been more manifestos and declarations in recent years, especially around the turn of the millennium. Paul Kurtz put his effort into a major book-length document sponsored by the Center for Inquiry, entitled *Humanist Manifesto 2000: A Call for a New Planetary Humanism*. The American Humanist Association issued "Humanism and Its Aspirations: Humanist Manifesto III" in 2003. Even the International Humanist and Ethical Union produced an updated "Amsterdam Declaration" in 2002, marking fifty years since the international group was founded. Finally, in 2006, the Center for Inquiry published a second statement: "Declaration in Defense of Science and Secularism." Each of these documents affirms basic humanist ideals with different emphases. These declarations usually come with a familiar list of signatories, well-known public figures that include prominent scientists and academics.[3]

What I have called religious humanism in these pages has remained a vibrant subset within the movement, growing in ways that incorporate a greater emphasis on secularism. This may seem contradictory, but we have seen this before. Now, however, it seems that even more people whose ideas and behavior place them easily into the category that we have been calling religious humanism no longer identify themselves as religious at all. Instead, they use terms such as *life stance* to describe how they incorporate humanism in their lives. These groups, ranging from people affiliated with explicitly secularist societies to student organizations like the Harvard Humanist Chaplaincy, continue to meet and celebrate important life moments, such as births, weddings, and funerals, in affirmative and nontheistic ways.[4]

In sharp contrast to those congregational and life stance humanists, the twenty-first century also witnessed a remarkable efflorescence of out-and-out atheism that shook the American religious scene for a number of years, making news, inspiring a large group of followers, and also drawing heavy fire from people across the religious spectrum. It was a movement spawned by the largely coincidental publication of several books by four strident public intellectuals who argued that, especially in the post-9/11 world, religion needed to be called out for its many sins against humanity. These four men, Richard Dawkins, Sam Harris, Daniel Dennett, and Christopher Hitchens, wrote sharp and unapologetic attacks on supernatural religion in all its forms, calling all such belief dangerous and outdated. The positions of the new atheists were so strident and negative that even secular humanists like Paul Kurtz considered the new atheists extreme by overemphasizing the antagonism against religion at the expense of a positive and forward-looking secular alternative.[5]

With regard to science, humanism has grown and developed in new ways. On the one hand, the generally optimistic and future-oriented idealization of science was occasionally countered by more critical concerns. After Asimov died in 1992, for instance, the AHA selected Kurt Vonnegut to be their honorary president, an especially interesting choice given Vonnegut's deeply cynical perspective on mankind and on modern technology and science, views that were in many respects diametrically opposed to Asimov's views. Also, it is notable that since the 1990s over half of the Humanist of the Year awardees have been not trained scientists but rather politicians, lawyers, humanist leaders, and writers. That said, however, many of the fourteen scientists who were selected as Humanist of the Year were noted for the strong critical positivist mode that I have discussed here. These include people like evolutionary scientist Richard Dawkins, the biologist P. Z. Myers, the myth buster Adam Savage, and the cognitive psychologist Steven Pinker. (One needs to be careful here, however. Several of those who have been sharply critical of organized religion are far from simply negative. Dawkins, for example, has been enormously successful in popularizing science in a mode not that different from Sagan, extoling the wonders of nature and the delight he finds in knowing about the universe. Nor is Dawkins averse to tying his ethical views to his science.) Outside of the AHA, many humanists continued to fret about dangerous implications of postmodernism and antihumanism in the academy, especially during the 1990s and early 2000s. Indeed, the formation of secular student groups on college campuses during this period reflects this larger concern that irrationalism and antiscience were invading the academy.[6]

The controversy over evolution, of course, did not abate with the conclusion of *Edwards v. Aguillard* in 1987. Indeed, the antievolution movement returned with a vengeance in an even more aggressive form based on a new type of argument: intelligent design, or ID for short. Despite its claim to be different from traditional creationism, the ID challenge mirrored the long-standing antievolutionist crusades: evangelical Christianity was the powerhouse behind ID activism, and these Christians continued to try to push for changes in public school science curricula. They extended their efforts, however, beyond the public schools, pushing hard against the scientific and academic establishment in order to gain influence and credibility in areas of the culture that held the greatest authority. The ID threat remained atop the humanists' agenda throughout this period and was clearly a major consideration in the AHA's naming of P. Z. Myers, an aggressive crusader for evolution, their Humanist of the Year in 2010.[7]

The ever-growing concerns about the environment and climate have challenged humanists, as they have humanity as a whole. What, indeed, is the future of humanity on this planet? If there were one other topic that I would have included in this book, it would be this one. There is an important story to be told about how humanists have grappled with the nature of the earth and mankind's environmental impact on it. As our global climate begins to step beyond its historical limits, humanists have become ever more engaged with the problem.

Finally, the skeptics movement has flourished over the years, and there are now two separate popular magazines promoting this agenda. With climate change deniers and anti-vaxers having entered into the mainstream in ways not unlike New Age beliefs during the 1970s and 1980s, skeptics have extended their agenda to deal with these new threats. These areas of health and environment join old worries over occult and supernatural belief, the psychology of human error, and the need to defend the scientific method.

From the rationalist skeptics to the life stance humanists, the wider humanist movement remains a home for defenders of a naturalistic and rationalistic outlook that they see as critical to the long-term survival of humanity. Things have changed in many ways, but the humanists remain among the country's most ardent defenders of a liberal Enlightenment-inspired outlook, a framework that has been an integral part of the American cultural landscape since the country's inception.

Introduction · The Scientific Spirit of American Humanism

Epigraph: Carl Sagan, *The Varieties of Scientific Experience: A Personal View of the Search for God* (New York: Penguin, 2006), 31. The statement here is from Sagan's first of nine Gifford Lectures held at the University of Glasgow in 1985.

1. William J. Harnack, "Carl Sagan: Cosmic Evolution vs. the Creationist Myth," *Humanist* 41, no. 4 (1981): 5–6; Jeremy Byman, *Carl Sagan: In Contact with the Cosmos* (Greensboro, NC: Morgan Reynolds, 2001), 8–26; Keay Davidson and Carl Sagan, *Carl Sagan: A Life* (New York: John Wiley & Sons, 1999).

2. Byman, *Carl Sagan*; Davidson and Sagan, *Carl Sagan*.

3. Richard A. Baer first pointed this symbolism out to me in a class at Cornell; see Richard A. Baer, "TV: Carl Sagan's Narrow View of the Cosmos," *Wall Street Journal*, October 24, 1980. See also Thomas M. Lessl, "The Priestly Voice," *Quarterly Journal of Speech* 75 (1989): 183–97; Thomas M. Lessl, "Science and the Sacred Cosmos: The Ideological Rhetoric of Carl Sagan," *Quarterly Journal of Speech* 71 (1985): 175–87.

4. Among the many works that have helped me come to grips with the conservative-liberal divide in American religion, I have found the following especially useful: Ferenc Morton Szasz, *The Divided Mind of Protestant America, 1880–1930* (Tuscaloosa: University of Alabama Press, 1982); Michael Ruse, *The Evolution-Creation Struggle* (Cambridge, MA: Harvard University Press, 2006); Robert Wuthnow, *The Restructuring of American Religion: Society and Faith since World War II* (Princeton, NJ: Princeton University Press, 1988); Mark A. Noll, ed., *Religion and American Politics: From the Colonial Period to the 1980s* (New York: Oxford University Press, 1990); James W. Fraser, *Between Church and State: Religion and Public Education in a Multicultural America* (New York: St. Martin's, 1999); Barry A. Kosmin and Seymour Lachman, *One Nation under God: Religion in Contemporary American Society* (New York: Harmony Books, 1993); Martin E. Marty, *Modern American Religion*, 3 vols. (Chicago: University of Chicago Press, 1986–1999); Martin E. Marty and R. Scott Appleby, eds., *Fundamentalisms Observed*, vol. 1, *The Fundamentalism Project* (Chicago: University of Chicago Press, 1991), as well as essays in other volumes of that series.

5. There is a huge literature on liberalism and progressivism in America, and my work has focused on the development of progressive thought within a religious context; the following works are among those that were especially useful in my thinking: David Ciepley, *Liberalism in the Shadow of Totalitarianism* (Cambridge, MA: Harvard University Press, 2006); James T. Kloppenberg, *Uncertain Victory: Social Democracy and Progressivism in European and American Thought, 1870–1920* (New York: Oxford University Press, 1986); Richard Wightman Fox, "The Culture of Liberal Protestant Progressivism, 1875–1925," *Journal of Interdisciplinary History* 23, no. 3 (1993): 639–60; Gary Gerstle, "The Protean Character of American Liberalism," *American Historical Review* 99, no. 4 (1994): 1043–73; Alan Ryan, *John Dewey and the High Tide of American Liberalism* (New York: W. W. Norton, 1995); Charles D. Cashdollar, *The Transformation of Theology, 1830–1890: Positivism and Protestant Thought in Britain and America* (Princeton, NJ: Princeton University Press, 1989). With regard to science and political thought, I have been aided by the following: George H. Daniels, "The Pure Science Ideal and Democratic Culture," *Science* 156, no. 3783 (1967): 1699–1705; Peter J. Kuznick, *Beyond the Laboratory: Scientists as Political Activists in 1930s America* (Chicago: University of Chicago Press, 1987); Albert E. Moyer, *A Scientist's Voice in American Culture: Simon Newcomb and the Rhetoric of Scientific Method* (Berkeley: University of California Press, 1992); Edward A. Purcell, *The Crisis of Democratic Theory: Scientific Naturalism and the Problem of Value* (Lexington: University Press of Kentucky, 1973); Andrew Jewett, *Science, Democracy, and the American University: From the Civil*

War to the Cold War (New York: Cambridge University Press, 2012); Robert B. Westbrook, *John Dewey and American Democracy* (Ithaca, NY: Cornell University Press, 1993); David A. Hollinger, *Science, Jews, and Secular Culture: Studies in Mid-twentieth-century American Intellectual History* (Princeton, NJ: Princeton University Press, 1996); Thomas L. Haskell, *The Authority of Experts: Studies in History and Theory* (Bloomington: Indiana University Press, 1984).

6. In understanding the larger implications of a naturalistic worldview in the late nineteenth and twentieth centuries, I found the following works especially useful: Edward Arthur White, *Science and Religion in American Thought: The Impact of Naturalism* (Stanford, CA: Stanford University Press, 1952); James Turner, *Without God, Without Creed: The Origins of Unbelief in America* (Baltimore: Johns Hopkins University Press, 1985); Bernard V. Lightman, *The Origins of Agnosticism: Victorian Unbelief and the Limits of Knowledge* (Baltimore: Johns Hopkins University Press, 1987); Purcell, *Crisis of Democratic Theory*; Jewett, *Science, Democracy, and the American University*; Westbrook, *John Dewey and American Democracy*; Paul F. Boller, *American Thought in Transition: The Impact of Evolutionary Naturalism, 1865–1900* (Chicago: Rand McNally, 1969); Merle Eugene Curti, *Human Nature in American Thought: A History* (Madison: University of Wisconsin Press, 1980); Yervant Hovhannes Krikorian, ed., *Naturalism and the Human Spirit* (New York: Columbia University Press, 1944).

7. Several works have helped me sort out the philosophical debates over the century: Westbrook, *John Dewey and American Democracy*; Jewett, *Science, Democracy, and the American University*; George A. Reisch, *How the Cold War Transformed Philosophy of Science: To the Icy Slopes of Logic* (Cambridge: Cambridge University Press, 2005); Gary L. Hardcastle and Alan W. Richardson, *Logical Empiricism in North America*, Minnesota Studies in the Philosophy of Science, vol. 18 (Minneapolis: University of Minnesota Press, 2003); Daniel J. Wilson, *Science, Community, and the Transformation of American Philosophy, 1860–1930* (Chicago: University of Chicago Press, 1990).

8. Roy Wood Sellars, *The Next Step in Religion: An Essay toward the Coming Renaissance* (New York: Macmillan, 1918).

9. Corliss Lamont, "Humanism Is a Way of Life," *Humanist* 2, no. 3 (1942): 106–7.

10. See chaps. 10 and 11 for the full story on this.

11. The term "sacred canopy" was popularized by the sociologist Peter Berger, who helped to characterize religion in its broad function as an enterprise that gave meaning to the cosmos. See Peter L. Berger, *The Sacred Canopy: Elements of a Sociological Theory of Religion* (1967; repr., New York: Anchor, 1990).

12. Mason Olds, *American Religious Humanism*, rev. ed. (Minneapolis: Fellowship of Religious Humanists, 1996); Edwin H. Wilson and Teresa Maciocha, *The Genesis of a Humanist Manifesto* (Amherst, NY: Humanist, 1995); Howard B. Radest, *The Devil and Secular Humanism: The Children of the Enlightenment* (New York: Praeger, 1990); Howard B. Radest, *Toward Common Ground: The Story of the Ethical Societies in the United States* (New York: Ungar, 1969); David Robinson, *The Unitarians and the Universalists* (Westport, CT: Greenwood, 1985); William F. Schulz, *Making the Manifesto: The Birth of Religious Humanism* (Boston: Skinner House Books, 2002). There have been a number of dissertations focusing on the humanist movement over the years. See William F. Schulz, "Making the Manifesto: A History of Early Religious Humanism" (PhD diss., Meadville-Lombard Theological School, 1975); Michael Anthony Schuler, "Religious Humanism in Twentieth-Century American Thought" (PhD diss., Florida State University, 1982); Andrew Robert Krieger, "Structural Ambiguity in a Social Movement Organization: A Case Study of the American Humanist Association" (PhD diss., Georgetown University, 1983); Beverley Margaret Earles, "The Faith Dimension of Humanism" (PhD diss., Victoria University of Wellington, 1989); Stephen Prugh Weldon, "The Humanist Enterprise from John Dewey to Carl Sagan: A Study of Science and Religion in American Culture" (PhD diss., University of Wisconsin–Madison, 1997).

13. Among the many works on the secularist and atheist histories, one finds the following: Fred Whitehead and Verle Muhrer, eds., *Freethought on the American Frontier* (Buffalo, NY: Prometheus Books, 1992); Sidney Warren, *American Freethought, 1860–1914* (New York: Gordian, 1966); Albert Post, *Popular Freethought in America, 1825–1850* (New York: Columbia University Press, 1943); Susan Jacoby, *Freethinkers: A History of American Secularism* (New York: Henry Holt, 2004); Jennifer Michael Hecht, *Doubt: A History: The Great Doubters and Their Legacy of Innovation from Socrates and Jesus to Thomas Jefferson and Emily Dickinson* (New York: Harper-Collins, 2003).

14. The efforts to understand secularism have yielded some extremely rich literature. The debate among sociologists over the nature and even existence of secularization provides fascinating reading and many insights into the modern condition. See Colin Campbell, *Toward a Sociology of Irreligion* (New York: Macmillan, 1971); Steve Bruce, ed., *Religion and Modernization: Sociologists and Historians Debate the Secularization Thesis* (Oxford: Clarendon, 1992); Steve Bruce, *Secularization: In Defence of an Unfashionable Theory* (Oxford: Oxford University Press, 2011). My own work has been more influenced by historical and theoretical treatments that have taken the discussion in quite different directions than the sociologists: Owen Chadwick, *The Secularization of the European Mind in the Nineteenth Century*, The Gifford Lectures in the University of Edinburgh for 1973–1974 (Cambridge: Cambridge University Press, 1975); Frank M. Turner, *Between Science and Religion: The Reaction to Scientific Naturalism in Late Victorian England* (New Haven, CT: Yale University Press, 1974); Michael J. Buckley, *At the Origins of Modern Atheism* (New Haven, CT: Yale University Press, 1987); Turner, *Without God, Without Creed*; Lightman, *Origins of Agnosticism*; Christian Smith, ed., *The Secular Revolution: Power, Interests, and Conflict in the Secularization of American Public Life* (Berkeley: University of California Press, 2003); Charles Taylor, *A Secular Age* (Cambridge, MA: Belknap Press of Harvard University Press, 2007).

15. I have been especially influenced by the following works in attempting to understand the social lives of intellectuals: Robert J. Brym, "The Political Sociology of Intellectuals: A Critique and a Proposal," in *Intellectuals in Liberal Democracies: Political Influence and Social Involvement*, ed. Alain Gagnon (New York: Praeger, 1987), 199–209; Thomas Bender, *Intellect and Public Life: Essays on the Social History of Academic Intellectuals in the United States* (Baltimore: Johns Hopkins University Press, 1997); Thomas Bender, *New York Intellect: A History of Intellectual Life in New York City, from 1750 to the Beginnings of Our Own Time* (New York: Alfred A. Knopf, 1987); Terry A. Cooney, *The Rise of the New York Intellectuals: Partisan Review and Its Circle*, History of American Thought and Culture (Madison: University of Wisconsin Press, 1986); Robert Booth Fowler, *Believing Skeptics: American Political Intellectuals, 1945–1964* (Westport, CT: Greenwood, 1978); Steve Fuller, *The Intellectual* (London: Icon Books, 2005); Alain Gagnon, ed., *Intellectuals in Liberal Democracies: Political Influence and Social Involvement* (New York: Praeger, 1987); Alvin Ward Gouldner, *The Future of Intellectuals and the Rise of the New Class: A Frame of Reference, Theses, Conjectures, Arguments, and an Historical Perspective on the Role of Intellectuals and Intelligentsia in the International Class Contest of the Modern Era* (New York: Continuum, 1979); Jeffrey Hart, *The Making of the American Conservative Mind: National Review and Its Times*, ed. (Wilmington, DE: Intercollegiate Studies Institute, 2006); Richard Hofstadter, *Anti-intellectualism in American Life* (New York: Knopf, 1963); Neil Jumonville, *Critical Crossings: The New York Intellectuals in Postwar America* (Berkeley: University of California Press, 1991); Michael James Lacey, ed., *Religion and Twentieth-Century American Intellectual Life*, Woodrow Wilson Center Series (Washington, DC: Woodrow Wilson International Center for Scholars, 1989); George H. Nash, *The Conservative Intellectual Movement in America since 1945*, 2nd ed. (Wilmington, DE: Intercollegiate Studies Institute, 2006); Harold Perkin, *The Third Revolution: Professional Elites in the Modern World* (London: Routledge, 1996); Richard A. Posner, *Public Intellectuals: A Study of Decline, with a New Preface and Epi-*

logue (Cambridge, MA: Harvard University Press, 2003); Steven C. Rockefeller, *John Dewey: Religious Faith and Democratic Humanism* (New York: Columbia University Press, 1991); George M. Marsden, *The Soul of the American University: From Protestant Establishment to Established Nonbelief* (New York: Oxford University Press, 1994); Shmuel Feiner, "Seductive Science and the Emergence of the Secular Jewish Intellectual," *Science in Context* 15 (2002): 121–35. The following works helped me think about the nature of science and its authority in American culture: Daniel Patrick Thurs, *Science Talk: Changing Notions of Science in American Popular Culture* (New Brunswick, NJ: Rutgers University Press, 2007); Haskell, *Authority of Experts*; John P. Diggins, *The Promise of Pragmatism: Modernism and the Crisis of Knowledge and Authority* (Chicago: University of Chicago Press, 1994); Lessl, "Priestly Voice"; David J. Hess, *Science in the New Age: The Paranormal, Its Defenders and Debunkers, and American Culture*, Science and Literature (Madison: University of Wisconsin Press, 1993).

16. Thomas Jefferson, "No. 622. Duty on Books: Communicated to the House of Representatives, December 10, 1821. (17th Congress. 1st Session)," in *American State Papers: Documents, Legislative and Executive of the Congress of the United States . . .*, pt. 3, vol. 3 (Washington, DC: Gales & Seaton, 1834), 681, https://books.google.com/books?id=2I4bAQAAMAAJ&pg=PA681&lpg.

Chapter 1 · *Liberal Christianity and the Frontiers of American Belief*

Epigraph: Thomas Paine, *The Age of Reason: Being an Investigation of True and Fabulous Theology* (Barrois, 1794), 63, Eighteenth Century Collections Online: Text Creation Partnership, http://name.umdl.umich.edu/004781203.0001.000.

1. The following discussion relies on many sources discussing freethought and deism in American history. See Paine, *Age of Reason*; Susan Jacoby, *Freethinkers: A History of American Secularism* (New York: Henry Holt, 2004); Stephen P. Weldon, "Deism," in *The History of Science and Religion in the Western Tradition: An Encyclopedia*, ed. Gary B. Ferngren, Edward J. Larson, and Darrel W. Amundsen (New York: Garland, 2000), 158–60; Peter Byrne, *Natural Religion and the Nature of Religion: The Legacy of Deism* (New York: Routledge, 1989).

2. Paine, *Age of Reason*; Jacoby, *Freethinkers*; Weldon, "Deism"; Byrne, *Natural Religion*.

3. Paine, *Age of Reason*, 55.

4. Paine, *Age of Reason*, 107.

5. Paine, *Age of Reason*, 12.

6. Paine, *Age of Reason*, 3.

7. Jacoby, *Freethinkers*, 66–103; Fred Whitehead and Verle Muhrer, eds., *Freethought on the American Frontier* (Buffalo, NY: Prometheus Books, 1992); Sidney Warren, *American Freethought, 1860–1914* (New York: Gordian, 1966); Albert Post, *Popular Freethought in America, 1825–1850* (New York: Columbia University Press, 1943); Leigh Eric Schmidt, *Village Atheists: How America's Unbelievers Made Their Way in a Godly Nation* (Princeton, NJ: Princeton University Press, 2016).

8. Jonathan A. Glickstein, "Garrison, William Lloyd," in *A Companion to American Thought*, ed. Richard Wightman Fox and James T. Kloppenberg (Oxford: Blackwell, 1995), 263–65.

9. Quoted in Annie Laurie Gaylor, *Women without Superstition: "No Gods—No Masters": The Collected Writings of Women Freethinkers of the Nineteenth and Twentieth Centuries* (Madison, WI: Freedom From Religion Foundation, 1997), 142.

10. Jacoby, *Freethinkers*, 198–205; Ellen Carol Dubois, "Stanton, Elizabeth Cady," in *A Companion to American Thought*, ed. Richard Wightman Fox and James T. Kloppenberg (Oxford: Blackwell, 1995), 651–53.

11. Kimberly A. Hamlin, *From Eve to Evolution: Darwin, Science, and Women's Rights in Gilded Age America* (Chicago: University of Chicago Press, 2014), 46–48.

12. David Robinson, *The Unitarians and the Universalists* (Westport, CT: Greenwood, 1985);

David E. Bumbaugh, *Unitarian Universalism: A Narrative History* (Chicago: Meadville Lombard, 2000); Earl Morse Wilbur, *A History of Unitarianism: In Transylvania, England, and America* (Cambridge, MA: Harvard University Press, 1952); Conrad Wright, *The Unitarian Controversy: Essays on American Unitarian History* (Boston: Skinner House Books, 1994).

13. Robinson, *Unitarians and the Universalists*, chap. 2; William Ellery Channing, *Unitarian Christianity and Other Essays* (New York: Liberal Arts Press, 1957), 9; see also the introduction by Irving H. Bartlett, x–xi.

14. Quoted in Channing, *Unitarian Christianity*, 8.

15. Robinson, *Unitarians and the Universalists*, chaps. 3–4.

16. Barry A. Kosmin and Seymour Lachman, *One Nation under God: Religion in Contemporary American Society* (New York: Harmony Books, 1993), esp. 258–62; Roger Finke and Rodney Stark, *The Churching of America, 1776–1990: Winners and Losers in Our Religious Economy* (New Brunswick, NJ: Rutgers University Press, 1992), 83.

17. A classic of Unitarian history is Wilbur, *History of Unitarianism*; see also Robinson, *Unitarians and the Universalists*, chap. 2; Channing, *Unitarian Christianity*, xi.

18. Robinson, *Unitarians and the Universalists*, chap. 2; Conrad Wright, "Institutional Reconstruction in the Unitarian Controversy," in *Unitarian Controversy*, 83–110; Conrad Wright, "Unitarian Beginnings in Western Massachusetts," in *Unitarian Controversy*, 137–54.

19. Conrad Wright, "Ministers, Churches, and the Boston Elite, 1791–1815," in *Unitarian Controversy*, 37–58. Even at the time Unitarianism was seen by its conservative foes to be only a religion suited to the "educated, rich and fashionable" (Channing quoted in Wright, *Unitarian Controversy*, 37–38).

20. Gary J. Dorrien, *The Making of American Liberal Theology: Imagining Progressive Religion, 1805–1900* (Louisville, KY: Westminster John Knox, 2001), 58–110; Robinson, *Unitarians and the Universalists*, 75–79.

21. Robinson, *Unitarians and the Universalists*, chaps. 8–9; Mason Olds, *American Religious Humanism*, rev. ed. (Minneapolis: Fellowship of Religious Humanists, 1996), 30–33; Wilbur, *History of Unitarianism*, 467–73.

22. Wilbur, *History of Unitarianism*, 472.

23. Many books talk about these events. See Robinson, *Unitarians and the Universalists*, chap. 9; Stow Persons, *Free Religion, an American Faith* (New Haven, CT: Yale University Press, 1947); Sydney E. Ahlstrom and Robert Bruce Mullin, *The Scientific Theist: A Life of Francis Ellingwood Abbot* (Macon, GA: Mercer University Press, 1987); "Appendix 2: Meetings of Unbelievers," in *The Encyclopedia of Unbelief*, ed. Gordon Stein (Buffalo, NY: Prometheus Books, 1985), 757–77; Whitehead and Muhrer, *Freethought on the American Frontier*; Warren, *American Freethought*; Post, *Popular Freethought in America*.

24. Quoted in Ahlstrom and Mullin, *Scientific Theist*, 65.

25. Ahlstrom and Mullin, *Scientific Theist*, 42–60, 84–87.

26. Robinson, *Unitarians and the Universalists*, 122; see also 116–22.

27. The quotations in this and the next paragraph are all taken from F. E. Abbot, "The God of Science," *Index* 3, no. 113 (February 24, 1872): 57–60.

28. John Fiske, one of America's most prominent philosophers, was developing an evolutionary philosophy at Harvard at about this time. See Edward Arthur White, *Science and Religion in American Thought: The Impact of Naturalism* (Stanford, CA: Stanford University Press, 1952), chap. 3; Charles H. Lyttle, *Freedom Moves West: A History of the Western Unitarian Conference, 1852–1952* (Boston: Beacon, 1952); Warren, *American Freethought*, esp. 56, 65–72. Warren is another good source for understanding radical religion in an evolutionary perspective in the late nineteenth century.

29. This point simply confirms the findings of historians like James Turner who have written about the origins of unbelief in theology.

30. D. G. Hart, "Nineteenth-Century Biblical Criticism," in *The History of Science and Religion in the Western Tradition: An Encyclopedia*, ed. Gary B. Ferngren (New York: Garland, 2000), 79–82; Dorrien, *Making of American Liberal Theology*; Claude Welch, *Protestant Thought in the Nineteenth Century* (New Haven, CT: Yale University Press, 1972). Richard Tison's dissertation "The Lords of Creation" includes a discussion of how this older philological approach to biblical studies was used by evangelicals and traditionalists; Richard Perry Tison, "Lords of Creation: American Scriptural Geology and the Lord Brothers' Assault on 'Intellectual Atheism'" (PhD diss., University of Oklahoma, 2008).

31. Ferenc Morton Szasz, *The Divided Mind of Protestant America, 1880–1930* (Tuscaloosa: University of Alabama Press, 1982); William R. Hutchison, *The Modernist Impulse in American Protestantism* (Durham, DC: Duke University Press, 1992); Susan Curtis, *A Consuming Faith: The Social Gospel and Modern American Culture* (Baltimore: Johns Hopkins University Press, 1991).

32. Martin E. Marty, *Varieties of Unbelief* (New York: Holt, Rinehart & Winston, 1964), 40–41; James P. Wind, *The Bible and the University: The Messianic Vision of William Rainey Harper* (Atlanta: Scholars, 1987), 59–66.

33. Wind, *Bible and the University*, 60–61.

34. "Guide to the Baptist Theological Union and Baptist Union Theological Seminary Records 1866–1922" (University of Chicago Library, 2006), https://www.lib.uchicago.edu/ead/rlg/ICU.SPCL.BUTS.pdf; Charles Harvey Arnold, *Near the Edge of Battle: A Short History of the Divinity School and the Chicago School of Theology, 1866–1966* (Chicago: Divinity School Association, University of Chicago, 1966), 11–21, 26–27; Bernard E. Meland, "Reflections on the Early Chicago School of Modernism," *American Journal of Theology and Philosophy* 5, no. 1 (1984): 6; Hutchison, *Modernist Impulse*, chap. 4.

35. Hart, "Nineteenth-Century Biblical Criticism"; Dorrien, *Making of American Liberal Theology*; Welch, *Protestant Thought*.

36. Shailer Mathews, *New Faith for Old: An Autobiography* (New York: Macmillan, 1936); J. Gordon Melton, *Religious Leaders of America* (Detroit: Gale Research, 1991); Meland, "Reflections," 5–8.

37. Shailer Mathews, *The Faith of Modernism* (New York: Macmillan, 1924), 108.

38. For the bigger story of this change, see James Turner, *Without God, Without Creed: The Origins of Unbelief in America* (Baltimore: Johns Hopkins University Press, 1985); Albert E. Moyer, *A Scientist's Voice in American Culture: Simon Newcomb and the Rhetoric of Scientific Method* (Berkeley: University of California Press, 1992); David A. Hollinger, "Inquiry and Uplift: Late Nineteenth-Century American Academics and the Moral Efficacy of Scientific Practice," in *The Authority of Experts: Studies in History and Theory*, ed. Thomas L. Haskell (Bloomington: Indiana University Press, 1984), 142–56.

39. Mathews, *Faith of Modernism*, 108, 115. See also Jon Roberts, *Darwinism and the Divine in America: Protestant Intellectuals and Organic Evolution, 1859–1900* (Madison: University of Wisconsin Press, 1988).

40. Ronald L. Numbers, *The Creationists: From Scientific Creationism to Intelligent Design*, exp. ed. (Cambridge, MA: Harvard University Press, 2006), 15–32.

41. Nicolaas A. Rupke, "Geology and Paleontology from 1700 to 1900," in *The History of Science and Religion in the Western Tradition: An Encyclopedia*, ed. Gary B. Ferngren, Edward J. Larson, and Darrel W. Amundsen (New York: Garland, 2000), 401–8.

42. A good discussion of the Social Gospel can be found in Martin E. Marty, *Modern American Religion*, vol. 1, *The Irony of It All: 1893–1919* (Chicago: University of Chicago Press, 1986), 283–96; Mark A. Noll, *A History of Christianity in the United States and Canada* (Grand Rapids, MI: William B. Eerdmans, 1992), 304–7. Richard Wightman Fox in "Liberal Protestantism," in

A Companion to American Thought, ed. Richard Wightman Fox and James T. Kloppenberg (Oxford: Blackwell, 1995), 394–97, explains the ties between liberalism and the Social Gospel.

43. Curtis, *Consuming Faith*; Hutchison, *Modernist Impulse*, chap. 5.

44. Arnold, *Near the Edge of Battle*, 11–21, 26–27; Meland, "Reflections," 6; Hutchison, *Modernist Impulse*, chap. 4.

45. Bernard V. Lightman, *The Origins of Agnosticism: Victorian Unbelief and the Limits of Knowledge* (Baltimore: Johns Hopkins University Press, 1987); Turner, *Without God, Without Creed*; Michael J. Buckley, *At the Origins of Modern Atheism* (New Haven, CT: Yale University Press, 1987); Charles Taylor, *A Secular Age* (Cambridge, MA: Belknap Press of Harvard University Press, 2007); Donald H. Meyer, "American Intellectuals and the Victorian Crisis of Faith," *American Quarterly* 27 (1976): 585–603; Richard J. Helmstadter and Bernard Lightman, *Victorian Faith in Crisis: Essays on Continuity and Change in Nineteenth-Century Religious Belief* (Basingstoke: Palgrave Macmillan, 1990).

46. Lightman, *Origins of Agnosticism*; Turner, *Without God, Without Creed*.

47. Matthew Stanley, *Huxley's Church and Maxwell's Demon: From Theistic Science to Naturalistic Science* (Chicago: University of Chicago Press, 2015).

Chapter 2 · The Birth of Religious Humanism

Epigraph: John H. Dietrich, "Who Are These Agnostic Humanists?," in *The Humanist Pulpit* (Minneapolis: First Unitarian Society, 1927), 1:6–7.

1. Wesley Mason Olds, "Three Pioneers of Religious Humanism: A Study of 'Religion Without God' in the Thought of John H. Dietrich, Curtis W. Reese, and Charles Francis Potter" (PhD diss., Brown University, 1973), 53–57; Carleton Winston, *This Circle of Earth: The Story of John H. Dietrich* (New York: G. P. Putnam's Sons, 1942), 38–82.

2. The biographical material about Dietrich's life comes from Winston, *This Circle of Earth*; Mason Olds, *Religious Humanism in America: Dietrich, Reese, and Potter* (Washington, DC: University Press of America, 1977); William F. Schulz, "Making the Manifesto: A History of Early Religious Humanism" (PhD diss., Meadville-Lombard Theological School, 1975); William F. Schulz, "Making the Manifesto," *Religious Humanism* 17 (1983): 88–97, 102. A good discussion of Dietrich's place in Unitarian history can be found in David Robinson, *The Unitarians and the Universalists* (Westport, CT: Greenwood, 1985), 143–57.

3. Schulz, "Making the Manifesto" (1975); Robinson, *Unitarians and the Universalists*, 143–57; Edwin H. Wilson, "The Origins of Modern Humanism," *Humanist* 51, no. 1 (1991): 9–11, 28.

4. Olds, *Religious Humanism in America*, 120–23.

5. Schulz, "Making the Manifesto" (1975), 120; Olds, *Religious Humanism in America*, 123–24.

6. Robinson, *Unitarians and the Universalists*, 142–57; Schulz, "Making the Manifesto" (1975); Mason Olds, *American Religious Humanism*, rev. ed. (Minneapolis: Fellowship of Religious Humanists, 2006), 53–150.

7. Ferenc Morton Szasz, *The Divided Mind of Protestant America, 1880–1930* (Tuscaloosa: University of Alabama Press, 1982), 1–29; Gary J. Dorrien, *The Making of American Liberal Theology: Imagining Progressive Religion, 1805–1900* (Louisville, KY: Westminster John Knox, 2001), 335–412.

8. Szasz, *Divided Mind*, chap. 2, esp. 26–27.

9. George Burman Foster, *The Finality of the Christian Religion* (Chicago: University of Chicago Press, 1906), 9–22; Olds, *American Religious Humanism*, 99–107, esp. 100.

10. Olds, *American Religious Humanism*, 33–34.

11. John H. Dietrich, "Who Are These Unitarians?," in *The Humanist Pulpit* (Minneapolis: First Unitarian Society, 1927), 1:17–18.

12. Dietrich, "Who Are These Agnostic Humanists?," 1:12.

13. Roy Wood Sellars, foreword to *The Next Step in Religion: An Essay toward the Coming Renaissance* (New York: Macmillan, 1918). Schulz in "Making the Manifesto" (1975), 336, calls Sellars "the man in whose thought early religious humanism is most sophisticatedly capsulized."

14. J. Ronald Engel, "American Religious Humanism (1916–1936) and Its Leading Ideas Functioning as Metaphors of Ultimate Reality and Meaning," *Ultimate Reality and Meaning* 8 (1985): 262; Olds, *American Religious Humanism*, 33–34.

15. On Potter, see Olds, *American Religious Humanism*, 125–33.

16. Alan Seaburg, "Curtis Reese," *Dictionary of Unitarian and Universalist Biography*, August 15, 2005, http://www25.uua.org/uuhs/duub/articles/curtiswillifordreese.html; Olds, *American Religious Humanism*, 99–105.

17. Olds, *American Religious Humanism*, 53–59.

18. Dietrich, "Who Are These Agnostic Humanists?," 17.

19. Dietrich, "Who Are These Agnostic Humanists?," 6–7.

20. John H. Dietrich, "What I Believe," in *The Humanist Pulpit* (Minneapolis: First Unitarian Society, 1934), 7:161–76.

21. Creighton Peden, "The Chicago School (1906–1926) in American Religious Thought and Its Contribution to a Better Understanding of Ultimate Reality and Meaning of Human Experience," *Ultimate Reality and Meaning* 6, no. 1 (1983): 51–79; Hutchison, *Modernist Impulse*.

22. Shailer Mathews, *The Faith of Modernism* (New York: Macmillan, 1924), 108.

23. John H. Dietrich, "The Religion of Experience," *Christian Register*, March 13, 1919, 241.

24. Dietrich, "Religion of Experience," 241.

25. Curtis W. Reese, "Do You Believe What He Believes?," *Christian Register*, September 9, 1920, 884; Olds, *Religious Humanism in America*, 35–36.

26. John H. Dietrich, "Who Are These Fundamentalists?," in *The Humanist Pulpit* (Minneapolis: First Unitarian Society, 1927), 1:12.

27. Dietrich, "Who Are These Agnostic Humanists?," 5.

28. Dietrich, "What I Believe."

29. Schulz, "Making the Manifesto" (1975); Olds, *American Religious Humanism*, 99–124.

30. Olds, *Religious Humanism in America*, 53–57; Winston, *This Circle of Earth*, 38–82.

31. Olds, *Religious Humanism in America*, 35–36; Reese, "Do You Believe What He Believes?," 883; Robinson, *Unitarians and the Universalists*, 147.

32. Olds, *Religious Humanism in America*, 35–36.

33. Robinson, *Unitarians and the Universalists*, 144–45.

34. Olds, *Religious Humanism in America*, 42.

35. Robinson, *Unitarians and the Universalists*, 147; Reese, "Do You Believe What He Believes?," 883–84.

Chapter 3 · *Manifesto for an Age of Science*

Epigraph: "A Humanist Manifesto," *New Humanist* 6 (May/June 1933): 1–2.

1. "Humanist Manifesto," 1–2.

2. "Prospectus of the Meadville Theological School," *Meadville Theological School Quarterly Bulletin* 15, no. 4 (June 1921): 1–28, https://books.google.com/books?id=NppVAAAAYAAJ&lpg=RA2-PA51&ots=r5xn387cH8&dq=prospectus&pg=RA2-PA51#v=onepage&q&f=false. See also Celian Ufford, "AHA! A Root!," *Humanist* 22, no. 6 (1962): 178–81.

3. Edwin H. Wilson, interview by Beverley Margaret Earles, March 6, 1987, tape 1, American Humanist Association Archives; Ufford, "AHA! A Root!"

4. "Prospectus 1921–1922," 16–18.

5. Alan Seaburg, "Curtis Reese," *Dictionary of Unitarian and Universalist Biography*, August 15, 2005, http://www25.uua.org/uuhs/duub/articles/curtiswillifordreese.html.

6. "Prospectus 1921–1922," 10.

7. Edwin H. Wilson, "The Origins of Modern Humanism," *Humanist* 51, no. 1 (1991): 9–11, 28.

8. H. G. C., "[Untitled]," *New Humanist* 1 (April 1928): 1; H. G. Creel, "The Program of the Humanist Fellowship," *New Humanist* 1 (May 1928): 1–2; Beverley Margaret Earles, "The Faith Dimension of Humanism" (PhD diss., Victoria University of Wellington, 1989), 194–95; Wilson, "Origins of Modern Humanism," 11. On Creel, see David T. Roy, "Herrlee Glessner Creel (1905–1994)," *Journal of Asian Studies* 53, no. 4 (November 1994): 1356–57. On the editor, see Philip Bock and Harry Basehart, "Stanley S. Newman (1905–1984)," *American Anthropologist* 88 (1986): 151–53.

9. Charles Harvey Arnold, "Pioneer Humanist: Albert Eustace Haydon," *Religious Humanism* 14 (Winter 1980): 15–20; Creighton Peden, "A. Eustace Haydon's Humanistic Religion," *Religious Humanism* 16 (Spring 1982): 71; "Dr. Haydon Made Honorary Member," *New Humanist* 1 (May 1928): 2; Wilson, interview by Earles, tapes 1 and 3; Creighton Peden, *A Good Life in a World Made Good: Albert Eustace Haydon, 1880–1975* (New York: Peter Lang, 2006), 55–56.

10. Peden, *Good Life*, 55–56.

11. A. Eustace Haydon, *The Quest of the Ages* (New York: Harper & Brothers, 1929), ix.

12. Haydon, *Quest of the Ages*, 118.

13. H. G. C., "[Untitled]," 1.

14. "Preamble to the Constitution of the Humanist Fellowship," *New Humanist* 1, no. 2 (May 1928): 4.

15. For an understanding of this mode of thinking about science, see Ronald L. Numbers and Daniel P. Thurs, "Science, Pseudoscience, and Science Falsely So-Called," in *Wrestling with Nature: From Omens to Science*, ed. Peter Harrison, Ronald L. Numbers, and Michael H. Shank (Chicago: University of Chicago Press, 2011), 281–306.

16. Edwin H. Wilson and Teresa Maciocha, *The Genesis of a Humanist Manifesto* (Amherst, NY: Humanist, 1995), 16–22.

17. Sellars had recently requested that Bragg write an article on humanism in the Soviet Union (never published) because Bragg had been to the Soviet Union in 1932. He was planning to lead a tour there in 1933 that promised to reveal the effectiveness of the Communist program. See *New Humanist* 6 (May/June 1933): inside front cover; William F. Schulz, "Making the Manifesto," *Religious Humanism* 17 (1983): 92n.

18. Schulz, "Making the Manifesto"; William F. Schulz, "Making the Manifesto: A History of Early Religious Humanism" (PhD diss., Meadville-Lombard Theological School, 1975); Wilson, "Origins of Modern Humanism." The list of signers of the manifesto is as follows: J. A. C. Fagginger Auer (Professor of Church History and Theology, Harvard and Tufts), E. Burdette Backus (Unitarian Minister), Harry Elmer Barnes (Editorial Assistant, Scripps-Howard Newspapers), L. M. Birkhead (Unitarian Minister), Raymond B. Bragg, Edwin Arthur Burtt (Professor of Philosophy, Cornell University), Ernest Caldecott (Unitarian Minister), A. J. Carlson (Professor of Physiology, University of Chicago), John Dewey, Albert C. Dieffenbach (Former Editor, *Christian Register*), John H. Dietrich, Bernard Fantus (Professor of Therapeutics, University of Illinois), William Floyd (Editor, *Arbitrator*), F. H. Hankins (Professor of Economics and Sociology, Smith College), A. Eustace Haydon, Llewellyn Jones (Author), Robert Morss Lovett (Editor, *New Republic*, and Professor of English, University of Chicago), Harold P. Marley (Unitarian Minister), R. Lester Mondale (Unitarian Minister), Charles Francis Potter, John Herman Randall Jr. (Professor of Philosophy, Columbia University), Curtis W. Reese, Oliver L. Reiser (Professor of Philosophy, University of Pittsburgh), Roy Wood Sellars, Clinton Lee Scott (Universalist Minister), Maynard Shipley (Author), W. Frank Swift (Ethical Society Leader), V. T. Thayer (Director, Ethical Culture Schools), Eldred C. Vanderlaan (Leader, Free Fellowship, Berkeley), Joseph Walker (Attorney), Jacob J. Weinstein (Rabbi), Frank S. C. Wicks (Unitarian Minister), David Rhys Williams (Unitarian Minister), and Edwin H. Wilson.

19. Most were Unitarians, but the list included Ethical Culture leaders, a rabbi, and other liberals.

20. Max Carl Otto et al., "Comment on the Humanist Manifesto," *New Humanist* 6 (1933): 29–33.

21. "Humanist Manifesto," 2.

22. "Preamble to the Constitution of the Humanist Fellowship."

23. "Humanist Manifesto," 2.

24. Roy Wood Sellars, *Evolutionary Naturalism* (Chicago: Open Court, 1922).

25. "Humanist Manifesto," 3.

26. "Humanist Manifesto," 2.

27. "Humanist Manifesto," 4.

28. A concise discussion of various types of naturalism can be found in Henry Samuel Levinson, "Naturalism," in *A Companion to American Thought*, ed. Richard Wightman Fox and James T. Kloppenberg (Oxford: Blackwell, 1995), 480–83.

29. Sellars, *Evolutionary Naturalism*.

30. "Humanist Manifesto," 2.

31. "Humanist Manifesto," 2.

32. "Humanist Manifesto," 4.

33. Roy Wood Sellars, *The Next Step in Democracy* (New York: Macmillan, 1916).

34. "[Endorsement]," *New Humanist* 5, no. 5 (1932): 31.

35. "Humanist Manifesto," 4.

36. Max Carl Otto, *Natural Laws and Human Hopes* (New York: Henry Holt, 1926), 89–93. See also Gary Gerstle, "The Protean Character of American Liberalism, " *American Historical Review* 99, no. 4 (1994): 1043–73, which discusses the liberal intellectuals' lack of surprise at the Depression, owing to their previous views on the society.

37. On the history of Universalism, see David Robinson, *The Unitarians and the Universalists* (Westport, CT: Greenwood, 1985).

38. Naomi Cohen, *Jews in Christian America: The Pursuit of Religious Equality* (New York: Oxford University Press, 1992); Naomi Wiener Cohen, *Encounter with Emancipation: The German Jews in the United States, 1830–1914* (Skokie, IL: Varda Books, 2001).

39. Horace Leland Friess, *Felix Adler and Ethical Culture: Memories and Studies*, ed. Fannia Weingartner (New York: Columbia University Press, 1981); Edward L. Ericson, "Ethical Culture since Felix Adler: An Afterword," in Friess, *Felix Adler and Ethical Culture*, 257–63; Howard B. Radest, *Toward Common Ground: The Story of the Ethical Societies in the United States* (New York: Ungar, 1969).

Chapter 4 · Philosophers in the Pulpit

Epigraph: John Dewey, *A Common Faith* (New Haven, CT: Yale University Press, 1934), 57.

1. David A. Hollinger, "Inquiry and Uplift: Late Nineteenth-Century American Academics and the Moral Efficacy of Scientific Practice," in *The Authority of Experts: Studies in History and Theory*, ed. Thomas L. Haskell (Bloomington: Indiana University Press, 1984), 142–56; Susan Curtis, *A Consuming Faith: The Social Gospel and Modern American Culture* (Baltimore: Johns Hopkins University Press, 1991); Dorothy Ross, "American Social Science and the Idea of Progress," in *The Authority of Experts: Studies in History and Theory*, ed. Thomas L. Haskell (Bloomington: Indiana University Press, 1984), 157–75; Dorothy Ross, "Professionalism and the Transformation of American Social Thought," *Journal of Economic History* 38, no. 2 (1978): 494.

2. Daniel J. Wilson, *Science, Community, and the Transformation of American Philosophy, 1860–1930* (Chicago: University of Chicago Press, 1990), esp. 1–11; Bruce Kuklick, *The Rise of American Philosophy, Cambridge, Massachusetts, 1860–1930* (New Haven, CT: Yale University

Press, 1977); Thomas Bender, *New York Intellect: A History of Intellectual Life in New York City, from 1750 to the Beginnings of Our Own Time* (New York: Alfred A. Knopf, 1987).

3. James Campbell, *A Thoughtful Profession: The Early Years of the American Philosophical Association* (Chicago: Open Court, 2006), 10–13.

4. James Turner, *Language, Religion, Knowledge: Past and Present* (Notre Dame, IN: University of Notre Dame Press, 2003). This ideology was expounded in a famous and influential 1917 article by Max Weber, "Science as a Vocation"; see its English translation in *From Max Weber: Essays in Sociology* (New York: Oxford University Press, 1946), 129–56.

5. Wilson, *Science, Community, and the Transformation of American Philosophy*, xv–xxvii, 565–67, 581–89. See also Campbell, *Thoughtful Profession*; Kuklick, *Rise of American Philosophy*; Bender, *New York Intellect*.

6. Jacques Barzun, ed., *A History of the Faculty of Philosophy, Columbia University* (New York: Columbia University Press, 1957); Horace Leland Friess, "The Department of Religion," in *A History of the Faculty of Philosophy, Columbia University*, ed. Jacques Barzun (New York: Columbia University Press, 1957), 146–67; John Herman Randall, introduction to *A History of the Faculty of Philosophy, Columbia University*, ed. Jacques Barzun (New York: Columbia University Press, 1957), 3–57; John Herman Randall, "The Department of Philosophy," in *A History of the Faculty of Philosophy, Columbia University*, ed. Jacques Barzun (New York: Columbia University Press, 1957), 102–45.

7. Quoted in George B. Pegram, John M. Nelson, and Leslie C. Dunn, "Frederick James Eugene Woodbridge," 4, Randall Papers, Columbia University Archives, box 19.

8. John Herman Randall Jr., "The Department of Philosophy," in *A History of the Faculty of Philosophy, Columbia University*, ed. Jacques Barzun (New York: Columbia University Press, 1957), 116–17; Pegram, Nelson, and Dunn, "Frederick James Eugene Woodbridge," 6.

9. Horace L. Friess, *Felix Adler and Ethical Culture: Memories and Studies*, ed. Fannia Weingartner (New York: Columbia University Press, 1981), 36–38; Howard B. Radest, "Ethical Culture and Humanism: A Cautionary Tale," *Religious Humanism* 16 (Spring 1982): 59-70.

10. Friess, *Felix Adler and Ethical Culture*; Radest, "Ethical Culture and Humanism," 36–44.

11. Friess, *Felix Adler and Ethical Culture*, 174–82; James Franklin Hornback, "The Philosophic Sources and Sanctions of the Founders of Ethical Culture" (PhD diss., Columbia University, 1983).

12. Randall, "Department of Philosophy," 122; Friess, *Felix Adler and Ethical Culture*, 214–16; Herbert W. Schneider to John H. Randall Jr., February 24, 1937, Randall Papers, box 2.

13. Kuklick, *Rise of American Philosophy*, 408–11; Bender, *New York Intellect*, 296–300; Friess, *Felix Adler and Ethical Culture*, 1.

14. Corliss Lamont, *Yes to Life: Memoirs of Corliss Lamont*, rev. ed. (New York City: Crossroad/Continuum, 1991), 4–5.

15. Lamont, *Yes to Life*, 4–5.

16. Radest, "Ethical Culture and Humanism"; see also Howard B. Radest, *Toward Common Ground: The Story of the Ethical Societies in the United States* (New York: Ungar, 1969).

17. This discussion of Dewey relies primarily on the following works: Steven C. Rockefeller, *John Dewey: Religious Faith and Democratic Humanism* (New York: Columbia University Press, 1991); Robert B. Westbrook, *John Dewey and American Democracy* (Ithaca, NY: Cornell University Press, 1993); Alan Ryan, *John Dewey and the High Tide of American Liberalism*, 1st ed. (New York: W. W. Norton, 1995). A good discussion of Dewey's thought can be found in John P. Diggins, *The Promise of Pragmatism: Modernism and the Crisis of Knowledge and Authority* (Chicago: University of Chicago Press, 1994), 205–49.

18. Rockefeller, *John Dewey*; Westbrook, *John Dewey and American Democracy*; Robert B. Westbrook, "Dewey, John," in *A Companion to American Thought*, ed. Richard Wightman Fox and James T. Kloppenberg (Oxford: Blackwell, 1995), 177–79.

19. See, e.g., Rockefeller, *John Dewey*, 36–38.

20. Rockefeller, *John Dewey*, 125–68.

21. Quoted in Rockefeller, *John Dewey*, 68. Ryan dismisses this story, told by Max Eastman, because of certain discrepancies and Eastman's tendency to embellish, but Rockefeller presents a cogent argument for taking the story seriously.

22. Quoted in Rockefeller, *John Dewey*, 530. (From Phil. 4:7.)

23. John Dewey, *Reconstruction in Philosophy* (Boston: Beacon, 1957).

24. John Dewey, *The Quest for Certainty: A Study of the Relation of Knowledge and Action* (New York: G. P. Putnam's Sons, 1929), 14; see also 72.

25. Dewey, *Reconstruction in Philosophy*, 212–13.

26. Dewey, *Common Faith*, 1–10.

27. Paul Oskar Kristeller, "John Herman Randall, Jr.: In Memoriam: Randall and the History of Philosophy," *Journal of the History of Ideas* 42, no. 3 (1981): 489–501.

28. John Herman Randall and John Herman Randall Jr., *Religion and the Modern World* (New York: Frederick A. Stokes, 1929), 185.

29. This aphorism of Randall's was widely reprinted in many daily newspapers throughout the country. Here, quoted in Eva G. Taylor, "By the Way," *Los Angeles Times*, July 7, 1929, 16 (from ProQuest Historical Newspapers).

30. Randall and Randall Jr., *Religion and the Modern World*, 176.

31. Kristeller, "John Herman Randall, Jr."

32. Randall and Randall Jr., *Religion and the Modern World*, 51.

33. From miscellaneous brochures in the Randall Papers, Columbia University.

34. Among the Columbia philosophy faculty members were Herbert Schneider, Horace Friess, Irwin Edman, and James Gutmann. In addition, there were other faculty members at Columbia with ties to the department, like V. T. Thayer at Columbia Teacher's College (a protégé of Dewey and Max Otto at Wisconsin), as well as committed humanist graduate students like Corliss Lamont, all of whom formed a humanist nexus centered on the department.

35. The biographical material here comes from two sources: W. Preston Warren, *Roy Wood Sellars* (Boston: Twayne, 1975), 19–32; and Warren's biographical sketch in Roy Wood Sellars, *Neglected Alternatives: Critical Essays by Roy Wood Sellars*, ed. W. Preston Warren (Lewisburg, PA: Bucknell University Press, 1973).

36. On his socialism, see the typescript of an untitled and undated article about Sellars by Grenell in the Judson Grenell Papers, University of Michigan, Ann Arbor. See also Warren, *Roy Wood Sellars*, 93.

37. Sellars, *Neglected Alternatives*, 20; William F. Schulz, "Making the Manifesto: A History of Early Religious Humanism" (PhD diss., Meadville-Lombard Theological School, 1975), 132ff.; see also William F. Schulz, "Making the Manifesto," *Religious Humanism* 17 (1983): 88–97, 102.

38. Roy Wood Sellars, *Religion Coming of Age* (New York: Macmillan, 1928).

39. Edwin H. Wilson and Teresa Maciocha, *The Genesis of a Humanist Manifesto* (Amherst, NY: Humanist, 1995), 23–30.

40. Roy Wood Sellars, foreword to the *The Next Step in Religion: An Essay toward the Coming Renaissance* (New York: Macmillan, 1918), n.p.

41. William Preston Warren, *Roy Wood Sellars* (Boston: Twayne, 1975).

42. Sellars, *Religion Coming of Age*, 51 (emphasis in the original).

43. Roy Wood Sellars, *Evolutionary Naturalism* (Chicago: Open Court, 1922), chap. 1.

44. Sellars, *Evolutionary Naturalism*, 19.

45. Sellars, *Evolutionary Naturalism*, preface and chap. 1.

46. Sellars, *Evolutionary Naturalism*, 20.

47. Sellars, *Evolutionary Naturalism*, 317–18.

48. Warren, *Roy Wood Sellars*, 46; Sellars, *Religion Coming of Age*, 51.

49. Otto wrote at least three autobiographical pieces in which he recounts much of his earlier life. Max Carl Otto, "Living Down the Hyphen," *Dial*, 1919, 401–5; Max Carl Otto, "Untitled Autobiography 'Dear L.G.W.' [Lewis G. Westgate]," October 17, 1945, M. C. Otto Papers, box 12, Wisconsin State Historical Society, a thirty-page "story" begun in October of 1945; Max Carl Otto, "He Came to Himself," M. C. Otto Papers, Wisconsin State Historical Society, a typescript probably written after 1950.

50. Otto, "Dear L.G.W.," 26.

51. I want to thank David A. Sandmire for sharing his unpublished manuscript with me, "Max Otto, the University Pastors, and the Holy 'Chapple': The Philosophical, Social and Political Implications of Evolution in the Early 20th Century" (December 1993). See also G. C. Sellery, "Max Otto: Unitarian Humanist, 1876–1968: A Biographical Note," Notable American Unitarians, http://www.harvardsquarelibrary.org/unitarians/otto.html. One of Otto's best arguments to this effect is in Max C. Otto, *The Human Enterprise: An Attempt to Relate Philosophy to Daily Life* (New York: F. S. Crofts, 1940), 1–29.

52. Max C. Otto to John Dewey, October 22, 1928, Otto Papers, Wisconsin State Historical Society.

53. The correspondence between Otto and Dewey as found in Otto's archive tells much of this story. In his letters Otto provides complements, asks small favors, and engages Dewey in philosophical topics, all in a tone of sincere reverence. See esp. Max C. Otto to John Dewey, October 22, 1928, Otto Papers, box 1, Wisconsin State Historical Society; Max C. Otto to John Dewey, July 8, 1941, Otto Papers, Wisconsin State Historical Society; Max C. Otto to John Dewey, [December 1941?], Otto Papers, Wisconsin State Historical Society.

54. Otto, *Human Enterprise*, 1.

55. Otto, *Human Enterprise*, 2.

Chapter 5 · Humanists at War

Epigraph: Sidney Hook, "Theological Tom-Tom and Metaphysical Bagpipe," *Humanist* 2 (1942): 96.

1. David Ciepley, *Liberalism in the Shadow of Totalitarianism* (Cambridge, MA: Harvard University Press, 2006); Benjamin L. Alpers, *Dictators, Democracy, and American Public Culture* (Chapel Hill: University of North Carolina Press, 2003). These works provide an understanding of how fascism and totalitarianism were viewed in America during the years leading up to World War II.

2. Jerome Nathanson, preface to *The Authoritarian Attempt to Capture Education*, vol. 2, *Papers from the 2d Conference on the Scientific Spirit and Democratic Faith*, ed. Jerome Nathanson (New York: King's Crown, 1945).

3. Two especially insightful sources dealing with many of the individuals and events related here are Edward A. Purcell, *The Crisis of Democratic Theory: Scientific Naturalism and the Problem of Value* (Lexington: University Press of Kentucky, 1973); and Andrew Jewett, *Science, Democracy, and the American University: From the Civil War to the Cold War* (New York: Cambridge University Press, 2012).

4. Edwin H. Wilson, "Introducing the H.P.A. News Letter," *Humanist Press Association News Letter* 1, no. 1 (February 1936): 1; [unknown], "[Untitled]," *Humanist Press Association Bulletin*, no. 1 (January 18, 1935). The "popularize and expand" language came from Edwin H. Wilson, "Dear Friend," October 1934, Meadville Theological School Library; and the "study and extend" quotation came from "By Laws of the Humanist Press Association, Inc.," *Humanist Press Association News Letter* 1, nos. 1–2 (February 1936): 3.

5. See esp. Fred W. Beuttler, "Organizing an American Conscience: The Conference on Science, Philosophy and Religion, 1940–1968" (PhD diss., University of Chicago, 1995); Jewett, *Science, Democracy, and the American University*, 320–24; Purcell, *Crisis of Democratic Theory*, 218–31.

6. Quoted in Fred W. Beuttler, "For the World at Large: Intergroup Activities at the Jewish Theological Seminary," in *Tradition Renewed: A History of the Jewish Theological Seminary*, vol. 2, *Beyond the Academy*, ed. Jack Wertheimer (New York: Jewish Theological Seminary of America, 1997), 8.

7. Quoted in Beuttler, "For the World at Large," 669. For interesting material on the Conference on Science, Philosophy and Religion, see JoAnn Palmeri, "An Astronomer beyond the Observatory: Harlow Shapley as Prophet of Science" (PhD diss., University of Oklahoma, 2000), 90–104. She also discusses the wider concerns about the fate of civilization by people involved in the conference.

8. Beuttler, "For the World at Large," esp. 681.

9. Van Wyck Brooks, "Conference on Science, Philosophy and Religion in Their Relation to the Democratic Way of Life," in *Science, Philosophy and Religion: A Symposium* (New York: Conference on Science, Philosophy and Religion in Their Relation to the Democratic Way of Life, 1941), 1:1–2.

10. Mortimer J. Adler, "God and the Professors," in *Science, Philosophy and Religion*, 1:128.

11. George A. Reisch, *How the Cold War Transformed Philosophy of Science: To the Icy Slopes of Logic* (Cambridge: Cambridge University Press, 2005).

12. Adler, "God and the Professors," 128.

13. Adler, "God and the Professors," 136–37.

14. Beuttler, "Organizing an American Conscience," 4 (on Finkelstein's attempt to stop Hook), 161–75 (on a detailed discussion of the intellectual exchange); Sidney Hook, *Out of Step: An Unquiet Life in the 20th Century* (New York: Harper & Row, 1987), 337.

15. This is Beuttler's main point throughout his dissertation. He addresses the issue directly in Beuttler, "Organizing an American Conscience," 111–12.

16. Beuttler, "Organizing an American Conscience," x–xi. Buettler goes further, however, claiming that the main thrust of the conference was a call to toleration and indicating that he thinks that Dewey and his followers held a quite narrow and uncompromising rejection of supernaturalism and orthodoxy.

17. Reinhold Niebuhr, *Moral Man and Immoral Society: A Study in Ethics and Politics* (New York: Scribner's Sons, 1960); see also Richard Wightman Fox, *Reinhold Niebuhr: A Biography* (New York: Pantheon Books, 1985).

18. For a general discussion of American Catholicism, see William M. Halsey, *The Survival of American Innocence: Catholicism in an Era of Disillusionment, 1920–1940* (Notre Dame, IN: University of Notre Dame Press, 1980); Philip Gleason, *Keeping the Faith: American Catholicism, Past and Present* (Notre Dame, IN: University of Notre Dame Press, 1987).

19. Purcell, *Crisis of Democratic Theory*, 68–169.

20. Gleason, *Keeping the Faith*, esp. chaps. 7 and 8; see also Martin E. Marty, *Modern American Religion*, vol. 3, *Under God, Indivisible: 1941–1960* (Chicago: University of Chicago Press, 1996), 89–95, 157–79; Mark A. Noll, *A History of Christianity in the United States and Canada* (Grand Rapids, MI: William B. Eerdmans, 1992), 433–34.

21. Mark Silk, "Notes on the Judeo-Christian Tradition in America," *American Quarterly* 36, no. 1 (1984): 65–85. See also the introduction to Charles L. Cohen and Ronald L. Numbers, eds., *Gods in America: Religious Pluralism in the United States* (New York: Oxford University Press, 2013), 6–7.

22. George M. Marsden, *The Soul of the American University: From Protestant Establishment to Established Nonbelief* (New York: Oxford University Press, 1994), 381; Beuttler, "Organizing an American Conscience." Indeed, Beuttler's main thesis is that the conference planners' goal was an extension of the optimistic pluralist agenda to bring together many dissonant voices and ideas to build a common democratic culture. Insofar as the religious agenda goes, he seems

to see it as a typical effort of mainline American religionists, but he also cautions about seeing the conference as overly focused on its religious mission.

23. Max C. Otto to Sidney Hook, November 24, 1942, Otto Papers, Wisconsin State Historical Society.

24. Robert B. Westbrook, "Hook, Sidney," in *A Companion to American Thought*, ed. Richard Wightman Fox and James T. Kloppenberg (Oxford: Blackwell, 1995), 312–13.

25. Wikipedia, s.v. "Sidney Hook," last modified April 29, 2018, https://en.wikipedia.org/w/index.php?title=Sidney_Hook&oldid=838746506; David Sidorsky, "Sidney Hook," in *The Stanford Encyclopedia of Philosophy*, ed. Edward N. Zalta (Stanford University, 2015), https://plato.stanford.edu/archives/win2015/entries/sidney-hook/.

26. Hook, *Out of Step*; Westbrook, "Hook, Sidney"; John P. Diggins, *The Promise of Pragmatism: Modernism and the Crisis of Knowledge and Authority* (Chicago: University of Chicago Press, 1994), 394–96.

27. Sidney Hook, "The New Medievalism," *New Republic* 103 (October 28, 1940): 602–6.

28. Hook, "Theological Tom-Tom," 96.

29. Gilbert Murray, *Five Stages of Greek Religion* (New York: Clarendon, 1925). This book was a revision of his original work entitled *Four Stages of Greek Religion*, published in 1912. Francis West, *Gilbert Murray: A Life* (New York: St. Martin's, 1984), 139, notes Murray's popularity and the readership of this book in 1912, and she points out Murray's desire to reach ever-broader audiences.

30. Sidney Hook, "The New Failure of Nerve," *Partisan Review* 10, no. 1 (1943): 2.

31. Hook, "New Failure of Nerve," 2–3.

32. Hook, "New Failure of Nerve," 2–4, 17.

33. J. Douglas Brown et al., "The Spiritual Basis of Democracy," in *Science, Philosophy and Religion: A Symposium* (New York: Science, Philosophy and Religion in Their Relation to the Democratic Way of Life, 1942), 2:256–57; Max C. Otto, "Take It Away!," *Humanist* 2, no. 1 (Spring 1942): 13.

34. Erwin R. Goodenough, "Scientific Living," *Humanist* 2, no. 1 (Spring 1942): 8–10.

35. "Minutes of the Dinner Meeting to Consider the Holding of a Conference of Educators, Scientists and Religionists, Hotel Ten Park Avenue," January 25, 1943, Wilson Papers, box 7, Special Collections, Morris Library, Southern Illinois University, Carbondale, Illinois.

36. [Edwin H. Wilson?], "Letter to Mrs. Alfred Alschuler," June 8, 1943, box 7, Special Collections, Morris Library, Southern Illinois University, Carbondale, Illinois.

37. Marsden, *Soul of the American University*; David A. Hollinger, "Science as a Weapon in Kulturkaempfe in the United States during and after World War II," in *Science, Jews, and Secular Culture: Studies in Mid-twentieth-century American Intellectual History* (Princeton, NJ: Princeton University Press, 1996), 155–74; "Conference on the Scientific Spirit and Democratic Faith [Program]," May 29, 1943, Wilson Papers, box 7, Morris Library, Southern Illinois University.

38. Eduard C. Lindeman, introduction to *Papers from the Conference on the Scientific Spirit and Democratic Faith* (New York: King's Crown, 1944), 1:ix.

39. Diggins, *Promise of Pragmatism*, 305–21; Hook, *Out of Step*, 335–55.

40. Robert Maynard Hutchins, *The Higher Learning in America* (New Haven, CT: Yale University Press, 1936), http://www.ditext.com/hutchins/hutcho.html.

41. [Edwin H. Wilson?], "Form Letter Unaddressed," May 24, 1944, Wilson Papers, box 39, Special Collections, Morris Library, Southern Illinois University, Carbondale, Illinois.

42. Arthur E. Murphy, "Tradition and Traditionalists," in Nathanson, *Authoritarian Attempt to Capture Education*, 2:13–25, 14 (quote).

43. "Conference on the Scientific Spirit and Democratic Faith [Program]," May 27, 1944, New York Society for Ethical Culture Archives.

44. Nathanson, *Authoritarian Attempt to Capture Education.*

45. Jerome Nathanson, ed., *Science for Democracy*, vol. 3, *Papers from the Conference on the Scientific Spirit and Democratic Faith* (New York: King's Crown, 1946).

46. AHA By-Laws 1943–1944, American Humanist Association Archives.

47. Harry Overstreet, "When Religion Becomes Mature," in *Science, Philosophy and Religion*, 184ff.; Hudson Hoagland, "Some Comments on Science and Faith," in *Science, Philosophy and Religion*, vol. 2, 33ff.

48. E. Burdette Backus, "Science and Democracy," *Humanist* 3, no. 2 (Summer 1943): 79.

49. Archie J. Bahm, "Humanism and Sect Membership," *Humanist* 1, no. 2 (Summer 1941): 54–56.

50. Quoted in James H. Leuba, "Humanism Is Religious," *Humanist* 2, no. 3 (Autumn 1942): 110. See also Archie J. Bahm, "Humanism Is More Than a Philosophy," *Humanist* 2, no. 3 (1942), 108.

51. Corliss Lamont, "Humanism Is a Way of Life," *Humanist* 2, no. 3 (Autumn 1942): 107.

52. Corliss Lamont to John Herman Randall Jr., April 10, 1934, 1–2, 4–5, Columbia University Library, J. H. Randall Jr. Papers, box 2.

53. Quote from Lamont to Randall Jr., 2 (emphasis in the original); see also Lamont, "Humanism Is a Way of Life," 107.

54. John H. Randall, "Letter to E. H. Wilson," January 3, 1934, Wilson Papers, box 34, Morris Library, Southern Illinois University.

55. Wilson tells a humorous story about his efforts to try to keep peace between the humanist organization and the Unitarians when a typo occurred in one of his humanist brochures in which the word "supplant" was mistakenly used instead of the word "supplement." Edwin H. Wilson, interview by Beverley Margaret Earles, March 6, 1987, tape 1, American Humanist Association Archives.

Chapter 6 · Scientists on the World Stage

Epigraph: "The Amsterdam Declaration 1952," *IHEU* (blog), https://humanists.international /policy/amsterdam-declaration-1952/.

1. There has been much research on science and the Cold War. See, e.g., John Krige, "Atoms for Peace, Scientific Internationalism, and Scientific Intelligence," *Osiris* 21, no. 1 (January 1, 2006): 161–81; Jamie Cohen-Cole, *The Open Mind: Cold War Politics and the Sciences of Human Nature* (Chicago: University of Chicago Press, 2014); Jessica Wang, "Scientists and the Problem of the Public in Cold War America, 1945–1960," *Osiris* 17 (2002): 323–47; David K. Hecht, "The Atomic Hero: Robert Oppenheimer and the Making of Scientific Icons in the Early Cold War," *Technology and Culture* 49, no. 4 (December 17, 2008): 943–66.

2. "Resolutions of the American Humanist Association [Typescript]," n.d., 16–21, American Humanist Association Archives.

3. Edwin H. Wilson, "News and Notes," *Humanist* 11, no. 2 (1951): 96.

4. Julian Huxley, "An Interview with Julian Huxley," *Humanist* 11, no. 4 (1951): 174.

5. J. van Praag, "How Dutch Humanism Supersedes Free Thought," *Humanist* 10, no. 3 (1950): 92-94.

6. *The International Humanist and Ethical Union and Its Member Organizations* (Utrecht: IHEU, 1959).

7. These papers were by Unitarian minister Curtis Reese, Columbia University philosopher Horace Friess, University of London social scientist Barbara Wootton, and Dutch philosopher Libbe van der Wal. (The transcripts of these papers, presented at the First International Congress on Humanism and Ethical Culture, Amsterdam, August 21–26, 1952, are in the American Humanist Association Archives.) "Summary of IHEU Congresses 1952–1992," *International Humanist*, July 1992, 34, states that Jacob Bronowski gave a paper at this meeting as well, al-

though I have no record of it. He also is supposed to have given a paper at the 1957 international meeting.

8. Curtis W. Reese, "Scientific Religion and Human Economy," 2, 6, and Barbara Wootton, "Science and Democracy," 1–9 (First International Congress on Humanism and Ethical Culture, Amsterdam, 1952) (transcripts in the American Humanist Association Archives).

9. Horace Friess, "Needs of a Greater Humanity" (First International Congress on Humanism and Ethical Culture, Amsterdam, 1952), 4 (transcript in the American Humanist Association Archives).

10. Eric Nguyen, email message to author, September 28, 2011.

11. "Descriptions of AHA Awards," n.d., American Humanist Association Archives.

12. Mildred McCallister and Lloyd Kumley, *The Humanist of the Year Book* (Amherst, NY: Humanist, 1992).

13. "Guide to the Bernard Fantus Collection 1874–2009," http://www.lib.uchicago.edu/e/scrc/findingaids/view.php?eadid=ICU.SPCL.FANTUSB; "Dr. B. M. Fantus, 'Blood Bank's' Originator, Dies," *Chicago Daily Tribune*, April 15, 1940.

14. "Guide to the Bernard Fantus Collection"; "Dr. B. M. Fantus"; Edwin H. Wilson, "Dr. Bernard Fantus: A Life of Kindness Guided by Intelligence: Funeral Address" (1940). Only one year before he launched the new AHA journal the *Humanist*, Wilson wrote a long obituary for Fantus and promoted his ideas for his revisions to the Humanist Manifesto.

15. "Anton J. Carlson, Physiology," The University of Chicago Faculty A Centennial View, University of Chicago Centennial Catalogues, https://www.lib.uchicago.edu/projects/centcat/fac/facch20_01.html.

16. Anton J. Carlson, "Biology and the Future of Man," *Humanist* 5, no. 1 (1945): 16–20.

17. "Maurice B. Visscher," *Dictionary of Unitarian & Universalist Biography*, http://www25.uua.org/uuhs/duub/articles/mauricevisscher.html.

18. Maurice B. Visscher, "Science for Humanity," *Humanist* 11, no. 1 (1951): 43. As early as 1941, the *Humanist* published articles advocating a strong scientific internationalist position. The philosopher Oliver Reiser, for example, proposed a worldwide New Deal that would promote global humanism. See Oliver L. Reiser, "Scientific Humanism: A Total Totalitarianism," *Humanist* 1, no. 3 (1941): 96–98.

19. Paul G. Roofe, "Herrick, Charles Judson," in *Complete Dictionary of Scientific Biography* (2008), www.encyclopedia.com; George W. Bartelmez, "Charles Judson Herrick, October 6, 1868–January 29, 1960," *National Academy of Sciences Biographical Memoir* (1973), 77–107.

20. C. Judson Herrick, *The Evolution of Human Nature* (Austin: University of Texas Press, 1956), 1, http://www.questia.com/read/101994230.

21. Herrick, *Evolution of Human Nature*, 3.

22. On scientific internationalism before and after the Second World War, see Elisabeth Crawford, "The Universe of International Science, 1880-1939," in *Solomon's House Revisited: The Organization and Institutionalization of Science*, ed. Tore Frängsmyr, Nobel Symposium (Canton, MA: Science History Publications, 1990), 251-69; Elisabeth Crawford, Terry Shinn, and Sverker Sörlin, "The Nationalization and Denationalization of the Sciences: An Introductory Essay," in *Denationalizing Science: The Contexts of International Scientific Practice*, ed. Elisabeth Crawford, Terry Shinn, and Sverker Sörlin (Boston: Kluwer, 1993), 1-42. See also John Beatty, "Scientific Collaboration, Internationalism, and Diplomacy: The Case of the Atomic Bomb Casualty Commission," *Journal of the History of Biology* 26 (1993): 205-31.

23. Sir Richard Gregory, "Science as International Ethics," in *The Scientific Spirit and Democratic Faith* (New York: King's Crown, 1944), 52.

24. Gregory, "Science as International Ethics," 52.

25. This outlook seems extreme these days, but one must place Gregory's comments in context. He was writing before the revelations of the crimes committed by Nazi doctors and

before the atomic era. The moral dilemmas created by these scientific and technological devices helped to change people's minds after the war. Before that, however, someone like Gregory could tout the virtues of scientific progress without sounding entirely divorced from reality.

26. Brock Chisholm, "The Future of Psychiatry and the Human Race," *Humanist* 7, no. 4 (1948): 157–62; see also G. B. Chisholm, *Psychiatry of Enduring Peace and Social Progress* (1946).

27. M. Cardwell, "Dr. Brock Chisholm: Canada's Most Famously Articulate Angry Man," *Medical Post* 34 (April 7, 1998): 13; see also John Farley, *Brock Chisholm, the World Health Organization, and the Cold War* (Vancouver: UBC Press, 2008), 27–47.

28. Farley, *Brock Chisholm*, 41–44, 199–202.

29. Chisholm, "Future of Psychiatry"; see also Chisholm, *Psychiatry of Enduring Peace*.

30. Quoted in Julian Huxley, *Memories* (London: Allen & Unwin, 1970), 153.

31. Most of the strict biographical information here is taken from Huxley's autobiography *Memories* and from C. Kenneth Waters, "Introduction: Revising Our Picture of Julian Huxley," in *Julian Huxley: Biologist and Statesman of Science*, ed. C. Kenneth Waters and Albert Van Helden (Houston: Rice University Press, 1992), 1–27.

32. Huxley, *Memories*, 55.

33. "Constitution of the United Nations Educational, Scientific and Cultural Organization," in *Basic Texts* (United Nations Educational, Scientific and Cultural Organization, 2004), 7, http://unesdoc.unesco.org/images/0013/001337/133729e.pdf.

34. Julian Huxley, *UNESCO: Its Purpose and Its Philosophy* (Washington, DC: Public Affairs Press, 1947), 37.

35. Huxley, *UNESCO*, 37.

36. Huxley, *UNESCO*, 6–7.

37. Huxley, *UNESCO*, 7.

38. Huxley, *UNESCO*, 11–13.

39. Lloyd Morain, "Reminiscences of IHEU's Founding from the USA," *International Humanism*, July 1992, 6; James Marshall, review of *UNESCO: Its Purpose and Its Philosophy*, by Julian Huxley, *Saturday Review of Literature* 30 (July 5, 1947): 16–17.

40. Biographical material about Muller is from Elof Axel Carlson, *Genes, Radiation, and Society: The Life and Work of H. J. Muller* (Ithaca, NY: Cornell University Press, 1981); Daniel J. Kevles, *In the Name of Eugenics: Genetics and the Uses of Human Heredity* (New York: Knopf, 1985). For an interesting unpublished essay on Muller, see Scott Mullen, "Mutation and Eugenics in the Thought of H. J. Muller" (master's thesis, University of Wisconsin–Madison, 1989).

41. Carlson, *Genes, Radiation, and Society*, 175–92.

42. Joseph Rotblat, *Pugwash—The First Ten Years: History of the Conferences of Science and World Affairs* (New York: Humanities, 1967); Peter H. Denton, *The ABC of Armageddon: Bertrand Russell on Science, Religion, and the Next War, 1919–1938* (Albany: State University of New York Press, 2001).

43. Rotblat, *Pugwash*, 12.

44. In addition, Einstein had sensationally argued that people needed to abandon the idea of a personal deity in his speech at the Conference on Science, Philosophy and Religion. This was soon after the Adler-Hook confrontation at the same meeting.

45. Edwin H. Wilson, "The Editor's Interview: The Other Cyrus Eaton," *Humanist* 16, no. 2 (1956): 86.

46. Wilson, "Editor's Interview," 87.

47. "Notes from the Annual Meeting," *Free Mind* 6 (March 1958): 2; Edwin H. Wilson, "About Humanists," *Free Mind* 6 (June 1958): 7.

48. Eugene Rabinowitch, "Science and Humanities in Education," *Humanist* 18 (September/October 1958): 279, 285.

49. H. J. Muller to Cyrus Eaton, November 15, 1957, Muller Papers, Lilly Library.

Chapter 7 · Eugenics and the Question of Race

Epigraph: H. J. Muller, "Man's Place in Living Nature: Part 2," *Humanist* 17, no. 2 (1957): 101–2.

1. John P. Diggins, *The Promise of Pragmatism: Modernism and the Crisis of Knowledge and Authority* (Chicago: University of Chicago Press, 1994), 39; Andrew Jewett, *Science, Democracy, and the American University: From the Civil War to the Cold War* (New York: Cambridge University Press, 2012), 169; Roy Wood Sellars, *Evolutionary Naturalism* (Chicago: Open Court, 1922).

2. See chaps. 3 and 4. Sellars's views on levels are spelled out most explicitly in Sellars, *Evolutionary Naturalism*. See also Max C. Otto, *The Human Enterprise: An Attempt to Relate Philosophy to Daily Life* (New York: F. S. Crofts, 1940).

3. Bernard Fantus, "A Humanist Affirmation" (1940), University of Chicago Library, Fantus Papers, printed in Edwin H. Wilson, "Dr. Bernard Fantus: A Life of Kindness Guided by Intelligence: Funeral Address" (1940).

4. Nadine Weidman, "Psychobiology, Progressivism, and the Anti-Progressive Tradition," *Journal of the History of Biology* 29, no. 2 (1996): esp. 271–76.

5. A. J. Carlson, review of *Out of the Night: A Biologist's View of the Future*, by H. J. Muller, *American Journal of Sociology* 42, no. 1 (1936): 134–35.

6. David Loth, "Planned Parenthood and the Modern Inquisition," *Humanist* 7, no. 3 (Autumn 1947): 64–68; Clarence Senior, "Women, Democracy, and Birth Control," *Humanist* 12, no. 5 (1952): 221–24; Maurice B. Visscher, "Science for Humanity," *Humanist* 12, no. 5 (1952): 220; Mildred Gilman, "Margaret Sanger, Birth Control, and World Peace," *Humanist* 13, no. 1 (1953): 9–12; Edwin H. Wilson, "Editorial Interview: Margaret Sanger," *Humanist* 16, no. 5 (1956): 234–36; "Resolutions on Planned Parenthood," *Humanist* 19, no. 4 (1959): 244.

7. "Margaret Sanger Honored as 'Humanist of the Year,'" *Free Mind* 5, no. 3 (March 1957): 1; Wilson, "Editorial Interview"; Gilman, "Margaret Sanger"; R. Marie Griffith, "Crossing the Catholic Divide: Gender, Sexuality, and Historiography," in *Catholics in the American Century: Recasting Narratives of U.S. History*, ed. R. Scott Appleby and Kathleen Sprows Cummings (Ithaca, NY: Cornell University Press, 2012), 81–108.

8. Jennifer Michael Hecht, *Doubt: A History: The Great Doubters and Their Legacy of Innovation from Socrates and Jesus to Thomas Jefferson and Emily Dickinson* (New York: HarperCollins, 2003), 440–41; Susan Jacoby, *Freethinkers: A History of American Secularism* (New York: Henry Holt, 2004), 235–38, 259–61.

9. Griffith, "Crossing the Catholic Divide"; Kathleen Tobin-Schlesinger, "Population and Power: The Religious Debate over Contraception, 1916–1936. (Volumes I and II)" (PhD diss., University of Chicago, 1994), http://search.proquest.com/docview/304136294/abstract/79BC142 E38B24C71PQ/1; Paul Blanshard, *American Freedom and Catholic Power*, 2nd ed. (Boston: Beacon, 1958).

10. Senior, "Women, Democracy, and Birth Control," 221.

11. "Resolutions of the American Humanist Association [Typescript]," n.d., 7–9, American Humanist Association Archives.

12. "Resolutions of the American Humanist Association [Typescript]," 8–9.

13. Alison Bashford, *Global Population: History, Geopolitics, and Life on Earth* (New York: Columbia University Press, 2014); Edmund Ramsden, "Eugenics from the New Deal to the Great Society: Genetics, Demography and Population Quality," *Studies in History and Philosophy of Science Part C: Studies in History and Philosophy of Biological and Biomedical Sciences* 39, no. 4 (December 1, 2008): 391–406, https://doi.org/10.1016/j.shpsc.2008.09.005; Tobin-Schlesinger, "Population and Power"; Griffith, "Crossing the Catholic Divide"; Susanne Klausen and Alison Bashford, "Fertility Control: Eugenics, Neo-Malthusianism, and Feminism," in *The*

Oxford Handbook of the History of Eugenics, ed. Alison Bashford and Philippa Levine (Oxford: Oxford University Press, 2010), 98–115; Nancy Ordover, *American Eugenics: Race, Queer Anatomy, and the Science of Nationalism* (Minneapolis: University of Minnesota Press, 2003).

14. "Margaret Sanger Honored," 2.

15. One of the few pieces that deals directly with the reasons why the Dewey circle of philosophy never embraced eugenic ideas is Timothy McCune, "Dewey's Dilemma: Eugenics, Education, and the Art of Living," *Pluralist* 7, no. 3 (2012): esp. 98, https://doi.org/10.5406/pluralist.7.3.0096.

16. H. J. Muller, *Out of the Night: A Biologist's View of the Future* (New York: Vanguard, 1935).

17. H. J. Muller, "Life," *Humanist* 15, no. 6 (1955): 258.

18. Muller, *Out of the Night*; Muller, "Life"; H. J. Muller, "Man's Place in Living Nature: Part 1," *Humanist* 17, no. 1 (1957): 3–13; Muller, "Man's Place in Living Nature: Part 2," 93–102.

19. Muller, *Out of the Night*, 125, 111; Muller, "Life," 259–61.

20. Quoted in Mark Swetlitz, "Julian Huxley and the End of Evolution," *Journal of the History of Biology* 28 (1995): 181.

21. Julian Huxley, *UNESCO: Its Purpose and Its Philosophy* (Washington, DC: Public Affairs Press, 1947), 12.

22. Huxley, *UNESCO*, 18–21, 31.

23. Huxley, *UNESCO*, 31–32.

24. "Ethical Forum: IQ and Race," *Humanist* 32, no. 1 (1972): 4–18.

25. Raymond E. Fancher, *The Intelligence Men: Makers of the IQ Controversy* (New York: W. W. Norton, 1987), 197–201.

26. H. J. Eysenck, "Maverick Humanist," *Humanist* 29, no. 4 (July/August 1969): 19–21; quotation on 19.

27. P. E. Vernon et al., "Testing Negro Intelligence: Comments on Eysenck," *Humanist* 30, no. 1 (January/February 1970): 35.

28. William Shockley, "The Apple-of-God's-Eye Obsession," *Humanist* 32, no. 1 (January/February 1972): 16–17; "Ethical Forum: IQ and Race."

29. Students for a Democratic Society, "Readers Forum: More on IQ and Race," *Humanist* 32, no. 5 (September/October 1972): 45–46.

30. Joseph P. DeMarco, "Readers Forum: I.Q.," *Humanist* 32, no. 4 (July/August 1972): 44–45.

31. "Ethical Forum: IQ and Race," 4.

32. Eysenck, "Maverick Humanist."

33. Sidney Hook, "Democracy and Genetic Variation," *Humanist* 32, no. 2 (1972): 7.

34. "Lewis A. McGee Joins AHA Staff," *Free Mind* 1, no. 2 (September 1953): 1; "Finding Aid for Lewis A. McGee Papers," Meadville Lombard Theological School Archives, https://www.meadville.edu/files/resources/lewis-a-mcgee-papers-56.pdf.

35. See, e.g., "Apartheid in S. Africa Protested," *Free Mind* 6, no. 1 (January 1958): 5; "Humanist Student Defies Florida University Segregation Ruling," *Free Mind* 5, no. 7 (August 1957): 3.

36. "Resolutions of the American Humanist Association [Typescript]."

37. Dan L. Garrett Jr., "Synanon: The Communiversity," *Humanist* 25, no. 5 (September/October 1965): 184–89; Tolbert McCarroll to Stephen Weldon, "Subject: First," February 15, 2000; Tolbert McCarroll to Stephen Weldon, "Subject: More," February 15, 2000; Arthur Jackson, interview by Stephen Weldon, February 2, 1997; Arthur Jackson, interview by Stephen Weldon, January 22, 2000; Lloyd Morain, interview by Stephen Weldon, January 21, 2000.

38. Richard Herbert and Erwin Pollack, "Inner-City Education: A Humanistic Alternative," *Humanist* 32, no. 2 (1972): 22–23.

39. See "The Messenger," Spartacus Educational, http://spartacus-educational.com/USAC

messenger.htm; Mildred McCallister and Lloyd Kumley, *The Humanist of the Year Book* (Amherst, NY: Humanist, 1992), 18.

Chapter 8 · *Inside the Humanist Counterculture*

Epigraph: Priscilla Robertson, "What Shall I Tell My Children?," *Harper's Magazine*, August 1952, 25.

1. "Priscilla Robertson, 1910–1989," TheHumanist.com, April 15, 2009, https://thehumanist .com/magazine/may-june-2009/commentary/priscilla-robertson-1910-1989; Edwin H. Wilson, "Introducing Our New Editor," *Humanist* 16, no. 3 (1956): 107–8.

2. Wilson, "Introducing Our New Editor."

3. Wikipedia, s.v. "Henry Preserved Smith," last modified May 31, 2018, https://en.wikipedia .org/w/index.php?title=Henry_Preserved_Smith&oldid=843857911.

4. Robertson, "What Shall I Tell My Children?," 21.

5. Priscilla Robertson, "The Humanist Looks Ahead," *Humanist* 16, no. 3 (1956): 108; Priscilla Robertson, "On a Scientific Standard of Personal Ethics," *Humanist* 16, no. 5 (1956): 217–23; Priscilla Robertson, "On Getting Values out of Science," *Humanist* 16, no. 4 (1956): 169–74.

6. Robertson, "What Shall I Tell My Children?," 22–23.

7. Robertson, "On Getting Values out of Science"; Robertson, "On a Scientific Standard of Personal Ethics."

8. Priscilla Robertson, "Thinking Out Loud . . . ," *Humanist* 18, no. 6 (1958): 322.

9. There are two folders devoted to the Robertson controversy in the George Axtelle Papers at the Special Collections, Morris Library, Southern Illinois University, Carbondale, Illinois. See box 12, folder 5 and box 15, folder 21.

10. McCarroll was executive director for six years and editor of the *Humanist* for three.

11. Todd Gitlin, *The Sixties: Years of Hope, Days of Rage* (New York: Bantam Books, 1993); James Livingston, *The World Turned Inside Out: American Thought and Culture at the End of the 20th Century* (Lanham, MD: Rowman & Littlefield, 2011).

12. *Humanist* 25, no. 1 (January/February 1965): front cover, and inside front cover.

13. "Chart of 'AHA Membership Data, All Categories, 1952–1963,'" n.d., American Humanist Association Archives.

14. Cliff Collins, "The Lawyer as Monk," *Oregon State Bar Bulletin*, January 2008, www.osbar .org/publications/bulletin/08jan/profiles.html; Arthur Jackson, interview by Stephen Weldon, January 22, 2000.

15. Tolbert McCarroll to Stephen Weldon, "Subject: First," February 15, 2000.

16. Collins, "Lawyer as Monk"; Edwin H. Wilson, interview by Beverley Margaret Earles, March 6, 1987, tape 2, American Humanist Association Archives; Jackson, interview, January 22, 2000.

17. Special Issue, *Humanist* 25, no. 2S (Spring 1965).

18. Tolbert McCarroll, "When Is Sex a Crime?," *Humanist* 25, no. 2S (Spring 1965): 86–87; Lester A. Kirkendall, "Sex Education: A Reappraisal," *Humanist* 25, no. 2S (Spring 1965): 77–82.

19. Charles S. Binderman, "The Playboy Paradox," *Humanist* 25, no. 6 (November/December 1965): 262–65; Ethem M. Nash, "More on the Sex Issue," *Humanist* 25, no. 4 (July/August 1965): 174.

20. Rudolf Dreikurs, "Humanist Youth," *Humanist* 25, no. 1 (January/February 1965): 32.

21. Howard Lischeron, "Conference Report: AHA in 1974," *Humanist* 25, no. 1 (January/ February 1965): 23.

22. Ward Tabler, "Lessons from Berkeley," *Humanist* 25, no. 2 (March/April 1965): 53.

23. Leslie A. Fiedler, "In Defense of Youth," *Humanist* 27 (1967): 117–19, 134–36. The educator Leslie A. Fiedler issued a strong defense of young people and their liberated spirit, Dreikurs

continued to preach the message that we needed to listen to the young people, and sometimes popular counterculture figures such as Joan Baez and Bob Dylan were discussed.

24. Harry Elmer Barnes, letter to the editor, *Humanist* 26 (1966): 64–65.

25. Tolbert McCarroll, "The Undead," *Humanist* 25, no. 3 (May/June 1965): 12.

26. Tolbert McCarroll and Hal Lenke, "A Humanist Attitude," *Humanist* 26 (March/April 1966): 46–50; Tolbert McCarroll, "Religions of the Future," *Humanist* 26 (November/December 1966): 190–92.

27. "Humanist Counseling," *Humanist* 23, no. 3 (1963): 96–98.

28. McCarroll to Weldon, "Subject: First"; Tolbert McCarroll to Stephen Weldon, "Subject: More," February 15, 2000.

29. "News from Humanist House," *Humanist* 25, no. 5 (September/October 1965): 214.

30. Roy José DeCarvalho, *The Founders of Humanistic Psychology* (New York: Praeger, 1991); Roy José DeCarvalho, "A History of the 'Third Force' in Psychology," *Journal of Humanistic Psychology* 30, no. 4 (September 1, 1990): 22–44, https://doi.org/10.1177/002216789003000403.

31. Mildred McCallister and Lloyd Kumley, introduction to *The Humanist of the Year Book* (Amherst, NY: Humanist, 1992); McCarroll to Weldon, "Subject: First."

32. Carl R. Rogers and B. F. Skinner, "Some Issues Concerning the Control of Human Behavior," *Science* 124, no. 3231 (1956): 1057–66.

33. Carl Rogers, "Freedom and Commitment," *Humanist* 24, no. 2 (1964): 37–40.

34. Rogers, "Freedom and Commitment," 40.

35. DeCarvalho, *Founders of Humanistic Psychology*, 19–21; Robert Frager, "The Influence of Abraham Maslow," in *Motivation and Personality*, by Abraham H. Maslow (New York: Harper & Row, 1987), xxxiii–xli.

36. Abraham Harold Maslow, *Religions, Values, and Peak-Experiences* (Columbus: Ohio State University Press, 1964), 4; Edward Hoffman, *The Right to Be Human: A Biography of Abraham Maslow*, rev. ed. (New York: McGraw-Hill, 1999), gives a good account of Maslow's irreligious views from his youth onward.

37. Maslow, *Religions, Values, and Peak-Experiences*, x.

38. Maslow, *Religions, Values, and Peak-Experiences*, 39.

39. Andrew Robert Krieger argues that "Maslow had in fact established relatively close collaborative contacts with the Association in San Francisco, and his thinking provided a further stimulus for the Institute's activities"; Andrew Robert Krieger, "Structural Ambiguity in a Social Movement Organization: A Case Study of the American Humanist Association" (PhD diss., Georgetown University, 1983), 105. However, Arthur Jackson stated that Maslow was not closely involved with any of the specific programs; Arthur Jackson, interview by Stephen Weldon, February 2, 1997.

40. McCarroll to Weldon, "Subject: First."

41. "Conference Report: AHA in 1974," *Humanist* 25, no. 1 (January/February 1965): 23; Jackson, interview, February 2, 1997; Jackson, interview, January 22, 2000.

42. Lloyd Morain, interview by Stephen Weldon, January 21, 2000; Jackson, interview, January 22, 2000.

43. Dan L. Garrett Jr., "Synanon: The Communiversity," *Humanist* 25, no. 5 (September/October 1965): 184–89; McCarroll to Weldon, "Subject: First"; McCarroll to Weldon, "Subject: More"; Jackson, interview, January 22, 2000; Jackson, interview, February 2, 1997; Morain, interview.

44. Krieger, "Structural Ambiguity"; McCarroll to Weldon, "Subject: First"; Jackson, interview, January 22, 2000; Jackson, interview, February 2, 1997.

45. Krieger, "Structural Ambiguity"; McCarroll to Weldon, "Subject: First"; Jackson, interview, January 22, 2000; Jackson, interview, February 2, 1997.

46. Morain, interview; Jackson, interview, January 22, 2000; Bette Chambers, interview by

Stephen Weldon, August 2, 1993; Bette Chambers, interview by Stephen Weldon, March 18, 1994.

47. McCarroll to Weldon, "Subject: First"; Jackson, interview, January 22, 2000; Morain, interview.

48. Morain, interview; Krieger, "Structural Ambiguity," 108–15.

49. Morain, interview; Krieger, "Structural Ambiguity," 108–15.

50. Rudolf Dreikurs, "The Scientific Revolution," *Humanist* 26 (1966): 8.

51. Dreikurs, "Scientific Revolution," 11.

52. H. J. Muller, letter to the editor, *Humanist* 26 (1966): 173–74.

53. Paul Jayes, "Paul Kurtz Loves a Good Argument," *Buffalo Currier-Express Sunday Magazine*, August 23, 1981; Miriam Berkley, "Prometheus Unbound: A Skeptical Man with a Mission Runs a Skeptical Publishing House," *Publishers Weekly*, January 16, 1987; Paul Kurtz, interview by Stephen P. Weldon, August 11, 1993.

54. Paul Kurtz to B. F. Skinner, September 25, 1967, Skinner Papers; B. F. Skinner to Paul Kurtz, October 5, 1967, Skinner Papers.

55. B. F. Skinner to Tolbert McCarroll, January 14, 1966, Skinner Papers.

56. B. F. Skinner, reply to a May 8, 1968, memo from Kurtz, Fairfield, and Gordon to Members of the Publications Committee, Skinner Papers.

57. Paul Kurtz to B. F. Skinner, May 24, 1968, Skinner Papers.

58. Floyd W. Matson, "Humanistic Theory: The Third Revolution in Psychology," *Humanist* 31, no. 2 (March/April 1971): 11.

59. Paul Kurtz to B. F. Skinner, April 28, 1970, Skinner Papers; B. F. Skinner to Paul Kurtz, May 20, 1970, Skinner Papers; Paul Kurtz to B. F. Skinner, March 1, 1971, Skinner Papers.

60. Kenneth MacCorquodale, "Behaviorism Is a Humanism," *Humanist* 31, no. 2 (March/April 1971): 12–13; Willard F. Day, "Humanistic Psychology and Contemporary Behaviorism," *Humanist* 31, no. 2 (March/April 1971): 13–16.

61. B. F. Skinner, "Humanistic Behaviorism," *Humanist* 31, no. 3 (May/June 1971): 35.

62. Daniel W. Bjork, *B. F. Skinner: A Life* (New York: Basic Books, 1993), 192.

63. B. F. Skinner, *Beyond Freedom and Dignity* (New York: Knopf, 1971); see 32–33 for his contrast of slavery and the wage system and the important difference between working for rewards and working out of fear of punishments.

64. Paul Kurtz, "Democracy and the Technology of Control," *Humanist* 31, no. 6 (November/December 1971): 33.

65. Kurtz, "Democracy and the Technology of Control," 34.

66. Robert Erdmann, "Awards Committee Choice—B. F. Skinner," *Free Mind*, June 1972, 3. For a biography of Erdmann, see "The Authors," *Human Factors* 10, no. 5 (1968): 553.

67. Erdmann, "Awards Committee Choice."

68. Erdmann, "Awards Committee Choice."

69. B. F. Skinner, "Humanism and Behaviorism," *Humanist* 32, no. 4 (July/August 1972): 20.

Chapter 9 · Skeptics in the Age of Aquarius

Epigraph: Broady Richardson and James Randi, "Uri Geller and the Chimera: An Interview with 'The Amazing' Randi," *Humanist* 36, no. 4 (July/August 1976): 16.

1. Richardson and Randi, "Uri Geller and the Chimera," 22.

2. There had long been collaboration between American pragmatists and logical positivists, and Kurtz's work can be seen as a belated version of this effort. See George A. Reisch, *How the Cold War Transformed Philosophy of Science: To the Icy Slopes of Logic* (Cambridge: Cambridge University Press, 2005).

3. David Kaiser, *How the Hippies Saved Physics: Science, Counterculture, and the Quantum Revival* (New York: W. W. Norton, 2012); Michael D. Gordin, *The Pseudoscience Wars: Imman-*

uel Velikovsky and the Birth of the Modern Fringe (Chicago: University of Chicago Press, 2013); Fred Turner, *From Counterculture to Cyberculture: Stewart Brand, the Whole Earth Network, and the Rise of Digital Utopianism* (Chicago: University of Chicago Press, 2010); Jeffrey J. Kripal, *Esalen: America and the Religion of No Religion* (Chicago: University of Chicago Press, 2008).

4. Paul Kurtz, ed., "The New Cults: A Critique," *Humanist* 34, no. 5 (September/October 1974): 4–33.

5. Kurtz, "New Cults," 5.

6. Kurtz, "New Cults," 5.

7. See chap. 5.

8. Gilbert Murray, *Five Stages of Greek Religion* (New York: Clarendon, 1925); Sidney Hook, "The New Failure of Nerve," *Partisan Review* 10, no. 1 (1943): 2

9. See the symposium on the Marxist-Humanist dialogue in issue 1 of the *Humanist* in 1971 and the Catholic-Humanist dialogue in issue 3 of that same year.

10. Ethel Grodzins Romm, "The Yinning of America," *Humanist* 34, no. 5 (September/October 1974): 6–7.

11. Joseph L. Blau, "American Religion: From Cosmos to Chaos," *Humanist* 34, no. 5 (September/October 1974): 33.

12. A. Theodore Kachel, "Myths, Madness, and Movements," *Humanist* 34, no. 5 (September/October 1974): 24.

13. For Hull's acknowledgment of Clay's assistance, see Richard T. Hull, "On Taking Causal Criteria to Be Ontologically Significant," *Behaviorism* 1, no. 2 (1973): 65.

14. Marjorie Clay, "The New Religious Cults and Rational Science," *Humanist* 34, no. 5 (September/October 1974): 29.

15. Clay, "New Religious Cults," 27.

16. Richard T. Hull, "Scientism, Occultism, and Rational Science," *Humanist* 34, no. 5 (September/October 1974): 30.

17. Clay, "New Religious Cults," 28–29; Hull, "Scientism, Occultism, and Rational Science," 30.

18. "[Prometheus Books Advertisement]," *Humanist* 34, no. 5 (September/October 1974): 27.

19. Harvey Lebrun et al., "Readers Forum: The New Cults," *Humanist* 34, no. 6 (November/December 1974): 4.

20. One can find historical discussion of astrology and the New Age movement in both David J. Hess, *Science in the New Age: The Paranormal, Its Defenders and Debunkers, and American Culture*, Science and Literature (Madison: University of Wisconsin Press, 1993), esp. the historical introduction, 3–40; and Wouter J. Hanegraaff, *New Age Religion and Western Culture: Esotericism in the Mirror of Secular Thought* (Leiden: E. J. Brill, 1996), 331–61.

21. David S. Heeschen, "Bart J. Bok," *Physics Today* 36, no. 12 (August 28, 2008): 73; Bart J. Bok, "A Critical Look at Astrology," *Humanist* 35, no. 5 (September/October 1975): 6–9.

22. Gordon Allport, "Psychologists State Their Views on Astrology," quoted in Bok, "Critical Look at Astrology," 9.

23. Bart J. Bok, Lawrence E. Jerome, and Paul Kurtz, "Objections to Astrology: A Statement by 186 Leading Scientists," *Humanist* 35, no. 5 (September/October 1975): 4; Bart J. Bok and Margaret W. Mayall, "Scientists Look at Astrology," *Scientific Monthly* 52, no. 3 (1941): 233–44.

24. Paul Kurtz, "A Note from the Editor," *Humanist* 35, no. 5 (September/October 1975): 3.

25. Bok, Jerome, and Kurtz, "Objections to Astrology," 4.

26. "Press Comment on 'Objections to Astrology,'" *Humanist* 35, no. 6 (November/December 1975): 21–23.

27. "The Astrologers Reply," *Humanist* 35, no. 6 (November/December 1975): 24–26.

28. Paul Kurtz, "Astrology and Gullibility," *Humanist* 35, no. 6 (November/December 1975): 20.

29. Martin Gardner, *In the Name of Science* (New York: G. P. Putnam's Sons, 1952); Hess, *Science in the New Age*; Ronald L. Numbers and Daniel P. Thurs, "Science, Pseudoscience, and Science Falsely So-Called," in *Wrestling with Nature: From Omens to Science*, ed. Peter Harrison, Ronald L. Numbers, and Michael H. Shank (Chicago: University of Chicago Press, 2011), 281–306; Henry Bauer, *Science or Pseudoscience: Magnetic Healing, Psychic Phenomena, and Other Heterodoxies* (Urbana: University of Illinois Press, 2001); Donald Goldsmith, ed., *Scientists Confront Velikovsky*, with a foreword by Isaac Asimov (Ithaca, NY: Cornell University Press, 1977).

30. Paul Kurtz, ed., "Antiscience and Pseudoscience," *Humanist* 36, no. 4 (July/August 1976): 4–37.

31. In 2006 the name and acronym changed, but the policing allusion remained: CSI—Committee for Skeptical Inquiry—was a not-so-subtle reference to popular twenty-first-century TV crime dramas that featured high-tech scientific forensics.

32. Paul Kurtz, "The Scientific Attitude vs. Antiscience and Pseudoscience," *Humanist* 36, no. 4 (July/August 1976): 27–28.

33. "Psychic Vibrations," *Skeptical Inquirer* 4, no. 2 (Winter 1979): 16–17; "Psychic Vibrations," *Skeptical Inquirer* 4, no. 4 (Summer 1980): 14; "Psychic Vibrations," *Skeptical Inquirer* 6, no. 3 (Spring 1982): 13.

34. Kurtz, "Scientific Attitude," 27–28; Ernest Nagel, "Philosophical Depreciations of Scientific Method," *Humanist* 36, no. 4 (July/August 1976): 34.

35. Marvin Zimmerman, "Subjective Thinking," *Humanist* 36, no. 4 (July/August 1976): 33.

36. Kurtz, "Scientific Attitude," 31.

37. Douglas Martin, "Marcello Truzzi, 67; Sociologist Who Studied the Supernatural," *New York Times*, February 9, 2003, http://www.nytimes.com/2003/02/09/us/marcello-truzzi-67-sociologist-who-studied-the-supernatural.html; Marcello Truzzi, "[Editor's Note]," *Explorations* 1, no. 1 (April 1972): 1; Marcello Truzzi, interview by Stephen P. Weldon, November 6, 1993.

38. Truzzi, "[Editor's Note]" (April 1972).

39. On Gardner, see Gardner, *In the Name of Science*. I am indebted to Marcello Truzzi for my understanding of Hyman. Also Truzzi, interview.

40. Marcello Truzzi, "[Editor's Note]," *Zetetic* 3, no. 2 (June 1975): 1–2.

41. Marcello Truzzi, "Reflections on 'Project Alpha': Scientific Experiment or Conjuror's Illusion?," *Zetetic Scholar*, nos. 12–13 (1987): 74.

42. Lawrence E. Jerome, "Astrology: Magic or Science?," *Humanist* 35, no. 5 (September/October 1975): 15–16.

43. Michel Gauquelin, *The Scientific Basis of Astrology: Myth or Reality*, trans. James Hughes (New York: Stein & Day, 1969). The *Humanist* printed a series of letters to the editor in the fall and winter of 1975/1976, and several further articles appeared by Jerome, the Gauquelins, the Belgian Comité Para that had initially studied the Gauquelin data, and a couple of other scientists who had been contacted by Paul Kurtz, George O. Abel, and Marvin Zelen.

44. Marvin Zelen, "Astrology and Statistics: A Challenge," *Humanist* 36, no. 1 (January/February 1976): 32–33; Truzzi, interview; Trevor Pinch and H. M. Collins, "Private Science and Public Knowledge: The Committee for the Scientific Investigation of the Claims of the Paranormal and Its Use of the Literature," *Social Studies of Science* 14, no. 521–46 (1984); H. M. Collins and T. J. Pinch, *Frames of Meaning: The Social Construction of Extraordinary Science* (London: Routledge & Kegan Paul, 1982).

45. Marvin Zelen, Paul Kurtz, and George Abell, "Is There a Mars Effect?," *Humanist* 37, no. 6 (November/December 1977): 36–39; Michel Gauquelin and Francoise Gauquelin, "The Zelen Test of the Mars Effect," *Humanist* 37, no. 6 (November/December 1977): 30–35.

46. Jerome Clark to Tom McIver, August 28, 1991; generously provided by Tom McIver.

47. Jim Lippard, " 'Mars Effect' Chronology, 22 January 1993 DRAFT" (unpublished, January 22, 1992); generously provided by Tom McIver. This document, though unfinished, is thirty-

two pages long and contains a list of publications, correspondence, and phone communications from 1962 to 1993.

48. George P. Hansen, "CSICOP and the Skeptics: An Overview," *Journal for the American Society for Psychical Research* 86 (1992): 20.

49. "The Committee for the Scientific Investigation of Claims of the Paranormal," *Skeptical Inquirer* 3, no. 1 (Fall 1978): back cover. See also *Skeptical Inquirer* 10, no. 3 (Spring 1986).

50. Norman R. King, "Review of The Creation-Evolution Controversy," *Zetetic* 1, no. 2 (Spring/Summer 1977): 80–85; Charles E. Spitz, "Creation-Evolution Controversy," *Zetetic* 2, no. 1 (Fall/Winter 1977): 127; R. L. Wysong and Norman R. King, "Creation-Evolution Controversy," *Zetetic* 2, no. 2 (Spring/Summer 1978): 135–40.

51. Laurie R. Godfrey, "Science and Evolution in the Public Eye," *Skeptical Inquirer* 4, no. 1 (Fall 1979): 22.

52. James E. Alcock, "Psychology and Near-Death Experiences," *Skeptical Inquirer* 3, no. 3 (Spring 1979): 25.

53. Isaac Asimov, "Science and the Mountain Peak," *Skeptical Inquirer* 5, no. 2 (Winter 1980–1981): 50–51.

54. Martin Gardner, *The Whys of a Philosophical Scrivener* (New York: Quill, 1983).

55. Irving Hexham and Karla Poewe-Hexham, "The Soul of the New Age," *Christianity Today*, September 22, 1988, 21.

Chapter 10 · The Fundamentalist Challenge

Epigraph: Paul Kurtz, "Secular Humanism under Attack: Max Rafferty, God and Country," *Humanist* 29, no. 5 (September/October 1969): 1.

1. "Guidelines for Moral Instruction in California Schools: A Report Accepted by the State Board of Education, May 9, 1969" (Sacramento: California State Department of Education, 1969), 42.

2. "Guidelines for Moral Instruction," 50–56.

3. "Guidelines for Moral Instruction," 6, 2.

4. "Guidelines for Moral Instruction," 8, 15, 42, 65.

5. "Guidelines for Moral Instruction," 14, 47.

6. J. H., "Max Rafferty, RIP," *National Review*, July 23, 1982, 878–80; "Guidelines for Moral Instruction," 7.

7. Natalia Mehlman Petrzela, *Classroom Wars: Language, Sex, and the Making of Modern Political Culture* (Oxford: Oxford University Press, 2017); Natalia Mehlman, "Sex Ed . . . and the Reds? Reconsidering the Anaheim Battle over Sex Education, 1962–1969," *History of Education Quarterly* 47, no. 2 (May 2007): 203–32; Jeffrey P. Moran, *Teaching Sex: The Shaping of Adolescence in the 20th Century* (Cambridge, MA: Harvard University Press, 2000); Andrew Hartman, *A War for the Soul of America: A History of the Culture Wars* (Chicago: University of Chicago Press, 2015).

8. On the general shift away from a Protestant establishment in America, especially in education, see George M. Marsden, *The Soul of the American University: From Protestant Establishment to Established Nonbelief* (New York: Oxford University Press, 1994). For a discussion of the way that Jewish intellectuals challenged this establishment, see David A. Hollinger, "Jewish Intellectuals and the De-Christianization of American Public Culture in the Twentieth Century," in *Science, Jews, and Secular Culture: Studies in Mid-twentieth-century American Intellectual History* (Princeton, NJ: Princeton University Press, 1996), 17–41. The classic discussion of the Protestant establishment in America is E. Digby Baltzell, *The Protestant Establishment: Aristocracy and Caste in America* (New York: Vintage Books, 1966).

9. It is important to differentiate secularization of American institutions from the general secularization of American society. Whereas the documentation that I provide here is designed

to highlight how the separation of explicit church influence on public institutions frightened Evangelicals and other conservative Christians, this was not the same as decline of religious belief. In fact, there has been relatively little decline in religious belief in America in the past one hundred years. Only recently are we beginning to see a rise in the number of people who claim to be atheist, agnostic, or otherwise nonreligious. For various perspectives on what it means to talk about secularization and secularism, see David A. Hollinger, "The 'Secularization' Question and the United States in the Twentieth Century," *Church History: Studies in Christianity and Culture* 70, no. 1 (March 2001): 132–43, https://doi.org/10.2307/3654413; C. John Sommerville, "Post-secularism Marginalizes the University: A Rejoinder to Hollinger," *Church History: Studies in Christianity and Culture* 71, no. 4 (December 2002): 848–57, https://doi.org/10.1017/S0009640700096323; David A. Hollinger, "Why Is There So Much Christianity in the United States? A Reply to Sommerville," *Church History: Studies in Christianity and Culture* 71, no. 4 (December 2002): 858–64, https://doi.org/10.1017/S0009640700096335; Steve Bruce, *Secularization: In Defence of an Unfashionable Theory* (Oxford: Oxford University Press, 2011); Charles Taylor, *A Secular Age* (Cambridge, MA: Belknap Press of Harvard University Press, 2007). In general, the views on secularism are split in understanding whether our current society is largely secular or religious. Historians have generally argued that American society as a whole is not secular, although certain pockets of it are highly secularized, which is the position I take throughout this book.

10. James W. Fraser, *Between Church and State: Religion and Public Education in a Multicultural America* (New York: St. Martin's, 1999), 137–40; Susan Jacoby, *Freethinkers: A History of American Secularism* (New York: Henry Holt, 2004), 292–98.

11. Jacoby, *Freethinkers*, 292–98; Douglas Martin, "Vashti McCollum, 93, Who Brought Landmark Church-State Suit, Is Dead," *New York Times*, August 26, 2006, http://www.nytimes.com/2006/08/26/obituaries/26mccullum.html?_r=1.

12. Fraser, *Between Church and State*, 139; Martin E. Marty, *Modern American Religion*, vol. 3, *Under God, Indivisible: 1941–1960* (Chicago: University of Chicago Press, 1996), 211–30.

13. Edwin H. Wilson, "The Sectarian Battlefront," *Humanist* 7, no. 4 (Spring 1948): 179.

14. Jacoby, *Freethinkers*, 292–98; Marty, *Under God, Indivisible*, 211–30.

15. Steven K. Green, *The Bible, the School, and the Constitution: The Clash That Shaped Modern Church-State Doctrine* (Oxford: Oxford University Press, 2012), 256–57; Mark A. Noll, *A History of Christianity in the United States and Canada* (Grand Rapids, MI: William B. Eerdmans, 1992), 452–53; Joan Delfattore, *The Fourth R: Conflicts over Religion in America's Public Schools* (New Haven, CT: Yale University Press, 2004), 67–81.

16. Robert S. Alley, *The Supreme Court on Church and State* (New York: Oxford University Press, 1988), 194–201, 204–24; Fraser, *Between Church and State*, 148–49; Delfattore, *Fourth R*, 67–76; Paul Blanshard, "A Day for Celebration—and Dedication," *Humanist* 22, nos. 2 and 3 (1962): 42–43; "The Schempp Murray Cases: The Humanist View," *Humanist* 23, no. 1 (1963): 3; Paul Blanshard, "The Church and State: Paul Blanshard's Commentary," *Humanist* 23, no. 3 (1963): 94; Paul Blanshard, "The Big Decision," *Humanist* 23, no. 4 (1963): 106–10.

17. Blanshard, "Big Decision," 110.

18. Alley, *Supreme Court on Church and State*, 410–14.

19. Tolbert H. McCarroll, "Brief of the American Humanist Association, as *Amicus Curiae*," United States of America, Petitioner, v. Daniel Andrew Seeger, in the Supreme Court of the United States, October Term, 1964, no. 50. (Note that C.O. status had been discussed earlier: see Rowland Watts, Legal Director, ACLU to Edwin Wilson, July 29, 1960, in Wilson Archives, box 45, COs.)

20. Quoted in Alley, *Supreme Court on Church and State*, 413–14; Joseph L. Blau, "Who First Used the Words 'Secular Humanism'?," *New York Times*, June 19, 1985.

21. Letter to Office of Commissioner of Internal Revenue from Joseph Cohen, Attorney for

the AHA, April 25, 1946; E. I. McLarney, Deputy Commissioner, Treasury Department, to American Humanist Association, June 19, 1946; Joseph M. Cullen, District Director, Internal Revenue Service, to American Humanist Association, July 3, 1968.

22. "Beginning a Membership-Wide Dialogue: Is Humanism a Religion, Philosophy or World-View? And Can It Ever Contain Supernatural Concepts?," *Free Mind* 31, no. 1 (January/ February 1988): 4; Corliss Lamont et al., "Continuing the Humanist Dialogue: Is Humanism a Religion, Life-Stance, Philosophy . . . ?," *Free Mind* 31, no. 2 (March/April 1988): 9–10; Bette Chambers et al., "Continuing the Humanist Dialogue: Is Humanism a Religion, Philosophy, World-View? All Three? And Can Humanism Ever Contain Supernatural Concepts?," *Free Mind* 31, no. 3 (May/June 1988): 2–8; "Opinion Poll: Nonscientific, 'Straw Poll' of Free Mind Readers," *Free Mind* 31, no. 4 (July/August 1988): 11; Bette Chambers, "Editorial," *Free Mind* 31, no. 5 (September/October 1988): 3; Bette Chambers, "And Now—the Nonscientific Results of Our Nonscientific Poll," *Free Mind* 31, no. 6 (November/December 1988): 16–17.

23. Fred Edwords, interview by Stephen P. Weldon, August 2, 1993.

24. US House of Representatives, "Conlan Kills Funding for 'Secular Humanism,'" *Congressional Record*, 94th Cong., 2d sess., vol. 122, no. 70, May 12, 1976.

25. Gordon V. Drake, *Is the School House the Proper Place to Teach Raw Sex?* (Tulsa, OK: Christian Crusade, 1968); Claire Chambers, *The SIECUS Circle: A Humanist Revolution* (Belmont, MA: Western Islands, 1977); Moran, *Teaching Sex*.

26. John W. Whitehead, *The Separation Illusion: A Lawyer Examines the First Amendment* (Milford, MI: Mott Media, 1977); John W. Whitehead and John Conlan, "The Establishment of the Religion of Secular Humanism and Its First Amendment Implications," *Texas Tech Law Review* 10, no. 1 (1978): 1–66.

27. Tim F. LaHaye, *The Battle for the Mind* (Old Tappan, NJ: F. H. Revell, 1980), 136.

28. LaHaye, *Battle for the Mind*, 183.

29. Nancy T. Ammerman, "North American Protestant Fundamentalism," in *Fundamentalisms Observed*, ed. Martin E Marty and R. Scott Appleby (Chicago: University of Chicago Press, 1991), 1:1–65. On Rushdoony, see William Edgar, "The Passing of R. J. Rushdoony," *First Things: A Monthly Journal of Religion and Public Life*, August 2001, 24–25; Rousas John Rushdoony, *By What Standard? An Analysis of the Philosophy of Cornelius Van Til* (Philadelphia: Presbyterian & Reformed, 1959); Rousas John Rushdoony, *The Messianic Character of American Education: Studies in the History of the Philosophy of Education* (Philadelphia: Presbyterian & Reformed, 1963); Rousas John Rushdoony and Herbert W. Titus, *The Institutes of Biblical Law: A Chalcedon Study* (Nutley, NJ: Craig, 1973); Molly Worthen, "The Chalcedon Problem: Rousas John Rushdoony and the Origins of Christian Reconstructionism," *Church History: Studies in Christianity and Culture* 77, no. 2 (June 2008): 399–437, https://doi.org/10.1017/S0009640708000590. On Whitehead, see Whitehead, *Separation Illusion*; Whitehead and Conlan, "Establishment of the Religion of Secular Humanism." On Schaeffer, see Forrest Baird, "Schaeffer's Intellectual Roots," in *Reflections on Francis Schaeffer*, ed. Ronald W Ruegsegger (Grand Rapids, MI: Academie Books, 1986), 45–67; Edith Schaeffer, *The Tapestry: The Life and Times of Francis and Edith Schaeffer* (Waco, TX: Word Books, 1981); Francis A. Schaeffer, *How Should We Then Live? The Rise and Decline of Western Thought and Culture* (Old Tappan, NJ: F. H. Revell, 1976).

30. Scott R. Burson and Jerry L. Walls, *C. S. Lewis and Francis Schaeffer: Lessons for a New Century from the Most Influential Apologists of Our Time* (Downers Grove, IL: InterVarsity, 1998), 34–44; Schaeffer, *How Should We Then Live?*; Francis A. Schaeffer, *Escape from Reason: A Penetrating Analysis of Trends in Modern Thought* (Downers Grove, IL: InterVarsity, 1968); Francis A. Schaeffer and C. Everett Koop, *Whatever Happened to the Human Race?* (Old Tappan, NJ: F. H. Revell, 1979).

31. Worthen, "Chalcedon Problem"; Rushdoony, *Messianic Character of American Education*.

32. Baird, "Schaeffer's Intellectual Roots"; Ammerman, "North American Protestant Fundamentalism"; Rushdoony, *By What Standard?*

33. Rushdoony, *Messianic Character of American Education*, 161.

34. Kurtz, "Secular Humanism under Attack," 1–2; Corliss Lamont et al., " 'A Most Immoral Statement': Responses to the 'Guidelines for Moral Instruction in California Schools,' " *Humanist* 29, no. 6 (November/December 1969): 11–14; Paul Kurtz, "A Defeat for California's Fundamentalists," *Humanist* 30, no. 2 (March/April 1970): 5, 41.

35. Kurtz, "Defeat for California's Fundamentalists."

36. Robert E. Jones, "Year in Review: Report from the Joint Washington Office for Social Concern (JWOSOC)," *Free Mind*, March 1971, 2.

37. Bart J. Bok, Lawrence E. Jerome, and Paul Kurtz, "Objections to Astrology: A Statement by 186 Leading Scientists," *Humanist* 35, no. 5 (September/October 1975): 4; "A New Bill of Sexual Rights and Responsibilities," *Humanist* 36, no. 1 (January/February 1976): 4–6; "A Statement Affirming Evolution as a Principle of Science," *Humanist* 37, no. 1 (January/February 1977): 4–6; Board of Directors American Humanist Association, "AHA's Statement on the Family," *Humanist* 40, no. 5 (September/October 1980): 40.

38. Eleanor Blau, "Humanist Manifesto II Offers a 'Survival' Philosophy," *New York Times*, August 26, 1973; Kenneth A. Briggs, "Secularists Attack 'Absolutist' Morals: 61 Scholars and Writers Denounce the Rise of Fundamentalism," *New York Times*, October 15, 1980.

39. The September/October 1976 issue of the *Humanist* had a fifteen-page special feature on "The Evangelical Right" with authors Paul Kurtz, Sidney Hook, James Luther Adams, Leo Pfeffer, William Van Alstyne, Albert Schanker, Jim Wallis, and Wes Michaelson, as well as the full text of the Conlan amendment. This was followed by a four-page article by Scott Edwards, "Jesus in the Now: The New Revivalism," *Humanist* 36 (September/October 1976): 16–19. The January/February 1977 issue contained two special features, one on "Evolution vs. Creationism in the Public Schools" (4–23) and the other on "The Resurgence of Fundamentalism" (37–43). The following issues contained more articles on the "The Resurgence of Fundamentalism," three articles in the March/April issue (38–43) and two in the May/June issue (46–51). Then, starting with the September/October issue of 1980 and going through November/December 1981, there were articles on the New Right's attacks on humanism, on "Born-Again Politics," and on the Moral Majority in nearly every issue.

40. Paul Kurtz, "The Resurgence of Fundamentalism: A Symposium," *Humanist* 37, no. 1 (January/February 1977): 36–43; Edward L. Ericson, "A Dynamic Religious Diversity," *Humanist* 37, no. 1 (January/February 1977): 37; Sidney Hook, "The New Religiosity," *Humanist* 37, no. 1 (January/February 1977): 38–39; Antony Flew, "Religion in America," *Humanist* 37, no. 1 (January/February 1977): 39–40; Morris B. Storer, "Humanism in a Time of Religious Revival," *Humanist* 37, no. 1 (January/February 1977): 40–41; Arthur C. Danto, "Religious Enthusiasm in the United States," *Humanist* 37, no. 1 (January/February 1977): 41–42; Joseph L. Blau, "Religion Today," *Humanist* 37, no. 1 (January/February 1977): 42–43; Paul Kurtz, "The Resurgence of Fundamentalism II," *Humanist* 37, no. 2 (March/April 1977): 36–44; Albert Ellis, "Religious Belief in the United States Today," *Humanist* 37, no. 2 (March/April 1977): 38–41; Walter Kaufmann, "Criticizing Religious Beliefs," *Humanist* 37, no. 2 (March/April 1977): 42–43; Lionel Abel, "Religion and Value," *Humanist* 37, no. 2 (March/April 1977): 43–44.

41. Ellis, "Religious Belief," 38.

42. Ellis, "Religious Belief," 39.

43. Flew, "Religion in America"; Storer, "Humanism in a Time of Religious Revival."

44. Ericson, "Dynamic Religious Diversity"; Blau, "Religion Today."

45. Hook, "New Religiosity."

46. Paul Kurtz, "The Attack on Secular Humanism," *Humanist* 36, no. 5 (1976): 4; Sidney Hook, "Is Secular Humanism a Religion?," *Humanist* 36, no. 5 (1976): 6.

47. Leo Pfeffer, "Is the Conlan Amendment Unconstitutional?," *Humanist* 36, no. 5 (1976): 9–10.

48. Kurtz, "Secular Humanism under Attack"; Lamont et al., " 'Most Immoral Statement.' "

49. Paul Kurtz, "The Attack on Secular Humanism"; Hook, "Is Secular Humanism a Religion?"

50. Maxine Negri, "Humanism under Fire," *Humanist* 41, no. 2 (March/April 1981): 5.

51. Paul Kurtz and Edwin H. Wilson, "Humanist Manifesto II," *Humanist* 33, no. 5 (1973): 4–9.

52. Blau, "Humanist Manifesto II"; "List of Manifesto Signers," *New York Times*, August 26, 1973; "Highlights of Manifesto II," *New York Times*, August 26, 1973.

53. Garry Wills, "Humanist Manifesto II Fails by Its Own Test," *Sunday Focus Section*, St. Paul, Minnesota, September 2, 1973, in the clippings files at CSICOP, and other clippings in the same file.

54. Kurtz and Wilson, "Humanist Manifesto II." Michael Anthony Schuler, "Religious Humanism in Twentieth-Century American Thought" (PhD, Florida State University, 1982), provides a detailed analysis of the difference between the manifestos.

55. Kurtz and Wilson, "Humanist Manifesto II," 5.

56. Kurtz and Wilson, "Humanist Manifesto II," 5.

57. Kurtz and Wilson, "Humanist Manifesto II," quoted here and the next five paragraphs.

58. Briggs, "Secularists Attack 'Absolutist' Morals."

59. Paul Kurtz, "A Secular Humanist Declaration," *Free Inquiry* 1, no. 1 (Winter 1980–1981): 3–7.

60. Kurtz, "Secular Humanist Declaration," quoted here and the next two paragraphs.

61. In the Humanist Manifesto II, religion and its cognates appear twenty-four times, whereas in the Secular Humanist Declaration, which is almost the same length, there are thirty-nine instances.

62. Kurtz, "Secular Humanist Declaration," quoted here and the next two paragraphs.

63. Quoted in Paul Kurtz to B. F. Skinner, March 21, 1979, Skinner Papers.

64. Paul Kurtz, "Announcing a New Magazine," *Free Inquiry* 1, no. 1 (Winter 1980–1981): front cover.

Chapter 11 · Battling Creationism and Christian Pseudoscience

Epigraph: Bette Chambers, "Bryan's Ghost," *Free Mind* 15, no. 6 (December 1972): 8.

1. Bette Chambers, "[Statement]," *Free Mind* 15, no. 5 (October 1972): 5.

2. Bette Chambers, "Why a Statement Affirming Evolution?," *Humanist* 37, no. 1 (January/February 1977): 23.

3. There is a vast amount of literature on this. Several good places to start are as follows: George E. Webb, *The Evolution Controversy in America* (Lexington: University Press of Kentucky, 1994); Edward J. Larson, *Trial and Error: The American Controversy over Creation and Evolution*, 3rd ed. (New York: Oxford University Press, 2003); Michael Ruse, *The Evolution-Creation Struggle* (Cambridge, MA: Harvard University Press, 2006); Adam R. Shapiro, *Trying Biology: The Scopes Trial, Textbooks, and the Antievolution Movement in American Schools* (Chicago: University of Chicago Press, 2013).

4. Max Carl Otto, "Untitled Autobiography 'Dear L.G.W.' [Lewis G. Westgate]," October 17, 1945, M. C. Otto Papers, box 12, Wisconsin State Historical Society.

5. Michael Lienesch, "Abandoning Evolution: The Forgotten History of Antievolution Activism and the Transformation of American Social Science," *Isis* 103, no. 4 (December 2012): 692–93, https://doi.org/10.1086/668963.

6. Otto, "Dear L.G.W." I must also thank David A. Sandmire for sharing his unpublished manuscript with me, "Max Otto, the University Pastors, and the Holy 'Chapple': The Philo-

sophical, Social and Political Implications of Evolution in the Early 20th Century" (December 1993), 3.

7. Mason Olds, *American Religious Humanism*, rev. ed. (Minneapolis: Fellowship of Religious Humanists, 1996), 125–29.

8. Michael Lienesch, *In the Beginning: Fundamentalism, the Scopes Trial, and the Making of the Antievolution Movement* (Chapel Hill: University of North Carolina Press, 2007), 178; Olds, *American Religious Humanism*, 127–29.

9. Lienesch, *In the Beginning*, 106.

10. Olds, *American Religious Humanism*, 128.

11. J. S. Huxley, "Evolution and Popular Thought," *The Nation and the Athenaeum (London)*, April 8, 1922 (proofsheet in Julian Sorell Huxley Papers, Woodson Research Center, Fondren Library, Rice University, Houston, Texas).

12. Hermann J. Muller to Julian Huxley, March 27, 1918, folder: Huxley, J. S. 1917–1918, Hermann J. Muller Papers, Lilly Library, Indiana University, Bloomington, Indiana.

13. In 1950, Huxley wrote to Muller, "I was glad to see your comments on the way in which evolution is neglected in general education, and am utilizing them in a campaign I am organizing to get something done about it (a) in ~~general~~ this country; (b) in general through Unesco." Julian Huxley to Hermann J. Muller, April 8, 1950, Julian Sorell Huxley Papers, Woodson Research Center, Fondren Library, Rice University, Houston, Texas.

14. See Webb, *Evolution Controversy in America*, 130–42. This standard picture is rebutted by Shapiro, *Trying Biology*. See Muller's original statement in H. J. Muller, "One Hundred Years without Darwinism Are Enough," *School Science and Mathematics* 59, no. 4 (April 1959): 304–16, https://doi.org/10.1111/j.1949-8594.1959.tb08235.x.

15. Carl Sagan, interview by Stephen Weldon, December 16, 1993; Richard C. Lewontin, "Billions and Billions of Demons," *New York Review of Books*, January 9, 1997, https://www.nybooks.com/articles/1997/01/09/billions-and-billions-of-demons/; Hermann J. Muller to Carl Sagan, October 12, 1960, Hermann J. Muller Papers, Lilly Library, Indiana University, Bloomington, Indiana; Hermann J. Muller, "Biologists' Statement on Teaching Evolution," *Bulletin of the Atomic Scientists* 23, no. 2 (February 1967): 39–40. The website http://wiki.creation.org/index.php/Debate gives the participants in the Little Rock debate as follows: Jack Wood Sears and James D. Bales, defending creationism, versus Carl Sagan, Ernan McMullin, R. C. Lewontin, and Thomas K. Shotwell, defending evolution. Neither Sagan nor Lewontin has indicated that the other two men participated.

16. Webb, *Evolution Controversy in America*, 151–52; Larson, *Trial and Error*.

17. The most thorough history of scientific creationism is Ronald L. Numbers, *The Creationists: From Scientific Creationism to Intelligent Design*, exp. ed. (Cambridge, MA: Harvard University Press, 2006). See also Larson, *Trial and Error*; Webb, *Evolution Controversy in America*.

18. Larson, *Trial and Error*, 122–24.

19. "Guidelines for Moral Instruction in California Schools: A Report Accepted by the State Board of Education, May 9, 1969" (Sacramento: California State Department of Education, 1969), 63.

20. "Guidelines for Moral Instruction," 62–64.

21. "Reagan Favors Creationism in the Public Schools," *Creation/Evolution* 1, no. 2 (Fall 1980): 45.

22. [Wendell Bird], "Freedom of Religion and Science Instruction in Public Schools," *Yale Law Journal* 87, no. 3 (January 1978): 5151970, https://doi.org/10.2307/795591. Bird's original agenda was to persuade public school boards, but a Catholic creationist by the name of Paul Ellwanger modified it so that it could be put before state legislatures. See Numbers, *Creationists*, 351–52.

23. Bette Chambers, interview by Stephen Weldon, August 2, 1993; Bette Chambers, "Who Is the American Scientific Affiliation?," *Free Mind* 16, no. 2 (May 1973): 10; Bette Chambers, "President's Message," *Free Mind* 19, no. 4 (August 1976): 7; anonymous, "Creationism Scores Again in California," *Free Mind* 19, no. 4 (August 1976): 6; Chambers, "Bryan's Ghost"; Chambers, "Why a Statement Affirming Evolution?"

24. "A.H.A. vs. Creationists," *Free Mind* 15, no. 6 (December 1972): 1; Chambers, "Bryan's Ghost"; Bette Chambers, "The President's Message," *Free Mind* 16, no. 1 (February 1973): 2; Chambers, "Who Is the American Scientific Affiliation?"

25. Sagan, interview; Lewontin, "Billions and Billions of Demons"; Muller, "Biologists' Statement on Teaching Evolution."

26. Chambers, interview; Chambers, "Why a Statement Affirming Evolution?," 24.

27. "A Statement Affirming Evolution as a Principle of Science," *Humanist* 37, no. 1 (January/February 1977): 4–5.

28. Chambers, "Why a Statement Affirming Evolution?," 24.

29. "Statement Affirming Evolution as a Principle of Science"; Chambers, "Why a Statement Affirming Evolution?," 24.

30. Preston Cloud, "'Scientific Creation'—a New Inquisition Brewing?," *Humanist* 37, no. 1 (January/February 1977): 15.

31. William V. Mayer, "Evolution: Yesterday, Today, Tomorrow," *Humanist* 37, no. 1 (January/February 1977): 22.

32. Sidney Hook, "The New Religiosity," *Humanist* 37, no. 1 (January/February 1977): 38–39; Joseph L. Blau, "Religion Today," *Humanist* 37, no. 1 (January/February 1977): 42–43; Antony Flew, "Religion in America," *Humanist* 37, no. 1 (January/February 1977): 39–40.

33. John Dart, "Scientists Defend Principle of Evolution: 'Creationism' Strictly Religious, Humanist Paper Says," *Los Angeles Times (1923–Current File)*, January 29, 1977, sec. Part One.

34. "Dear Reader," *Creation/Evolution* 1, no. 1 (Summer 1980): inside front cover.

35. "Dear Reader."

36. Stanley L. Weinberg, "Reactions to Creationism in Iowa," *Creation/Evolution*, no. 2 (Fall 1980): 5–6.

37. Paul Kurtz and Isaac Asimov, "An Interview with Isaac Asimov: On Science and the Bible," *Free Inquiry* 2, no. 2 (Spring 1982): 6–10.

38. Frederick Edwords to Isaac Asimov, March 27, 1986, Isaac Asimov Papers, Special Collections, Mugar Memorial Library, Boston University, Boston, Massachusetts.

39. Edd Doerr to Isaac Asimov, February 3, 1986, Isaac Asimov Papers, Special Collections, Mugar Memorial Library, Boston University, Boston, Massachusetts.

40. Paul Kurtz, "A Secular Humanist Declaration," *Free Inquiry* 1, no. 1 (Winter 1980–1981): 3–7.

41. H. James Birx, "The Creation/Evolution Controversy," *Free Inquiry* 1, no. 1 (Winter 1980): 24–26; Delos McKown, "'Scientific' Creationism: Axioms and Exegesis," *Free Inquiry* 1, no. 3 (Summer 1981): 23–28.

42. L. Sprague de Camp, "The Continuing Monkey War," *Free Inquiry* 2, no. 2 (Spring 1982): 12–17; A. J. Mattill Jr., "Three Cheers for the Creationists!," *Free Inquiry* 2, no. 2 (Spring 1982): 17–18.

43. Numbers, *Creationists*, 274.

44. de Camp, "Continuing Monkey War"; Mattill, "Three Cheers for the Creationists!"; Antony Flew, "The Erosion of Evolution: A Treason of the Intellectuals," *Free Inquiry* 2, no. 2 (Spring 1982): 19–23.

45. "Science, the Bible and Darwin: An International Symposium on Science, Religion and Ethics [Advertisement]," *Free Inquiry* 2, no. 2 (Spring 1982): 11; Paul Kurtz, "Introduction: Science, the Bible and Darwin," *Free Inquiry* 2, no. 3 (Summer 1982): 3.

46. Kurtz, "Introduction: Science, the Bible and Darwin."

47. Gerald Larue, "The Religion and Biblical Criticism Research Project," *Free Inquiry* 2, no. 4 (Fall 1982): 16.

48. "Call for the Critical Examination of the Bible and Religion," *Free Inquiry* 2, no. 2 (Spring 1982): front cover, back cover; Larue, "Religion and Biblical Criticism Research Project"; "New Committee Founded to Study Religion," *Secular Humanist Bulletin* 1, no. 2 (May 1985): 1.

49. Paul Kurtz, "Introduction: Is America a Judeo-Christian Republic," *Free Inquiry* 3, no. 3 (Summer 1983): 5–7.

50. Marvin M. Mueller, Walter McCrone, and Steven D. Schafersman, "Special Critique on the Shroud of Turin," *Skeptical Inquirer* 6, no. 3 (1982): 15–56; "News: Shroud Proved Medieval," *Skeptical Inquirer* 13, no. 2 (1989).

51. Steven D. Schafersman, "Raiders of the Lost Tracks: The Best Little Footprints in Texas," *Skeptical Inquirer* 7, no. 3 (1983): 2–6; "Supreme Court Hands Skeptics a Victory," *Skeptical Briefs* 3, no. 3 (August 1987): 3. I was one of those high school students. (Physics teacher Ronald Hastings led several trips to the site, where we all talked to creationists doing the work. The report was written up in an issue of *Creation/Evolution* and recounted in the *Skeptical Inquirer*.) See R. J. Hastings, "Tracking Those Incredible Creationists," *Creation/Evolution*, no. 15 (1985): 5–15. The movie production is mentioned in Ronnie J. Hastings, "The Rise and Fall of the Paluxy Mantracks," *Perspectives on Science and Christian Faith* 40 (September 1988): 144–54.

52. "New Committee Founded to Study Religion"; "Jesus Symposium Goes on Despite Disruptions," *Secular Humanist Bulletin* 1, no. 2 (May 1985): 1–4.

53. John Dart, "Evangelist Popoff off Air; Files Bankruptcy Petitions," *Los Angeles Times (1923–Current File)*, September 26, 1987; John Dart, "Skeptics' Revelations: Faith Healer Receives Heavenly Messages via Electronic Receiver, Debunkers Charge," *Los Angeles Times (1923–Current File)*, May 11, 1986; John Dart, "Skeptics Tune In on Source of Faith Healer's 'Divine' Messages," *Los Angeles Times (1923–Current File)*, May 12, 1986; Bruce Buursma, "Preacher Tuned to the Divine," *Chicago Tribune*, June 27, 1986.

54. Stuart Taylor Jr., "72 Nobelists Urge Court to Void Creationism Law," *New York Times*, August 19, 1986; Michael Brant Shermer, "Science Defended, Science Defined: The Louisiana Creationism Case," *Science, Technology, and Human Values* 16, no. 4 (October 1991): 517–39, https://doi.org/10.1177/016224399101600405; "Supreme Court Hands Skeptics a Victory"; Al Seckel, "Science, Creationism, and the U.S. Supreme Court," *Skeptical Inquirer* 11, no. 2 (Winter 1986–1987): 147–58.

Chapter 12 · The Humanist Ethos of Science in Modern America

Epigraph: Carl Sagan, *Cosmos* (New York: Random House, 1980), 4.

1. Isaac Asimov to Lyle L. Simpson, May 13, 1984, Isaac Asimov Papers, Special Collections, Mugar Memorial Library, Boston University, Boston, Massachusetts.

2. Carl Sagan, interview by Stephen Weldon, March 9, 1994.

3. On the early connections to the AHA, see Edwin H. Wilson to Isaac Asimov, July 17, 1959, box 1, folder: Asimov, Isaac 1985, Special Collections, Morris Library, Southern Illinois University, Carbondale, Illinois; Edwin H. Wilson to Isaac Asimov, March 7, 1960, box 1, folder: Asimov, Isaac 1985, Special Collections, Morris Library, Southern Illinois University, Carbondale, Illinois. There is a great deal of correspondence between Bette Chambers and Asimov in the Asimov Papers at Boston University. See, e.g., Chambers's discussion of Ericson in Bette Chambers to Isaac Asimov, November 12, 1986, Isaac Asimov Papers, Special Collections, Mugar Memorial Library, Boston University, Boston, Massachusetts. See also Edward L. Ericson, *The Humanist Way: An Introduction to Ethical Humanist Religion* (New York: Continuum, 1988).

4. Carl Sagan, *Contact* (New York: Pocket Books, 1985); Isaac Asimov, *The Foundation Trilogy: Three Classics of Science Fiction* (Garden City, NY: Doubleday, 1982).

5. Jared Scott Buss, "Willy Ley, the Science Writers, and the Popular Reenchantment of Science" (PhD diss., University of Oklahoma, 2014); Kendrick Oliver, *To Touch the Face of God: The Sacred, the Profane, and the American Space Program, 1957–1975* (Baltimore: Johns Hopkins University Press, 2012); David J. Tietge, *Flash Effect: Science and the Rhetorical Origins of Cold War America* (Athens: Ohio University Press, 2002); Martin Halliwell, *Romantic Science and the Experience of Self: Transatlantic Crosscurrents from William James to Oliver Sacks* (Aldershot: Ashgate, 1999); James William Gibson, *A Reenchanted World: The Quest for a New Kinship with Nature* (New York: Metropolitan Books, 2009); Anne Harrington, *Reenchanted Science* (Princeton, NJ: Princeton University Press, 1999). On Sagan, see esp. Thomas M. Lessl, "The Priestly Voice," *Quarterly Journal of Speech* 75 (1989): 183–97; Thomas M. Lessl, "Science and the Sacred Cosmos: The Ideological Rhetoric of Carl Sagan," *Quarterly Journal of Speech* 71 (1985): 175–87.

6. Isaac Asimov, "Popularizing Science," in *Past, Present, and Future* (Buffalo, NY: Prometheus Books, 1987), 91. This was originally published in *Nature* in 1983.

7. Sagan, interview.

8. Carl Sagan and Ann Druyan, "Real Patriots Ask Questions," *Parade Magazine*, September 8, 1991, 14.

9. Sagan, interview.

10. William J. Harnack, "Carl Sagan: Cosmic Evolution vs. the Creationist Myth," *Humanist* 41, no. 4 (1981): 5–6; Carl Sagan et al., *Cosmos* (Los Angeles: Carl Sagan Productions, 1980), and the companion book (Sagan, *Cosmos*).

11. Isaac Asimov, "Nightfall," in *Science Fiction: The Science Fiction Research Association Anthology*, ed. Patricia S. Warrick, Charles C. Waugh, and Martin H. Greenberg (New York: Pearson, 1997), 128–29.

12. Sagan et al., "Episode 13: Who Speaks for the Earth?," *Cosmos*, at 25–37 min. The narrative can also be found in Sagan, *Cosmos*, 335–37; David C. Lindberg, "Myth 1. That the Rise of Christianity Was Responsible for the Demise of Ancient Science," in *Galileo Goes to Jail: And Other Myths about Science and Religion*, ed. Ronald L. Numbers (Cambridge, MA: Harvard University Press, 2009), 8–18.

13. Sagan, *Contact*, 153–54.

14. Isaac Asimov to Taimi Saha, July 16, 1989, Isaac Asimov Papers, Special Collections, Mugar Memorial Library, Boston University, Boston, Massachusetts.

15. Robert Wiebe explores the American ideology of self-reliance and its political implications in the nineteenth and twentieth centuries. See Robert H. Wiebe, *Self-Rule: A Cultural History of American Democracy* (Chicago: University of Chicago Press, 1995), esp. 185–201.

16. Karl Giberson and Mariano Artigas, *Oracles of Science: Celebrity Scientists versus God and Religion* (Oxford: Oxford University Press, 2007).

17. "Announcing the Academy of Humanism," *Free Inquiry* 3, no. 4 (Fall 1983): 5–7.

18. "Announcing the Academy of Humanism," 6.

19. The classic study of religion and science interactions in the West is John Hedley Brooke, *Science and Religion: Some Historical Perspectives* (Cambridge: Cambridge University Press, 1991).

20. For a really clear essay exploring the development of what is now called methodological naturalism, see Ronald L. Numbers, "Science without God: Natural Laws and Christian Beliefs," in *When Science and Christianity Meet*, ed. Ronald L. Numbers and David C. Lindberg (Chicago: University of Chicago Press, 2003), 265–85. The general trend of all modern professions, however, has been to shift religious discussions into the private sphere. There is enormous literature on the nature of professionalization in the modern world; a few significant works are Samuel Haber, *The Quest for Authority and Honor in the American Professions, 1750–1900* (Chicago: University of Chicago Press, 1991); Harold Perkin, *The Third Revolution: Professional Elites in the Modern World* (London: Routledge, 1996); William Sullivan, *Work and Integrity: The Crisis*

and Promise of Professionalism in America, 2nd ed. (New York: Jossey-Bass, 2005). For an interesting study of science and professionalization, see Thomas F. Gieryn, George M. Bevins, and Stephen C. Zehr, "Professionalization of American Scientists: Public Science in the Creation/Evolution Trials," *American Sociological Review* 50, no. 3 (1985): 392–409. Sociological literature on secularization deals with this as well; see Steve Bruce, *Secularization: In Defence of an Unfashionable Theory* (Oxford: Oxford University Press, 2011); Steve Bruce, ed., *Religion and Modernization: Sociologists and Historians Debate the Secularization Thesis* (Oxford: Clarendon, 1992); Christian Smith, ed., *The Secular Revolution: Power, Interests, and Conflict in the Secularization of American Public Life* (Berkeley: University of California Press, 2003); Charles Taylor, *A Secular Age* (Cambridge, MA: Belknap Press of Harvard University Press, 2007).

21. Numbers from the Gallup polls go back to 1944. Since that time, the proportion of people who have answered "yes" to the question "Do you, personally, believe in God?" has ranged from a high of 98 percent in the 1950s and 1960s to a low of 87 percent in 2013. "Religion," Database, Gallup, September 26, 2013, http://www.gallup.com/poll/1690/religion.aspx#2; Edward J. Larson and Larry Witham, "Scientists Are Still Keeping the Faith," *Nature* 386, no. 6624 (1997): 435; James H. Leuba, *The Belief in God and Immortality*, 2nd ed. (Chicago: Open Court, 1921); James H. Leuba, "Religious Beliefs of American Scientists," *Harper's Monthly Magazine* 169 (August 1934): 291–300; Edward J. Larson and Larry Witham, "Leading Scientists Still Reject God," *Nature* 394 (1998): 313.

22. Elaine Howard Ecklund, *Science vs. Religion: What Scientists Really Think* (New York: Oxford University Press, 2010).

23. Ecklund, *Science vs. Religion*, esp. 61.

24. Norm R. Allen Jr., *African-American Humanism: An Anthology* (Buffalo, NY: Prometheus Books, 1991); Norm R. Allen Jr., "Humanism in Political Action," in *By These Hands: A Documentary History of African American Humanism*, ed. Anthony B. Pinn (New York: New York University Press, 2001), 147–60; Norm Allen Jr., "Humanism in the Black Community," *Sunrays* 1, no. 1 (December 1991): 11–13; Norm Allen Jr., "Introducing a Bold, Original Newsletter," *AAH Examiner* 1, no. 1 (Spring 1991): 1, 6; "AAH African Americans for Humanism" (undated brochure), Council for Democratic and Secular Humanism, Clippings files, Archives, Center for Inquiry Libraries, Amherst, New York; "About AAH," African Americans for Humanism, https://web.archive.org/web/20160727070405/http://aahumanism.net/info/about_aah (as of this writing the main location for this group is a Facebook page, https://www.facebook.com/pg/africanamericansforhumanism/about/?ref=page_internal); Paul Kurtz, "Why African Americans for Humanism?," *AAH Examiner* 1, no. 1 (Spring 1991): 3; Olga Bourlin, "Norman Robert Allen Jr. (1957–)," BlackPast.org, https://blackpast.org/aah/allen-norman-robert-jr-1957.

25. Anthony B. Pinn, "On Becoming Humanist: A Personal Journey," *Religious Humanism* 32, nos. 1 and 2 (1998): 22–23; Anthony B. Pinn, "Anybody There? Reflections on African American Humanism," *Religious Humanism* 31, nos. 3 and 4 (1997): 61–78; Anthony B. Pinn, ed., *By These Hands: A Documentary History of African American Humanism* (New York: New York University Press, 2001).

26. See discussion in chap. 7.

27. Allen, "Humanism in Political Action," 159.

28. Jennifer Bardi, "The HUMANIST Interview with Neil DeGrasse Tyson," TheHumanist.com, August 14, 2009, https://thehumanist.com/magazine/september-october-2009/features/the-humanist-interview-with-neil-degrasse-tyson.

29. Ann Druyan et al., *Cosmos: A Spacetime Odyssey* (Beverly Hills, CA: Twentieth Century Fox Home Entertainment, 2013–2014).

30. Fredric Jameson, "Postmodernism, or the Cultural Logic of Late Capitalism," *New Left Review* 146 (1984): 53–92; Jean-François Lyotard, *The Postmodern Condition: A Report on Knowledge*, trans. Brian Massumi and Geoffrey Bennington (Minneapolis: University of Minnesota

Press, 1984); John McGowan, *Postmodernism and Its Critics* (Ithaca, NY: Cornell University Press, 1991); Christopher Norris, *What's Wrong with Postmodernism: Critical Theory and the Ends of Philosophy* (Baltimore: Johns Hopkins University Press, 1990); Stephen P. Weldon, s.v. "Postmodernism," in *The History of Science and Religion in the Western Tradition: An Encyclopedia*, ed. Gary B. Ferngren, Edward J. Larson, and Darrel W. Amundsen (New York: Garland, 2000). One of the clearest examples of the use of the postmodern critique against rationality relates to the *Humanist* attack on astrology from 1975. The philosopher of science Paul Feyerabend took up the defense of the astrologers against the skeptics in this case. See Paul Feyerabend, "The Strange Case of Astrology," in *Philosophy of Science and Occult*, ed. Patrick Grim (Albany: State University of New York Press, 1982), 19–32.

31. Two books that have provided some good perspective on the overall path of American thought along these lines are Robert Genter, *Late Modernism: Art, Culture, and Politics in Cold War America* (Philadelphia: University of Pennsylvania Press, 2010); and Mark Greif, *The Age of the Crisis of Man: Thought and Fiction in America, 1933–1973* (Princeton, NJ: Princeton University Press, 2015).

Epilogue · Science and Millennial Humanism

1. Pew polls are the gold standard for measuring these things, and they show a currently fairly high level of secularity in the culture, nearing 30 percent; see Becka A. Alper, "From the Solidly Secular to Sunday Stalwarts, a Look at Our New Religious Typology," Pew Research Center, August 29, 2018, https://www.pewresearch.org/fact-tank/2018/08/29/religious-typology -overview/. This seems to be a pretty recent phenomenon and happening quickly; see Michael Lipka, "A Closer Look at America's Rapidly Growing Religious 'Nones,'" Pew Research Center, May 13, 2015, https://www.pewresearch.org/fact-tank/2015/05/13/a-closer-look-at-americas -rapidly-growing-religious-nones/.

2. Richard P. Cimino and Christopher Smith, *Atheist Awakening: Secular Activism and Community in America* (Oxford: Oxford University Press, 2014).

3. Paul Kurtz, *Humanist Manifesto 2000: A Call for a New Planetary Humanism* (Amherst, NY: Prometheus Books, 2000); "Humanism and Its Aspirations: Humanist Manifesto III, a Successor to the Humanist Manifesto of 1933," American Humanist Association, 2003, https:// americanhumanist.org/what-is-humanism/manifesto3/; "The Amsterdam Declaration," Humanists International, 2002, https://humanists.international/what-is-humanism/the-amsterdam -declaration/; "Declaration in Defense of Science and Secularism," Institute for Ethics and Emerging Technologies, 2006, https://ieet.org/index.php/IEET2/more/cfi200611.

4. Cimino and Smith, *Atheist Awakening*, 16–21. See also, the introduction by Niels De Nutte and Bert Gasenbeek (1–21) and my essay ("Organized Humanism in the United States," 75–93) in Niels De Nutte and Bert Gasenbeek, eds., *Looking Back to Look Forward: Organised Humanism in the World: Belgium, Great Britain, the Netherlands and the United States of America, 1945–2005* (Brussels: VUBPRESS, 2019).

5. Cimino and Smith, *Atheist Awakening*, 76.

6. Cimino and Smith, *Atheist Awakening*, 29–30.

7. On ID, see Ronald L. Numbers, *The Creationists: From Scientific Creationism to Intelligent Design*, exp. ed. (Cambridge, MA: Harvard University Press, 2006), chap. 17.

American Humanist Association. Organization Archives. American Humanist Association, Washington, DC.

Asimov, Isaac, Papers. Special Collections. Mugar Memorial Library, Boston University, Boston, Massachusetts.

Axtelle, George, Papers. Special Collections. Morris Library, Southern Illinois University, Carbondale, Illinois.

Committee for the Scientific Investigation of Claims of the Paranormal. Archives. Center for Inquiry Libraries. Amherst, New York.

Council for Democratic and Secular Humanism, Clippings files. Archives. Center for Inquiry Libraries. Amherst, New York.

Dietrich, John H., Papers. Andover-Harvard Theological Library, Harvard Divinity School, Cambridge, Massachusetts.

Huxley, Julian Sorell, Papers. Woodson Research Center, Fondren Library, Rice University, Houston, Texas.

McIver, Tom, Personal papers. Private collection.

Muller, Herman J., Papers. Lilly Library, Indiana University, Bloomington, Indiana.

New York Society for Ethical Culture, Archives. New York Society for Ethical Culture, New York, New York.

Otto, Max C., Papers. Wisconsin State Historical Society Archives, Madison, Wisconsin.

Randall, John Herman, Jr., Papers. Rare Book and Manuscript Library, Columbia University, New York, New York.

Skinner, B. F., Papers. Special Collections. Pusey Library, Harvard University, Cambridge, Massachusetts.

Truzzi, Marcello, Personal papers. Private collection.

Wilson, Edwin H., Papers. Special Collections. Morris Library, Southern Illinois University, Carbondale, Illinois.

The designations G1 and G2 refer to photographic galleries.

AAH Examiner, 222
Abbot, Francis Ellingwood, 22, 23
Abbott, Lyman, 24
Abel, George O., 259n43
abortion, 136, 178, 180, 183. *See also* women
Abraham Lincoln Center, 43, 60
academics/academia, 7, 8, 9, 67, 79, 85–94, 98, 204, 215–16, 227–28, 232; and church, 63, 74; and Humanist Manifestos, 185, 216; and hyperprofessionalization, 63; and irrationalism, 233; and New Age thought, 153; and postmodernism, 224–25; and Protestant modernism, 24; and radicalism, 23–28
Academy of Humanism, 215, 218–19
Adler, Felix, 60–61, 66–68, 71, 72, 102, G1
Adler, Mortimer J., 87, 89, 90, 91, 92, 225, 252n44; "God and the Professors," 85–86, 94
African Americans, 10, 124–25, 127–29, 221, 222–24, G1. *See also* people of color; race
African Americans for Humanism (AAH), 222, G2
agnosticism, 17, 29, 108, 174, 221, 231
Akeley, Carl, 194
Alcock, James E., 168
Allen, Norm R., Jr., 222, 223–24, G2
Allport, Gorden, 157
American Association for the Advancement of Science, 95, 160
American Astronomical Society, 158
American Civil Liberties Union (ACLU), 173, 177, 180, 211
American Ethical Union, 101
American Federation of Astrologers, 159
American Humanist Association (AHA), 7, 97, 98, 99, 104, 111, 127–29, 160, 186, 222, G1; and Asimov, 202–3, 210, 211; and birth and population control, 118, 119–20; board of directors of, 103, 127, 135, 143, 147–48; and *The Case of the Texas Footprints*, 207; and Chambers, 177, 191, 199, 203, 210, G2; and church-state issues, 173; counselor program of, 137–38; and CSICOP, 167; and disciplinary divisions, 217; and Dreikurs, 138; and Eaton, 113; and education, 176; and Edwords, 202; and evolution, 198–201; founding

of, 82, 95–96; and fundamentalism, 176, 201; "Humanism and Its Aspirations," 231–32; Humanist Youth Fund, 137; and IHEU, 101; and Kirkendall, 172; and Lamont, 5–6; and T. Mc-Carroll, 128, 135, 136, 137, 143; and McCollum, 173, 174, 175; and Morain, 143; and Muller, 110, 115, 195, 211; and Myers, 234; and O'Hair, 175; and Rabinowitch, 112; and Rafferty report, 171, 179; and Randi, 151; and Sagan, 1–2, 211; and San Francisco, 141–42; and Sanger, 118; "Statement Affirming Evolution as a Principle of Science," 200, 202; and *Torcaso v. Watkins*, 176; and Tyson, 224; and Vonnegut, 232–33; and E. H. Wilson, 132, 135–36
American Philosophical Society, 67
American Psychological Association, 139
Americans for Religious Liberty, 203
Americans United for Separation of Church and State, 180, 211
American Unitarian Association, 19, 38
Ames, Edward Scribner, 25
analytic philosophy, 4, 152, 161, 169, 228
Anfinson, Christian B., G2
anthropology, 58, 73, 132, 133, 167
Aquinas, Thomas, 84, 85, 88
Aristotelianism, 73, 85
Arkansas, 197, 198
Armenian genocide, 178
Asimov, Isaac, 168–69, 185, 199, 202–3, 210–14, 215, 232, 233, G2; Foundation series, 211; "Nightfall," 212–13
Association for Humanistic Psychology, 146
astrology, 10, 153, 157–59, 164–67, 168, 199, G2
astronomy, 83, 157, 158, 160, 167
atheism, 7, 15–16, 17, 81, 127, 172, 174–75, 193, 195, 214, 219–21, 231, 232; and African Americans, 222; and Asimov, 212; in Christian culture, 29; and Comte, 35; and Humanist Manifesto, 56; and Kurtz, 232; and Lamont, 97; and Protestant modernism, 26; and religious humanism, 46; and secular humanism, 186; and skepticism, 169
atomic energy, 106, 111–12, 251–52n25

authoritarianism, 81, 91, 94, 95, 184, 187, 188–89, 225
Ayala, Francisco, G2

Backus, E. Burdette, 96
Bahá'í faith, 72
Bahm, Archie J., 96–97
Baptists, 25, 28, 34–35, 50, 174, 181
Barnes, Harry Elmer, 54, 137
behaviorism, 10, 138, 139, 140, 143, 145–49, 155, 171, 179, 197
Bentley, A. F., 103
Berger, Peter, 6, 236n11
Bermingham, Carolyn, 159
Beuttler, Fred, 86–87, 248–49n22
Bible, 14–15, 17, 18, 21, 23, 66, 69, 132, 160, 161, 168, 200, 206; Creation in, 27–28, 55, 56; and creationism, 196, 197; and education, 173, 174–75, 197; factual inaccuracies in, 194; and fundamentalism, 186, 212; and higher criticism, 26; literal interpretation of, 25, 27, 181; and Protestant modernism, 24, 45; and religious humanism, 39, 45
biological determinism, 113, 117, 120, 123, 126, 129
biology, 22, 58, 96, 115, 116, 129, 199, 200, 204; and Carlson, 105; and Chambers, 191; and Herrick, 105–6, 117; and J. Huxley, 108, 110, 123; and Muller, 144, 195; and Otto, 78; and Robertson, 133; and Sellars, 76
biomedicine, 116, 165, 217, 218, 219
Bircher, John, 182
Bird, Wendell, 198, 204
Birkhead, Leon Milton, 53
birth control, 117–20, 122, 222
Black, Hugo, 174
Black Power, 155
Blanshard, Paul, 175; *American Freedom and Catholic Power*, 119; "The Sectarian Battlefront" column, 119
Blau, Joseph, 154–55
Bok, Bart J., 157; "Objections to Astrology," 158, 159, 160, 164, 180
Bragg, Raymond B., 53, 243n17
British Ethical Union, 101
Bronowski, Jacob, 250–51n7
Bryan, William Jennings, 191, 193, 194
Buddhism, 22, 35, 72
Bulletin of the Atomic Scientists, 112, 199
businessmen, 103, 135, 148, 215

Calderone, Mary Steichen, 180
California, 171, 172, 179, 197, 198, 199
Calvinism, 19, 69, 105
Camus, Albert, 134
Capra, Fritjof, *The Tao of Physics*, 169
Carlson, Anton J., 95, 103, 104–5, 106, 116–17; "Biology and the Future of Man," 105
Carruth, William H., 194
Carson, Johnny, 2, 207
Carter, Jimmy, 1, 188
Catholicism, 16, 35, 73, 85, 87, 89, 97–98, 107–8, 119, 142, 155, 187. *See also* Roman Catholic Church
Catholics, 81, 84, 88–89, 94, 145, 172, 174
Center for Inquiry, 231; "Declaration in Defense of Science and Secularism," 232. *See also* Council for Democratic and Secular Humanism (CODESH)
Chambers, Bette, 177, 191, 199–201, 202, 203, 210, G1, G2
Chambers, Claire, *The SIECUS Circle*, 176–77
Channing, William Ellery, 18, 20, 23, 44
Chautauqua Literary and Scientific Circle, 25
chemistry, 57, 76, 167, 228
Chicago, 61, 67, 78, 104
Chicago Ethical Society, 51
Chicago Sinai Congregation, 51
Chisholm, Brock, 103, 107–8, 110, 112, 113
Christian Century, 54
Christian Crusade, *Is the School House the Proper Place to Teach Raw Sex?*, 172–73
Christian fundamentalism, 1, 3, 31, 42, 83, 168, 172, 182, 189, 209, 212; and AHA, 176, 201; and CODESH, 203–5; and CSICOP, 203–5; and evolution, 38, 192, 194, 195, 196–203; and Harper, 25; and J. Huxley, 109; and Moral Majority, 179; and Protestant modernism, 36; and secular humanism, 6, 180, 181, 186, 188, 203, G2; and secularism, 187
Christianity, 1, 3, 4, 6, 9, 24, 29, 34, 168, 171, 172, 192, 228; and M. Adler, 85; and America, 30, 172, 196, G2; deism vs., 14–15; and Dewey, 69; and Emerson, 20; and Foster, 36, 43; and *Free Inquiry*, 188; and Free Religious Association, 21; and Garrison, 17; and Harper, 25; and Haydon, 51; and Hook, 91; and Humanist Manifesto, 55, 56, 60; and ID, 233; and Mathews, 27; and New Age thought, 169–70; and Otto, 77, 78; and Paine, 13–14; and Potter-Straton debates, 193–94; and

Protestant modernism, 26, 42; and Reese, 36; and religious humanism, 39, 40, 45–46; and *Skeptical Inquirer*, 170; and social problems, 28, 37; and Unitarianism, 18, 20–22; and Van Til, 178

Christianity Today, 169–70

Christian Register, 54

church-state issues, 132, 134, 172, 173–76, 178, 179–80, 182, 188, 203, 221–22

civilization, 2, 3, 38, 81, 91–92, 93, 94, 123, 154, 213

civil rights, 9, 124, 127, 128, 134, 183

Civil War, 16, 20, 21, 64–65

class, 17, 23, 45, 67, 120, 121, 221

Clay, Marjory, 155–56, 159, 165

clergy, 17, 23–28, 101, 103, 140

Cloud, Preston, 200

Cohen, Daniel, 160

Cold War, 1, 8, 101, 111, 122, 129

Cole, John and Pia Nicolini, *The Case of the Texas Footprints* (documentary film, 1983), 207

Columbia University, 61, 65–74, 90, 97

Commission to Defend Humanism, 182

Committee for Skeptical Inquiry, 7. *See also* Committee for the Scientific Investigation of Claims of the Paranormal (CSICOP)

Committee for the Scientific Investigation of Claims of the Paranormal (CSICOP), 7, 160–62, 164, 166, 167–70, 202–5, 207, 210, 211, G2

Communism, 84, 99, 110–11, 172, 173, 174, 188, 193, 243n17

Comte, Auguste, 35

"Conference on Science, Philosophy and Religion, The," 83–87, 88, 89, 90, 91, 92

"Conference on the Scientific Spirit and Democratic Faith," 92–95, 96

Congregationalism, 19, 20, 69

Congress, 47, 105

Conlan, John, 176, 177, 182

conscientious objectors, 175, 176

Constitution of the Humanist Fellowship, 52

Constitution of the United States, 212; Establishment Clause, 175, 197; First Amendment, 173, 174, 197

cosmopolitanism, 100, 106, 121, 183

Cosmos (1980 TV series), 1, 2, 209, 212, 213

Cosmos (2014 TV series), 224

Council for Democratic and Secular Humanism (CODESH), 7, 180, 186, 188, 203–5, 215, 218, 219; Faith Healing Investigation Project, 207

Council for Liberal Religious Thought, 179

Council for Secular Humanism, 7

counterculture, 9, 128, 137, 138–39, 142, 143, 148, 153, 157, 228, G2

Counter-Reformation, 88

Creation/Evolution, 201–2, 203

creationism, 161, 168, 197, 211, 222, 223, 233, 265n22, G2; and Chambers, 191, 199–201; and education, 187, 192, 196, 199–201, 204, 207, 208; and Edwords, 201–2; equal time for, 198, 200, 204, 207, 208; and Kurtz, 201, 203, 204; and "Science, the Bible, and Darwin" conference, 205; and Secular Humanist Declaration, 187–88. *See also* education

Creation Research Society, 197

creation science, 168, 196, 198, 204, 207, 208

creativity, 133, 182, 184, 185

creeds, 19–22, 23, 34, 43, 45, 49

Creel, Herrlee Glessner, 50

Crick, Francis, 185, 219

cultural revolutions, of 1960s, 8–9, 137, G2

culture wars, 9, 10, 153, 192

Curie, Marie, 218

Däniken, Erich von, 168

Darrow, Clarence, 54, 194

Darwin, Charles, 25, 195, 205, 218; *On the Origin of Species*, 22

Darwinism, 17, 41, 58, 70, 116, 193

Dawkins, Richard, 214, 232

deism, 13–15, 17, 18, 23, 26

DeMarco, Joseph P., 125

democracy, 2, 5, 52, 81, 84, 96, 119, 121, 129, 189, 212, 213, 227; and M. Adler, 85, 86; and Dewey, 69, 86; and Deweyans, 9; and Dreikurs, 138; and evolution, 192; and Foster, 36–37, 43; and fundamentalism, 181, 187; and Gregory, 107; and Hook, 86, 91, 92, 181; and Humanist Manifesto, 59; and Humanist Manifesto II, 184, 185; and IHEU, 101, 102; and Mathews, 28; and Otto, 92; and Reese, 34, 43; and Scientific Spirit conferences, 93, 94, 95; and Secular Humanist Declaration, 188; and Sellars, 38, 75; and Skinner, 147, 148; and Thomism, 88; and Unitarianism, 1

Dennett, Daniel, 232

Derrida, Jacques, 225

Descartes, René, 155

determinism, 4, 77, 78, 80, 116, 117, 120, 124, 135, 139, 146, 179; biological, 113, 117, 120, 123, 126, 129; and Humanist Manifesto, 54, 56, 57

Dewey, John, 4, 7, 8, 66, 68–71, 116, 148, 171, 246n34, G1; and M. Adler, 86; and American thought, 228; *A Common Faith*, 63, 71, 141; and education, 94; and eugenics, 120–21; and Herrick, 117; and Hook, 90, 91, 227; and Humanist Manifesto, 54; and Humanist Manifesto II, 183; and Kurtz, 152; and Maslow, 141; and Niebuhr, 87; and Otto, 78, 79; and Rushdoony, 179; and Scientific Spirit conferences, 93, 95; and technocracy, 107; and Thomism, 88

Deweyism, 9, 87, 100, 106, 123, 144, 152, 226

Dietrich, John H., 33–36, 37, 38, 39–40, 46, 51, 93, 175, 193, 214, G1; and Christianity, 34, 36, 42; heresy trial of, 33, 34, 43; and self-reliance, 41–42, 59; and Sullivan, 44, 49; and supernaturalism, 37, 40, 42; and Visscher, 105

Diggers, 142

Dill, John R., 124–25

Douglass, Frederick, 127, 222

Dreikurs, Rudolf, 138, 139, 145, G2; "The Scientific Revolution," 143–44

drugs, 128, 134, 142, 156, 171

dualism, 56, 57, 75, 92

DuBois, W. E. B., 222

Dutch Humanist League, 101

Eastern Philosophical Association, 74

Eastern religions, 20, 35, 153

Eaton, Cyrus, 112, 113, G1

Ecklund, Elaine, 221

Edman, Irwin, 246n34

education, 8, 17, 46, 64, 108, 117, 123, 128, 178, 185, 233, G2; and M. Adler, 85; and AHA, 176; and Bible reading, 173, 174–75, 197; and Chambers, 191, 199–201, G2; and creationism, 187, 192, 196, 199–201, 204, 207, 208; and CSICOP, 161; and Dewey, 68, 69, 94; and Ethical Culture movement, 67; and evolution, 196, 200; and IHEU, 102; and Kauffman, 181; and LeHaye, 177; and prayer, 174, 197; and Rafferty report, 171–72, 176, 179, 197; released time religious, 173–74; and religious humanists, 40; and Scientific Spirit conferences, 94, 95; and secular humanism, 176, 187–88, 204; sex, 136, 171–72, 176–77, 180, 222;

and Skinner, 148; and Unitarianism, 19; universal, 5. *See also* church-state issues

Edwards v. Aguillard, 207–8, 233

Edwords, Fred, 201, 202, 203

Einstein, Albert, 112, 252n44

election, presidential, of 1980, 186, 187, 188, 198

Ellis, Albert, 181

Ellis, Havelock, 136

Ellwanger, Paul, 265n22

Emerson, Ralph Waldo, 20, 21, 22, 23, 35

empiricism, 27, 68, 70, 133, 149, 152

Engel v. Vitale, 174–75

Enlightenment, 2, 5, 81, 89, 153, 162, 187, 188, 212, 223, 224, 225, 226–27, 234

environment and ecology, 116, 118, 147, 184, 233–34

Episcopalians, 19

Epperson v. Arkansas, 196–97, 198

Epstein, Jacob, 78

equality, 4, 16, 69, 121–22, 123, 149, 182, 185, 222, 223

Erdmann, Robert L., 148

Ericson, Edward, 211

ESP (extrasensory perception), 156, 160

Ethical Culture movement, 60–61, 73, 91, 93, 101, 102, 154, 174, 181, 211, G1; and F. Adler, 66, 67, 68, 71; and Rafferty report, 179; and *Torcaso v. Watkins*, 176

Ethical Culture School, 67, G1

Ethical Culture Society, 5, 53, 68, 82, 135

ethics and morality, 6, 29, 65, 67, 87, 92, 93, 134, 157, 214, 217, 228; and M. Adler, 85, 86, 89; as autonomous, 185; and Channing, 18; and Chisholm, 108; and Columbia University philosophers, 68; and "Conference on Science, Philosophy and Religion," 83, 84; and Dawkins, 233; and deism, 14, 15; and Dewey, 70, 94; and Deweyans, 9; and Dietrich, 39, 42; and evolution, 198; and Frankfurt School, 225; and freethought, 16, 17; and Harlow, 140; and humanistic psychology, 138, 139; and Humanist Manifesto, 54, 58–59; and Humanist Manifesto II, 184, 185; and Hutchins, 95; and IHEU, 101, 102; and Jefferson, 11; and Kurtz, 126, 144, 147, 182; and Maslow, 141; and naturalism, 4; and Niebuhr, 88; and postmodernism, 225; and Protestant modernism, 59; and Rafferty report, 171–72, 197; and Reese, 35; and religion, 3, 94, 102, 175, 184, 185, 188, 205; and religious humanists, 41; and Robertson, 131;

and Rogers, 139; and Sagan, 2, 3; and Schaeffer, 178; and science, 11, 102, 184; secular, 180; and secular humanism, 186, 188; and Sellars, 77; and World War II, 82

eugenics, 10, 102, 117, 120–23, 124, 129, G2

evangelicalism, 17, 168, 178, 179, 180–81, 182, 186, 193, 198, 212, 233, 260–61n9

evolution, 25, 88, 94, 115, 117, 129, 189, 191–205, 223, G2; and F. E. Abbot, 22, 23, 25; and AHA, 198–201; and contingency, 58–59; and education, 191, 192, 196, 200; and *Edwards v. Aguillard*, 207–8, 233; and Edwords, 201–2; and eugenics, 122–23; and fundamentalists, 38, 42, 192, 194, 195, 196–203; and Harper, 25, 27; and Humanist Manifesto, 55, 58; and Humanist Manifesto II, 184; and J. Huxley, 110, 122–23, 194–95, 196; and Mathews, 27, 41; and Muller, 194, 195, 196, 199, 200; and Otto, 78, 192–93; and Potter, 38; and Potter-Straton debates, 193–94; and religious humanists, 41; and Secular Humanist Declaration, 187–88; and Sellars, 56, 76, 100, 116

experience, 72, 75, 79, 92, 94, 128, 135, 140, 152, 156, 162, 169; and Dewey, 69, 70, 71, 79, 116, 141; and Deweyans, 152

experimental method, 9–10. *See also* scientific method

Eysenck, H. J., 124–25, 126, 129

fact-value distinction, 85, 92

Falwell, Jerry, 181

Fantus, Bernard, 104, 105, 106, 116–17, 183

Faraday, Michael, 159

fascism, 81, 84, 92

Fate, 166, G2

Federation of Scientific Astrologers, 159

Fellowship of Religious Humanists, 7

feminism, 131, 180, 191, G1. *See also* women

Feyerabend, Paul, 269–70n30

Fifth Dimension, "Aquarius," 157

Finkelstein, Louis, 83, 84, 85, 86, 88, 89, 90, 92, 93

First Humanist Society of New York, 193

First Unitarian Church, Dayton, Ohio, 53

First Unitarian Church, Madison, Wisconsin, 78

First Unitarian Church, Minneapolis, 39

First Unitarian Society, Madison, Wisconsin, 51

Fletcher, Joseph, 217; *Situation Ethics*, 180; "Why Ethics Should Avoid Religion," 205

Flew, Antony, 162, 186; "The Erosion of Evolution," 204

Fosdick, Harry Emerson, 83

Foster, George Burman, 25, 43; *The Finality of the Christian Religion*, 36–37, 50

Foucault, Michel, 225

Frankfurt School, 225

Franklin, Benjamin, 13

Frazier, Kendrick, 164

freedom, 21, 34, 36, 42, 116, 129, 149, 155, 172, 182, 187, 227; of conscience, 16, 18, 33, 43; and Fromm, 139; and humanistic psychology, 139; and Humanist Manifesto, 57; and Humanist Manifesto II, 185; and IHEU, 101; and mechanism, 4; and Muller, 121, 122; and Reese, 43; and Rogers, 133, 139, 140; and Sagan, 2, 212; and Schaeffer, 178; and Secular Humanist Declaration, 188; and Sellars, 75, 77; and Skinner, 146–47; and Unitarianism, 43, 44

Freedom From Religion Foundation, 211

Free Inquiry, 168, 180–81, 186, 188, 204, 205, 218, G2

Free Mind, 127, 180–81, 191, 199

Free Religious Association, 21

Free Religious Fellowship, 127

Free Speech Movement, 137

freethinkers, 7, 13, 16–17, 18, 23, 24, 28, 29, 35, 101, 118, 144, 188, 222

Freie Gemeinde leagues, 16

Freud, Sigmund, 139, 218

Friedan, Betty, *The Feminine Mystique*, 180

Friess, Horace, 68, 102, 246n34, 250n7

Fromm, Erich, 104, 139

Fuller, Buckminster, 104

Galilei, Galileo, 126, 159, 218, 220

Galton, Francis, 120

Gardner, Martin, 163; *In the Name of Science*, 159–60; "Notes of a Psi-Watcher," 169

Garrison, William Lloyd, 16, 17

Gauquelin, Françoise, 165–66, 259n43

Gauquelin, Michel, 164–66, 259n43

Geller, Uri, 151, 160

Gell-Mann, Murray, 207–8

genetics, 57, 111, 115, 121, 122, 123, 124, 126

geology, 27, 167, 197, 200, 207

Germany, 24, 26

Gladden, Washington, 24

God, 5, 17, 19, 29, 36, 73, 84, 88, 173, 174, 220; and
    F. E. Abbot, 22, 23; and F. Adler, 67; and Chan-
    ning, 18, 20, 23; and Dewey, 69, 71; and Dietrich,
    34, 37; and Emerson, 20, 22; and Foster, 43; and
    Gardner, 169; and higher criticism, 24; and
    Humanist Manifesto, 54, 55, 56, 58; and Lamont,
    97; and Mathews, 27, 41; and Otto, 78; and Paine,
    13; and Protestant modernism, 24, 26, 41, 45; and
    providence, 41, 42, 172; and Reese, 37; reliance
    on, 14, 15, 37, 40, 185; and religious humanists, 39,
    40, 41, 45; and Robertson, 132; and Russell, 111;
    and Schaeffer, 178; and Sellars, 77; and *Torcaso v.
    Watkins*, 175–76; and Unitarianism, 1, 18, 20–21.
    See also Jesus Christ
Godfrey, Laurie, 168
Goldstein, Kurt, 132
Gould, Stephen Jay, 208, 214, 219, G2
Great Depression, 47, 60, 82, 83, 87
Greco-Roman culture, 91, 182
Greece, ancient, 14, 154
Gregory, Sir Richard, 110, 251–52n25; "Science as
    International Ethics," 106–7
Gutmann, James, 68, 246n34

Haeckel, Ernst, 22, 76
Harlow, Harry, 140
Harper, William Rainey, 25, 27, 50
Harris, Sam, 232
Hartford Theological Seminary, 74
Harvard College, 19, 38, 44, 65, 66
Harvard Divinity School, 20, 43–44, 47
Harvard Educational Review, 124
Harvard Humanist Chaplaincy, 232
Haydon, A. Eustace, 50–51, 52, 53, 93, 95, G1; *The
    Quest of the Ages*, 51
Hefner, Hugh, 136
Heidelberg Catechism, 33
Herrick, C. Judson, 103, 116–17; *The Evolution of
    Human Nature*, 105–6
higher criticism, 24, 25, 26, 66. *See also* Protestant
    modernism
Hinduism, 20, 72
history, 24, 26, 29, 36, 58, 73, 154, 212
Hitchens, Christopher, 232
Hitler, Adolf, 86, 91, 125
Hitlerism, 81, 172
Hoagland, Hudson, 104, 200
holism, 57, 75, 76–77, 80, 116–17, 138, 148, 152

Holmes, John Haynes, 72
Hook, Sidney, 81, 86, 90–92, 93, 97–98, 137, 181, G2;
    and M. Adler, 86, 252n44; and IQ, 126–27, 129,
    223; and Kurtz, 123, 144, 154, 188; and secularism,
    182, 227; "Theological Tom-Tom and Metaphysi-
    cal Bagpipe," 90, 91, 92
Houdini, Harry, 159
House of Representatives, 1–2, 176
Hull, Richard T., 155, 156, 159, 165
humanism, as term, 4–8
Humanist, 92, 95–96, 98, 103, 105, 108, 112, 115,
    117–18, 148, 179; and astrology, 158, 159, 164, 165,
    259n43, 269–70n30; "Birth Control and World
    Peace," 118; changing presentation of, 134–35;
    and church-state separation, 175; and CSICOP,
    161; and Eaton, 112; and Edwords, 202; and
    eugenics, 123, 124; and evangelicalism, 180–81;
    and evolution, 199–200; and Free Speech Move-
    ment, 137; fundamentalism symposia in, 181; and
    humanistic psychology, 139; and Humanist
    Manifesto II, 183; "Inner-City Education," 128;
    and Kurtz, 123, 141, 143, 145, 151, 167, 168, 186, G2;
    and T. McCarroll, 134, 135, 136, 141, G2; and New
    Age thought, 153, 156; "New Bill of Sexual Rights
    and Responsibilities," 180; "Planned Parenthood
    and the Modern Inquisition," 118; and race, 125,
    127, 128, G2; "The Resurgence of Fundamental-
    ism," 181; and Robertson, 131, 132–34, 139; and
    Sanger, 119, 123; "Sex: Is There a Humanist
    View?," 136; "Statement Affirming Evolution as
    a Principle of Science," 180; "Statement on the
    Family," 180; "When Is Sex a Crime?," 136; and
    E. H. Wilson, 49, 82–83, 90, 132; "The Woman
    Rebel," 136; "Women, Democracy, and Birth
    Control," 119
Humanist Alternative, The (TV series), 145, 183
Humanist Fellowship, 50, 51–53, 78
Humanist House, San Francisco, 141–42, 145, G2
Humanist House, Yellow Springs, Ohio, 141, G1, G2
humanistic psychology (third force), 9, 138–41, 143,
    144, 147, 148, 157, 185, 225, G1; and Dreikurs, 138,
    145, G2; and T. McCarroll, 128, 138, 142, 145
Humanist League of Belgium, 101
Humanist Manifesto (1933), 53–61, 82, 90, 93, 95,
    97, 102, 152, 183, 184, G1; and Carlson, 105; and
    Christianity, 55, 56, 60; composition of, 47; and
    determinism, 54, 56, 57; and Dewey, 63, 71;
    and disciplinary divisions, 215; and ethics and

morality, 54, 58–59; and evolution, 55, 58; and
Fantus, 104, 117; and God, 54, 55, 56, 58; and
human nature, 54, 56, 57, 58; and morality, 54,
58–59; and naturalism, 55, 56, 57, 58; and Potter,
193; and Rafferty report, 171; and Randall, Jr.,
73; and scientific method, 55, 56, 60, 184; and
Secular Humanist Declaration, 187; signatories
of, 185–86, 215, 216; and Visscher, 105
Humanist Manifesto II (1973), 124, 145, 180, 183–87,
199, 215, 216
Humanist Manifesto III (2003), 233
Humanist of the Year award, 103–4, 105, 107, 112,
145; and Asimov, 203, 210; and Calderone, 180;
and disciplinary divisions, 215, 217–18; and
A. Ellis, 181; and Fletcher, 180, 205; and Friedan,
180; and Fromm, 139; and J. Huxley, 123, 196; and
Larue, 206; and Maslow, 139; and Morgentaler,
180; and Muller, 123, 196; and Myers, 233; and
Randolph, 128, 180; and Rogers, 139; and Sagan,
1–2, 196, 209, 210, G2; and Sanger, 118, 119, 120;
and scientists vs. non-scientists, 233; and
Skinner, 146, 147–48
Humanist Press Association, 53, 82, G1
Humanist Society, 112
Humanist Student Union of North America, 137
human sciences, 58, 131, 217, 218, 219
Hurston, Zora Neale, 222
Hutchins, Robert Maynard, *The Higher Learning in
America*, 94–95
Huxley, Aldous, *Brave New World*, 122
Huxley, Julian, 100–101, 103–4, 107, 108–10, 120, 121,
122–23, 129, 141, 194–95, 196; *Religion without
Revelation*, 108, 109; *UNESCO: Its Purpose and
Its Philosophy*, 109, 122–23
Huxley, T. H., 76, 100, 108, 194
Hyman, Ray, 163

Inherit the Wind (motion picture, 1960), 200
Institute for Creation Research, 198
intelligent design (ID), 233–34
International Humanist and Ethical Union
(IHEU), 101–2; "Amsterdam Declaration" (1952),
99; "Amsterdam Declaration" (2002), 232
internationalism, 98, 99–113, 132, 138, 184, 185, G2
IQ and race debates, 124–27, 129, 145–46, 149, 223,
G2
Isaac Asimov Science Award, 224
Islam, 1, 187, 188

James, William, 69
Jastrow, Robert, *God and the Astronomers*, 168–69
Jefferson, Thomas, 11, 13, 93, 212
Jensen, Arthur, 124, 125, 126, 128, 223
Jerome, Lawrence E., 158, 164–65, 259n43;
"Objections to Astrology," 158
Jesus Christ, 14, 21–22, 26, 28, 36, 206. *See also* God
Jewish Theological Seminary, 83
Jews, 21, 60, 67, 84, 85, 125, 127, 129, 174, 178, 226, 227
John Birch Society, 172
Joint Washington Office for Social Concern,
179–80
Jones, William R., 222–23, 224
Judaism, 66, 67, 87, 187
Judeo-Christian tradition, 30, 48, 85, 89, 92, 176,
182

Kachel, A. Theodore, 155
Kauffman, Walter, 181
Kepler, Johannes, 159, 220
King, Norman R., *The Creation-Evolution
Controversy*, 168
Kirkendall, Lester A., 172; *Premarital Intercourse
and Interpersonal Relationships*, 136
Klass, Philip, 160
Kurtz, Paul, 124, 155, 181, 189, 205, 227, 232, 257n2,
G2; and Academy of Humanism, 218, 219; and
astrology, 157, 158, 159, 166, 199, 259n43; and be-
haviorism, 145–46, 147, 148, 149; and CODESH,
180, 186, 203; and creationism, 201, 203, 204; and
CSICOP, 160, 161, 162, 163, 203; and ethics, 126,
144, 147, 182; and *Free Inquiry*, 168; and fundamentalism, 186, 188, 203, 204; and Hook, 123,
144, 154, 188; and *Humanist*, 123, 141, 143, 145, 151,
167, 168, 186, G2; and *Humanist Alternative*, 183;
Humanist Manifesto 2000, 231; and Humanist
Manifesto II, 180, 183, 199; and IQ and race, 125,
126–27, 145–46, 223; and T. McCarroll, 128, 143;
and New Age thought, 153–54, 156, 157; "Objections to Astrology," 158; and race and IQ, 129;
and Rafferty report, 171, 179; and reason, 144,
152–53, 162; and Secular Humanist Declaration,
180, 204; and skepticism, 160, 167, 169; and
Truzzi, 162–63, 164

La Barre, Weston, 132
LaHaye, Tim: *The Battle for the Mind*, 177, 186, G2;
Left Behind, 177

Lamont, Corliss, 5–6, 93, 97, 98, 134, 141, 246n34, G1

Larson, Edward, 220

Larue, Gerald, 205–6

lawyers, 8, 19, 103, 135

Leake, Chauncey, 200

LeBrun, Harvey, 156

Leuba, James H., 96, 97, 220

life stance humanists, 232, 234

Lightman, Bernard, 29

Lippmann, Walter, 54

logic, 9, 13, 152, 162

logical positivism, 85–86, 152, 160, 257n2

Los Angeles Times, 201

Louisiana, 198, 207

Lutheranism, 77

Lysenko, T. D., 111

Lyttle, Charles H., "Humanism, America's Real Religion," 38

Madison, James, 175

magic, 153, 158, 160, 163

Maimonides, 84

Maritain, Jacques, 83

Marxism, 87, 90, 97, 110, 126, 145

Maslow, Abraham, 7, 104, 133, 139, 140–41, 145, 228

Massachusetts Congregational Church, 18

materiality/materialism, 22, 24, 72, 73, 106, 178, 179, 193; and Dewey, 70; and Emerson, 23; and Humanist Manifesto, 56, 57; and Sellars, 75, 76, 77

mathematics, 9, 13, 23, 149, 152, 167

Mathews, Shailer, 25, 27, 41; *The Social Teachings of Jesus*, 28

Matson, Floyd, 146

Mayer, William, 200

McCarroll, Claire, 142–43

McCarroll, Tolbert, 134, 135–38, 141–44, 175, 176, G2; and humanistic psychology, 139, 145; and Kurtz, 128, 143; and Maslow, 140, 141; and Skinner, 145; "The Undead," 137

McCollum, Vashti, 173–74, 175, G1

McGee, Lewis A., 127, 222, G1

McGee, Marcella Walker, 127

Meadville Theological Seminary, 22, 48–50, 51, 52, 53, 61, 127, 132, G1

mechanism, 4, 10, 57, 76, 116, 117, 129, 138, 139, 140, 155

metaphysics, 9, 28, 79, 164, 168

middle class, 17, 67, 221

ministers, 1, 8, 23, 28, 37, 64, 68, 100, 106, 115, 227; and AHA, 103, 135; and *Humanist*, 96; and Humanist Manifestos, 185, 215

miracles, 13, 14, 15, 26, 36, 37, 45, 109, 160, 205

Missouri, 16

Money-Kyrle, Roger, 132

Moody Bible Institute, 199

Morain, Lloyd, 103, 143, 148, 179

Moral Guidelines Committee, 171

Moral Majority, 179, 181, 182

Morgan, T. H., 110

Morgentaler, Henry, 180

Mott, Lucretia, 16

Muller, Hermann, 104, 110–11, 112, 113, 125, 133, 144, 202, 211, 228; and eugenics, 120, 121–22, 123; and evolution, 194, 195, 196, 199, 200; "Man's Place in Living Nature," 115; *Out of the Night*, 117, 121

Mumford, Lewis, 54

Murray, Gilbert, 91, 154

Myers, P. Z., 233

mysticism, 10, 70, 79, 108, 109, 140, 154, 156, G2

Nagel, Ernest, 7, 160, 162

Nathanson, Jerome, 82, 93, G1

National Academy of Sciences, 105, 158

National Center for Science Education, 202, 203

National Conference of Unitarian Churches, 20–22, 23

National Council of Churches, 177

National Education Association, 177

National Science Foundation, 105

Natural History Museum in London, 204

naturalism, 3, 4, 9, 36, 59, 73, 74, 84, 89, 140, 158, 161; and F. E. Abbot, 23; and F. Adler, 67; and Alcock, 168; and Dewey, 63, 68, 70, 71, 94, 141; and Dietrich, 39; and Emerson, 23; and Hook, 92; and Humanist Manifesto, 55, 56, 57, 58; and J. Huxley, 141; and Niebuhr, 87, 88; and Protestant modernism, 24, 26, 40–41, 45; and Reese, 35; and Sellars, 56, 75–77, 100; and Truzzi, 163

natural law, 14, 233

natural philosophy, 13, 14

natural rights, 86

natural sciences, 9, 24, 27, 40, 58, 70, 100, 113, 216, 217, 218, 219, 226 natural theology, 13–14

nature, 3, 13–14, 15, 29, 41, 80, 100, 116, 146, 209, 212, 233; and F. E. Abbot, 22, 23; and Emerson, 20, 22, 23; and Humanist Manifestos, 56, 184; and J. Huxley, 108, 109

Nazis, 92, 121, 129, 171–72, 188, 251–52n25

Negri, Maxine, 182

neo-Pentecostalism, 155

neo-Thomism, 94

New Age thought, 2, 153–56, 157, 159, 163, 169–70, 188, 214, 228, 234, G2

New England, 20, 21

New Humanist, 52, 53, 54, 59, 82, 243n17

New Left, 126, 188, G2

New Republic, 54, 91

New Right, 182

Newton, Isaac, 155, 159, 220

New York City, 61, 65

New York Society for Ethical Culture, 67, G1

Niebuhr, Reinhold, *Moral Man and Immoral Society*, 87–88

nuclear energy, 111, 184

nuclear weapons, 1, 100, 105, 106, 110, 112

occultism, 151, 154, 155, 156, 157, 159–60, 163, 166, 170, 234

O'Hair, Madalyn Murray, 174–75

Otto, Max Carl, 54, 77–79, 90, 93, 192–93, 194, 246n34, G1; "He Came to Himself," 247n49; *The Human Enterprise*, 78–79; "Living Down the Hyphen," 247n49; *Natural Laws and Human Hopes*, 78; "Take It Away!," 92; *Things and Ideals*, 78; "Untitled Autobiography 'Dear L.G.W.,'" 247n49

Overstreet, Harry, 96

paganism, 22, 85, 91, 170

Paine, Thomas, 13, 16, 218, 226; *The Age of Reason*, 13–15, 16

paranormal beliefs, 151, 161, 162, 163, 164, 166, 167, 168, 188, 205

Partisan Review, 91

Pauling, Linus, 103, 199

Peirce, Charles S., 69

People for the American Way, 180

people of color, 124, 221, 222. *See also* African Americans; race

Pfeffer, Leo, 182

philosophers, 8, 64, 68, 100, 115, 120, 160, 185, 199, 216, 217, 227–28

philosophy, 3, 6, 7, 9, 24, 29, 80, 88, 106, 167, 175, 219, 225; and F. Adler, 67; and M. Adler, 85–86; at Columbia University, 65–74; and CSICOP, 162; and Dewey, 68, 70–71; and Humanist Manifesto, 56, 57; and J. Huxley, 109; and Judeo-Christian tradition, 89; and Kurtz, 144, 152, 162; and Otto, 77–79; professional, 64–65, 79; public, 64, 65, 66, 68, 74, 78, 79; and religious humanism, 41, 63, 64, 74, 79, 131; and Sellars, 75; technical, 4, 65, 75

physics/physicists, 57, 83, 144, 197, 228

Picasso, Pablo, *The Love of Jupiter and Semele*, 136

Pinker, Steven, 233

Pinn, Anthony, 222–23, 224

Planned Parenthood, 119–20, 135, 136, 143

Platonism, 40, 68

pluralism, 18, 72, 87, 89, 102, 248n22

Popoff, Peter, 207, G2

Popper, Karl, 204, 219

population control, 99, 117–20, 184

Portland Ethical Study Society, 135

positivism, 113, 123, 133, 139, 143, 149, 184, 203, 208, 214, 225, 228, 233; and T. McCarroll, 135; and postmodernism, 226; and pragmatism, 4, 10, 152; and scientific method, 9–10; and skeptics, 152, 169

postmodernism, 221, 224–26, 233, 269–70n30

Potter, Charles Francis, 38, 53, 61, 93, 112, 193–94

poverty/poor people, 4, 28, 35, 43, 59–60, 118, 119, 129, 184, 185

pragmatism, 4, 8, 9, 10, 68–69, 87, 90, 106, 123, 148, 152, 214, 228, 257n2

Presbyterians, 177

Princeton University, 66

Prometheus Press, 151, 156

Protestantism, 8, 29, 36, 81, 84, 87, 89, 97, 119, 172, 174

Protestant modernism, 24–28, 36, 37, 42, 49, 58, 72, 83, 85, 89, 226, 227; and Bible, 24, 45; and Dewey, 69, 70; and Dietrich, 42; and ethics, 59; and God, 41, 45; and naturalism, 24, 26, 40–41, 45; and Niebuhr, 87, 88; and religious humanists, 40; and Sellars, 74–75; and World War II, 82

pseudoscience, 2, 10, 151, 160, 161, 164, 206

psychiatry, 108

psychics, 2, 151, 159, 160, 161, 163

psychoanalysis, 138, 139

psychobiology, 105, 117

psychology, 9, 73, 83, 116, 131–32, 137–49, 157, 167, 168, 216, 217; and Chisholm, 107, 108; and CSICOP, 161; and Dreikurs, 138; and Eysenck, 124; and IHEU, 102; and Rafferty report, 171; and Robertson, 132, 133–34; and Sellars, 76. *See also* behaviorism; humanistic psychology (third force)

Pugwash Conferences, 100, 110–13, G1

Quakers, 21

quantum mechanics, 144, 169

Quine, Willard Van Orman, 162

Rabinowitch, Eugene, 112–13; "Atoms and Man," 112

race, 110, 111, 117, 120, 121, 124–29, 134, 145–46, 185, 223, 231, G2. *See also* African Americans; people of color

Radical Humanist Movement of India, 101

radical religion, 15–31, 214, 221, 222

Radical Religious Research Society, 48, 51–52

Rafferty, Max, 171, 172, 177, 197

Rafferty report, 171–72, 176, 179, 182, 197

Randall, John Herman, Jr., 68, 71–74, 92–93, 95, 97–98, 137; "The Ethical Challenge of Pluralistic Society," 73; "The Ethical Life and the Humanist Temper," 73; "The Experimental Attitude in Morals," 73; *The Making of the Modern Mind,* 72; *Religion and the Modern World,* 72

Randall, John Herman, Sr., 72–73; *Religion and the Modern World,* 72

Randi, James, 151, 152, 160, 163, 205, 207, G2

Randolph, A. Philip, 7, 128, 180

Rationalist Press Association, 111

Rawlins, Dennis, "sTarbaby," 166–67, G2

Reagan, Ronald, 171, 179, 187, 188, 206

reason/rationalism, 3, 5, 10, 11, 16, 40, 81, 87, 107, 157, 159, 224, G2; and F. E. Abbot, 22; and F. Adler, 68, 71; and Asimov, 211; and Channing, 18, 20, 23, 44; and Clay, 155–56; and deism, 14, 15; and Dill, 125; and A. Ellis, 181; and Emerson, 20; and Eysenck, 124; and *Free Inquiry,* 188; and fundamentalism, 187, 202; and Hook, 154; and Humanist Manifesto II, 185; and IHEU, 102; and Jefferson, 93; and Kachel, 155; and Kurtz, 144, 152–53, 162; and Mathews, 27; and T. McCarroll, 135, 143; and New Age thought, 2, 153; and Niebuhr, 88; and postmodernism, 225, 226;

269–70n30; and Protestantism, 29; and Protestant modernism, 26; and Rafferty report, 172; and Randi, 151, 152; and Robertson, 133; and skepticism, 169; and Unitarianism, 18, 19, 23; and Van Til, 178; and Zimmerman, 162

reductionism, 4, 10, 54, 56, 57, 76, 80, 115, 116, 117, 131, 140, 149, 152

Reese, Curtis, 34–35, 36, 37, 38–39, 42, 43–44, 45, 49, 53, 93, 250n7; Harvard Divinity School lecture, 47; and IHEU, 102; and Potter, 193; and religion, 175; and social reform, 59–60

Reformed Church, 105

Reform Judaism, 66

Reiser, Oliver, 251n18

Religion and Biblical Criticism Research Project, 205–6

"Religion in American Politics" conference, 206

Religion of Humanity, 35

religious humanism, 5, 33–46, 74, 106, 152, 232, G1; and AHA as religious organization, 176; and atheism, 46; and Christianity, 45–46; and Dewey circle, 148; and *Free Inquiry,* 188; and Humanist Manifesto, 47, 57; and Kurtz, 144; and Maslow, 141; and Meadville Theological Seminary, 49–50; and Muller, 115; and philosophy, 63, 64, 79, 131; and Potter, 193; and Sellars, 38, 77; and *Torcaso v. Watkins,* 176; and Unitarianism, 37, 38, 39; and E. H. Wilson, 98

Renaissance humanism, 4

Rhine, J. B., 156

Riddle, Oscar, 103

Robertson, Priscilla, 131–34, 136, 139, G1

Robinson, James Harvey, 54

Rockefeller, John D., 25, 28

Rogers, Carl, 104, 133, 139–40

Roman Catholic Church, 98, 118–19

Rome, ancient, 91, 154. *See also* Greco-Roman culture

Romm, Ethel, "The Yinning of America," 154

Roosevelt, Franklin D., 47, 59, 60, 83

Rushdoony, Rousas J., 177–79; *The Messianic Character of American Education,* 178

Russell, Bertrand, 111–12; "A Free Man's Worship," 111; *Religion and Science,* 111; *The Scientific Outlook,* 111; *Why I Am Not a Christian,* 111

"Russell-Einstein Manifesto," 112

Russia, 99

Russo, Kathleen, 159

Sagan, Carl, 1–3, 7, 160, 195–96, 209–12, 214, 219, 221, 228, G2; *Contact*, 211, 213; and *Cosmos*, 1, 2, 209, 212, 213; and Dawkins, 233; as popularizer, 2, G2; and religion, 2, 212, 213; and Tyson, 224

Sakharov, Andrei, 186, 219

San Francisco, 141

Sanger, Margaret, 7, 103, 118–19, 120, 121, 122, 123, 129, 136, 218

Savage, Adam, 233

Schaeffer, Francis, 177–79

School District of Abington Township v. Schempp, 174–75

"Science, the Bible, and Darwin" conference, 205

science fiction, 115, 163, 168, 185, 202, 210, 211, 213, G2

science popularizers, 2, 22, 54, 200, 202, 210, 212, 214–15, 224, G2

Scientific American, 159

scientific humanism, 106, 109, 115, 117, 120, 123, 129, 131, 135, 143, 144, 146, 148–49, 152

scientific method, 9, 36, 52, 85, 93–94, 102, 149, 152, 160, 163, 205, 208, 234; and Dewey, 94, 148; and Dietrich, 33, 39; and Dreikurs, 144; and Hook, 91; and Hull, 156; and Humanist Manifesto, 55, 56, 60, 184; and J. Huxley, 109; and religious humanists, 34, 40, 41; and Robertson, 134; and Rogers, 139; and Russell, 111. *See also* experimental method

scientism, 155, 156

scientists, 99–100, 106–7, 112, 113, 158, 169, 217, 218, 220–21, 228, 232; and AHA, 103, 135; and creationism, 197, 199, 200, 201; and CSICOP, 160; and Edwords, 201; and Humanist Manifesto II, 185; and occult, 151, 156, 157, 159, 160; and Potter, 194; and Randi, 151; and Truzzi, 164

Scopes "Monkey" Trial, 194, 196

secular humanism, 6, 171, 172, 176, 180, 181, 186, 189, 200, 203, G2

Secular Humanist Declaration, 180, 186–88, 204, 205, 215, 216

self-reliance, 14, 15, 34, 41–42, 59, 75, 213–14

Sellars, Roy Wood, 5, 37–38, 53, 56, 57, 58, 61, 74–77, 93, 100, 116, 243n17; *Evolutionary Naturalism*, 59; "Has Human Life Intrinsic Meaning?," 75; "Has Man a Cosmic Companion?," 75; "Humanism as a Religion," 96; "Is the Universe Friendly?," 75; *The Next Step in Democracy*, 59, 75; *The Next Step in Religion*, 38, 75, G1; *Religion Coming of Age*, 75

settlement houses, 28, 43, 67, 69, 102

Sevareid, Eric, 158

sex/sexuality, 132, 133, 134, 136, 138, 171–72, 177, 185, 197, 231

Sexuality Information and Education Council of the United States (SIECUS), 172, 176, 177

Shapley, Harlow, 54

Shelley, Percy Bysshe, 155

Shipley, Maynard, 54

Shockley, William, 125

Shroud of Turin, 161, 206

Simpson, George Gaylord, 200

Skeptical Inquirer, 160–61, 162, 164, 167, 168, 170, 206, G2

skeptics/skepticism, 151–70, 220, 222, 234, G2; and astrology, 157–59, 164–67; and New Age thought, 153–56; organization of, 159–64

Skinner, B. F., 7, 139, 140, 145–48, 155, 179, 228; *Beyond Freedom and Dignity*, 146–47, 149; and *Humanist Manifesto II*, 186; *Walden Two*, 147

slavery, 16, 147, 178

Social Gospel, 28, 45, 59, 67, 69, 75, 87

socialism, 28, 42, 54, 59–60, 75

social justice, 79, 185

social problems, 37, 64, 104, 105, 107, 108, 128

social reform, 28, 42–43, 59–60, 67, 106, 121

social sciences, 43, 58, 64, 105, 131–32

social scientists, 83, 100, 216

Society of the Psychological Study of Social Issues, 157

sociology, 163, 167

Socrates, 218

Southern Baptists, 34, 35

Southern Christian states' rights, 177

Soviet Union, 111, 243n17

Spencer, Herbert, 22, 76

Spinoza, Baruch, 218

spiritualism, 21, 109, 161

spirituality, 2, 68, 74, 77, 78, 96, 138, 139, 140, 141, 162, 178, 184

Spock, Benjamin, 104

Stalin, Joseph, 111, 178

Stalinism, 188

Stanley, Matthew, 29

Stanton, Elizabeth Cady, 16; *The Women's Bible*, 17

sTarbaby affair, 164–67, G2

St. Mark's Reformed Church, Pittsburgh, 33
Straton, John Roach, 193–94
Strauss, Friedrich, *The Life of Jesus*, 26
Students' Christian Association, 69
Students for a Democratic Society, 125
subjectivity, 139–40, 155, 156, 162
Sullivan, William Laurence, 44, 49
Supreme Court, 6, 172, 173–76, 177, 182, 196–97, 207–8
Synanon, 142
Szilard, Leo, 103

Tabler, Ward, 137
Taoism, 35, 169
technocracy, 81, 82, 102, 106, 107, 147
technology, 73, 104, 117, 154, 184, 187, 233
teleology, 41–42, 58, 59, 70, 75
Tennessee, 196
Texas Tech Law Review, 177
Thayer, V. T., 246n34
theism, 22, 57, 84, 85, 88, 92, 193
theology, 22, 24–28, 48, 64, 74, 87–89, 91, 109
Third Unitarian Church of Chicago, 105
Thomas, Norman, 59
Tillich, Paul, 83
Torcaso, Roy, 175
Torcaso v. Watkins, 175–76
totalitarianism, 81, 147, 148, 179, 184
Transcendentalist movement, 20
Transcendental Meditation, 155, 156
Trinitarianism, 18, 20, 44
Truzzi, Marcello, 162–64
Turner, James, 29
Tuskegee experiment, 223
Tyson, Neil deGrasse, 214, 224

UFOs, 2, 160
Union Theological Seminary, 66
Unitarian General Conference, 43, 44, 49
Unitarianism, 18–23, 44, 91, 98, 105, 205; and African Americans, 127, 222; and Bible, 21; and creeds, 19–22, 23, 34, 43, 45; and Dietrich, 34; and Ethical Culture movement, 61; and *Free Inquiry*, 188; and Humanist Fellowship, 53; and Humanist Manifesto, 60; and Humanist Press Association, 82; and professionals, 19, 23; and Protestant modernism, 26; and Reese, 35, 43; and religious humanists, 38, 39, 43–45; and

scientists, 100; and Sellars, 38, 74–75; seminaries of, 45
Unitarians, 1, 5, 7, 8, 22, 38, 39, 48, 136; and AHA, 135; and Bible reading in schools, 174–75; and Fantus, 104; and Kurtz, 144; and Mathews, 28; and T. McCarroll, 136; and Protestant modernism, 24, 28; and Rafferty report, 179; and religious humanists, 37; and school prayer, 174
Unitarian Universalism, 7, 22, 222
Unitarian Universalist Association (UUA), 60
United Nations, 99, 101, G2
United Nations Educational, Scientific, and Cultural Organization (UNESCO), 100, 101, 105, 109–10, 122–23
United States v. Seeger, 175, 176
Universalism, 21, 60, 61
University of Chicago, 36, 40, 61, 75, 94, 104, G1; Divinity School, 25–26, 27, 28, 49–50, 51, 66, 74–75
University of Michigan, 69, 74
University of Minnesota, 105
University of Wisconsin, 77–78, 192–93, G1

van der Wal, Libbe, 102, 250n7
Van Horne, Harriet, 158
Van Til, Cornelius, 178
Velikovsky, Immanuel, 160
Vienna Ethical Society, 101
Vietnam War, 134, 175, 176
Visscher, Maurice, 106, 116–17; "Science for Humanity" column, 105
Vonnegut, Kurt, 232–33

Warbasse, James P., 103
Wendt, Gerald, 95, G1
West, 5, 81, 94, 182
Western Philosophical Association, 74
Western Unitarian Conference, 21, 23, 37, 38, 43, 45, 49, 53
Whitehead, John W., 177
Will, George, 158
Willett, Herbert, 24
Wills, Garry, 183, 184
Wilson, Daniel, 65
Wilson, Edward O., 219
Wilson, Edwin H., 49, 53, 54, 92, 95, 98, 134, 250n55; and AHA, 132, 135–36; and Asimov, 211; and Chisholm, 108; and "Conference on Science,

Philosophy and Religion," 90; and Eaton, 113; and Fantus, 104; and *Humanist*, 49, 82–83, 90; and Humanist Manifesto II, 183; and J. Huxley, 100; and T. McCarroll, 135–36; and McGee, 127, 222, G1; and Sanger, 118, 119; and Visscher, 105

Witham, Larry, 220

women, 16, 17, 103, 118, 119, 121, 129, 131, 155, 180, 221–22. *See also* abortion; feminism

Woodbridge, Frederick J. E., 66, 68

Wootton, Barbara, 102, 250n7

World Health Organization (WHO), 100, 107, 108

World Population Conference in Geneva, Switzerland (1927), 120, 122

World War II, 8, 81, 82, 84, 86, 89, 101, 109, 154, 228; era following, 99, 106, 129; and Protestant modernism, 87; and Scientific Spirit conferences, 93

Zelen, Marvin, 165, 166, 259n43

Zetetic, 160, 168. See also *Skeptical Inquirer*

Zimmerman, Marvin, 161, 162